MICROBIOLOGY OF EXTREME ENVIRONMENTS AND ITS POTENTIAL FOR BIOTECHNOLOGY

*Proceedings of the Federation of
European Microbiological Societies Symposium held in
Troia, Portugal, 18–23 September 1988*

MICROBIOLOGY OF EXTREME ENVIRONMENTS AND ITS POTENTIAL FOR BIOTECHNOLOGY

Edited by

M. S. DA COSTA
Department of Zoology, University of Coimbra, Portugal

J. C. DUARTE
LNETI, Queluz, Portugal

and

R. A. D. WILLIAMS
Department of Biochemistry, The London Hospital Medical College, London, UK

ELSEVIER APPLIED SCIENCE
LONDON and NEW YORK

ELSEVIER SCIENCE PUBLISHERS LTD
Crown House, Linton Road, Barking, Essex IG11 8JU, England

Sole Distributor in the USA and Canada
ELSEVIER SCIENCE PUBLISHING CO., INC.
655 Avenue of the Americas, New York, NY 10010, USA

WITH 66 TABLES AND 77 ILLUSTRATIONS

© 1989 ELSEVIER SCIENCE PUBLISHERS LTD

British Library Cataloguing in Publication Data

Federation of European Microbiological Societies
(*Symposium; 1988; Troia, Portugal*)
Microbiology of extreme environments and its
potential for biotechnology.
1. Extreme environments. Adaptation of microorganisms
I. Title II. Costa, M. S. da III. Duarte, J. C. IV. Williams, R. A. D.
576'.15

ISBN 1-85166-361-4

Library of Congress CIP data applied for

No responsibility is assumed by the publisher for any injury and/or damage to persons or property as a matter of products liability, negligence or otherwise, or from any use or operation of any methods, products, instructions or ideas contained in the material herein.

Special regulations for readers in the USA

This publication has been registered with the Copyright Clearance Center Inc. (CCC), Salem, Massachusetts. Information can be obtained from the CCC about conditions under which photocopies of parts of this publication may be made in the USA. All other copyright questions, including photocopying outside the USA, should be referred to the publisher.

All rights reserved. No part of this publication may be reproduced, stored in a retrieval system, or transmitted in any form or by any means, electronic, mechanical, photocopying, recording, or otherwise, without the prior written permission of the publisher

Printed in Great Britain at the University Press, Cambridge

PREFACE

Many environments which, from the viewpoint of man, seem to be extremely inhospitable are colonised by microorganisms that are especially adapted to their ecological niches. Research in this area was stimulated by the description of *Thermus* by Brock in 1969. Since then, a surprising number and variety of isolates have been made from environments of extreme temperature, pH, salinity and pressure, as well as strains resistant to radiation and toxic chemicals. The diversity of microorganisms from such niches has shed new light on the diversity of life and evolution. The identification of the Archaebacteria as a taxon distinct from eubacteria and eukaryotes highlights the significance of the last 20 years' work on extreme environments.

The numbers of new microorganisms being described shows no signs of decreasing, yet our knowledge of those that have been identified for many years is still fragmentary. There is a need for further investment in the isolation, characterisation, physiology and biochemistry of these fascinating bacteria. An understanding of the taxonomic relationships and ecological interactions will arise out of this knowledge of their properties. The practical applications of microorganisms from extreme environments vary from the production of low-volume, high-value products such as DNA polymerase for DNA amplification in the polymerase chain reaction, to large-scale industrial processes in competition with more traditional technologies.

The Federation of European Microbiological Societies (FEMS) Symposium in Troia, Portugal, was conceived to bring together those interested in these microorganisms from all points of view. We wish to express our thanks to the Executive Council of FEMS for the financial support, encouragement and advice that made this meeting possible. Our gratitude is also due to the Commission of the European Communities for their financial support to some of the delegates, and for convening a European Laboratory Without Walls to consider further funding initiatives by the Community. This is outlined in a statement by Dr Alfredo Aguilar. Financial support for the meeting was also provided by Junta Nacional de Investigação Científica e Tecnológica (JNICT), Instituto Nacional de Investigação Científica (INIC), Fundação Luso-Americana, Fundação Calouste Gulbenkian, the Sociedade Portuguesa de Microbiologia and Novo Industri A/S.

The editors wish to thank Paula Morgado for help with the pre-meeting planning and Janet Guidal for retyping some of the contributions, as well as their colleagues and their respective institutions for their support.

Special gratitude is expressed to those who contributed as speakers, participated in discussions and contributed their manuscripts to this volume.

M. S. da Costa
J. C. Duarte
R. A. D. Williams

CONTENTS

Preface v

European Laboratory Without Walls on Extremophiles 1
    A. Aguilar

Physiology of Clostridial Homoacetogens: Autotrophy Energy Metabolism and Potential for Industrial Production of Acetate . . 6
    L. G. Ljungdahl, J. Hugenholtz, A. Das and J. Wiegel

Acidophilic, Mineral-Oxidizing Bacteria: The Utilization of Carbon Dioxide with Particular Reference to Autotrophy in *Sulfolobus* . . 24
    P. Norris, A. Nixon and A. Hart

Isolation and Characterisation of Plasmids Present in *Thermus* Strains from Yellowstone National Park 44
    N. D. H. Raven and R. A. D. Williams

Taxonomic and Genetic Studies of *Bacillus thermophilus* 62
    R. Sharp, M. Munster, A. Vivian, S. Ahmad and T. Atkinson

Biochemical Taxonomy of the Genus *Thermus* 82
    R. A. D. Williams

Inducible 'SOS' Repair in the Extreme Thermophile *Bacillus caldotenax* 98
    P. Riley, T. Atkinson, A. Vivian and R. Sharp

Development of Host–Vector Systems in the Extreme Thermophile *Thermus thermophilus* 103
    Y. Koyama, S. Okamoto and K. Furukawa

Biocatalysis Acting under Extreme Physico-Chemical Conditions . . 106
B. Heinritz, M. Thiem, C. Gwenner, J. Sawistowsky, R. Hedlich and M. Ringpfeil

Thermophilic Organisms in Submarine Freshwater Hot Springs in Iceland 109
S. Hjörleifsdottir, J. K. Kristjansson and G. A. Alfredsson

Catechol Production by *Bacillus stearothermophilus* 113
P. Oriel and G. Gurujeyalakshmi

Cell Envelopes of Archaebacteria 117
H. König

Lipids of Archaebacteria: Structural and Biosynthetic Aspects. . . 131
M. De Rosa, V. Lanzotti, B. Nicolaus, A. Trincone and A. Gambacorta

DNA Polymerases and Topoisomerases in Archaebacteria: Potential Applications 152
P. Forterre

Deep-Sea Hydrothermal Vents of the 13°N (East Pacific Rise): Preliminary Bacterial Survey and Biotechnological Potential . . . 159
D. Prieur

Diversity of Archaebacteria in Biotechnologically Important Environments Revealed by Antibody Probes 163
E. Conway de Macario and A. J. L. Macario

Lipid Structures in *Thermotoga maritima* 167
M. De Rosa, A. Gambacorta, R. Huber, V. Lanzotti, B. Nicolaus, K. O. Stetter and A. Trincone

Behaviour of the Thermostable *Sulfolobus solfataricus* Malic Enzyme in Water-Miscible Organic Solvents: Biotechnological Prospects . . . 174
A. Guagliardi, M. Rossi and S. Bartolucci

Mechanism of Cyclopentane Ring Formation in Tetraether Lipids of *Sulfolobus solfataricus* 180
A. Trincone, A. Gambacorta and M. De Rosa

The Microbiology of Methane Production in Landfills 187
    D. B. Archer and M. W. Peck

Methanogenesis and Reductive Dechlorinations in an Alkaline, Hypersaline Sediments and Groundwater 205
    D. R. Boone, R. L. Johnson, D. C. Chen, I. M. Mathrani and R. A. Mah

Properties of (NiFe) and (NiFeSe) Hydrogenases from Methanogenic Bacteria 216
    G. Fauque

Methanogenesis in Artificially Created Extreme Environments. . . 237
    A. C. Wilkie and P. H. Smith

Approaches to the Development of Gene Transfer Systems in *Methanobacterium thermoautotrophicum* 253
    L. Meile, T. Rechsteiner, U. Jenal, M. Jordan and T. Leisinger

In vivo NMR Studies of the Metabolism and Bioenergetics of *Methanosarcina barkeri* 258
    H. Santos, P. Fareleira, R. Toci, Y. Berlier, J. LeGall, H. D. Peck Jr and A. V. Xavier

Taxonomy of Halophilic Bacteria 262
    A. Ventosa

Halophilic Bacteria: Their Life In and Out of Salt. 280
    D. J. Kushner

Ecological Distribution and Biotechnological Potential of Halophilic Microorganisms 289
    A. Ramos-Cormenzana

Polyol Accumulation in Yeasts in Response to Water Stress . . . 310
    M. S. da Costa and M. F. Nobre

Denitrification of Concentrated Sodium Nitrate Solutions by the Moderate Halophilic Denitrifier *Bacillus halodenitrificans* 328
    G. Denariaz, W. J. Payne and J. LeGall

Alkaliphiles 346
    W. D. Grant and K. Horikoshi

Modification of Immune Response by Extreme Halophilic Bacteria . 367
    C. Ruiz, M. Monteoliva-Sanchez and A. Ramos-Cormenzana

Isolation and Characterization of Novel Anaerobic, Halophilic Eubacteria from Hypersaline Environments of Western America and Kenya. 371
    H. Shiba and K. Horikoshi

The Potential Use of Halophilic Eubacteria for the Production of Organic Chemicals and Enzyme Protective Agents. 375
    E. A. Galinski

Cyclodextrin Production by Extremophilic Bacilli 380
    E. G. Afrikian

POSTERS

Alkaline Phosphatase from *Thermus ruber* 384
    D. Cossar and R. J. Sharp

Loss of Pigmentation in *Thermus* sp. 385
    D. Cossar and R. J. Sharp

Characterisation of an Excision Repair Deficient Mutant of *Bacillus caldotenax* 386
    P. Riley, T. Atkinson, A. Vivian and R. J. Sharp

Preliminary Taxonomic Studies on 1,000 Isolates of Thermophilic Bacilli 387
    D. White, F. G. Priest and R. J. Sharp

Amino Acid Utilization by Strains of the Genus *Thermus* 388
    G. Holtom, D. Cossar, R. Sharp and R. A. D. Williams

A New Sequence-Specific Endonuclease from the Genus *Thermus* . 389
    N. D. H. Raven, R. Mullings, P. Eastlake and R. A. D. Williams

Isolation and Characterisation of Thermophilic Aerobic Bacteria, Producing Thermostable Enzymes 390
    M. Kambourova and E. Emanuilova

Metabolism of Carbamoylphosphate in Extreme Thermophiles . . 391
    M. Van De Casteele, C. Legrain, N. Glansdorff and A. Peirard

Cellulolytic Anaerobic Bacteria from Hot Springs 392
    B. E. Jones

Thermophilic Methylotrophic Bacilli: Possible Biotechnological Potential 393
    A. G. Brooke, M. M. Attwood and D. W. Tempest

Methanol Oxidising System of Thermophilic *Bacillus* 394
    A. Netrusov, M. Guettler and R. S. Hanson

Mn^{2+} Recovery from a Low Grade Ore by Microorganisms . . . 395
    B. Paponetti, L. Toro, C. Abbruzzese, A. Marabini and M. Y. Duarte

Acidophilic Bacteria 396
    M. E. Simas Marques, M. L. Quinta and C. M. Rangel

A New Solid Medium for the Isolation and Phenotypic Characterisation of *Thiobacillus ferrooxidans* 397
    E. Bianchi, P. Valenti, P. Visca and N. Orsi

Purification, Properties and Cloning in *E. coli* of the Structural Gene of the S-Layer Component from *Thermus thermophilus* 398
    J. Berenguer, M. L. Faraldo, J. R. Caston and M. A. de Pedro

Malate Dehydrogenase from *Chloroflexus aurantiacus* 399
    A. K. Rolstad, E. Howland, B. Synstad and R. Sirevåg

Stability of Thermophilic Plasmid DNA in Heterologous Hosts . . 400
    D. J. Hardman and M. F. Tuite

Thermotropic Behaviour of the Polar Lipids of *B. stearothermophilus* 401
    A. S. Jurado, M. S. Costa and V. M. C. Madeira

Bacillus schlegelii and *Hydrogenobacter* spp.: Compared Ecology and Taxonomy of Two Thermophilic, Hydrogen and/or Sulfur Oxidising Aerobes from Geothermal Areas. 403
    M. Aragno and F. Bonjour

Cold Adaptation of Biopolymer-Degrading Bacteria from Permanently
Cold Environments. Is There a Biotechnological Potential?. . . .   405
   W. Reichardt

Microbial Protein Thermostability Correlates Directly with Resistance
to Denaturation in Biphasic Organic:Aqueous Solvents   406
   R. K. Owusu and D. A. Cowan

Partial Purification of Intracellular Enzymes by Heat-Precipitation of
Cell Debris: A Model Study   407
   Y. Takesawa, D. Cowan, M. Hoare and J. Bonnerjea

Investigation of the Enzymes Related to the Energy Metabolism of
Sulfolobus acidocaldarius, a Thermoacidophilic Archaebacterium . .   408
   T. Wakagi, T. Yamauchi, H. Wakao, H. Eguchi and T. Oshima

Purification and Characterisation of the C0 Dehydrogenase from
Methanothrix soehngenii.   409
   M. S. M. Jetten, A. J. M. Stams and A. J. B. Zehnder

Growth of a Thermophilic Butyrate-Degrading Bacterium in Axenic
Culture with Butyrate as Sole Carbon and Energy Source   410
   B. K. Ahring, P. Westermann and R. A. Mah

Properties of Enzymes and Electron Carriers Isolated from the Thermophilic Sulfate-Reducing Bacterium *Desulfovibrio thermophilus* . . .   411
   G. Fauque, M. Czechowski, A. R. Lino, Y. Berlier, D. V. Dervartanian, I. Moura, P. A. Lespinat, L. Kang, J. Lampreia, A. V. Xavier, H. D. Peck Jr, J. J. G. Moura and J. LeGall

Biosynthesis of Acetate from Methanol by *Methanosarcina barkeri* as
Monitored by *in vivo* ^{13}C NMR   412
   H. Santos, P. Fareleira, R. Toci, J. LeGall and A. V. Xavier

Redox Properties of Two B_{12} Corrinoid Proteins Isolated from
Methanosarcina DSM 800 and DSM 2905 in the Methylated and Aquo
Forms.   414
   A. R. Lino, J. J. G. Moura, A. V. Xavier, J. LeGall and I. Moura

Isolation and Characterisation of Two Novel Methanogens, a New
Haloalkalophilic Methanogen and a New Alkalophilic *Methanosarcina*   415
   N. Nakatsugawa and K. Horikoshi

Isolation and Identification of Bacteria Living in Environments Severely Contaminated with Heavy Metals 416
    L. Diels, L. Hooyberghs, A. Ryngaert and M. Mergeay

Effect of External Salinity Changes on Aminoacids and Ions Composition of *Deleya halophila*. 417
    A. del Moral, M. J. Valderrama, M. R. Ferrer, E. Quesada, F. Peran and A. Ramos-Cormenzana

Does the High Mg^{2+} Content Inhibit the $CaCO_3$ Precipitation by *Deleya halophila*? 418
    A. Rivadeneyra, R. Delgado, E. Quesada and A. Ramos-Cormenzana

Haloadaptation and Membrane Lipid Composition in a Halophilic *Vibrio* sp. HX 419
    R. L. Adams and N. J. Russell

The Gene of a Putative DNA-Binding Protein from *Halobacterium halobium* 420
    G. Baldacci

Screening for Compatible Solutes of Halophilic Eubacteria. . . . 421
    A. Wohlfarth, J. Severin and E. A. Galinski

Enhancement of Proline Production of *Bacillus subtilis* During Salt Stress 422
    E. Müller and H. G. Truper

Antibiotic-Resistant Moderately Halophilic Gram-Negative Motile Rods from Hypersaline Waters 423
    J. Quevedo-Sarmiento, A. del Moral, M. R. Ferrer and A. Ramos-Cormenzana

Polyol Accumulation by Yeasts and a Yeast-like Fungus under Osmotic Stress 424
    M. F. Nobre and M. S. da Costa

Isolation and Characterisation of Bacteriophages Active on Moderately Halophilic Microorganisms 425
    A. M. Garcia de la Paz, A. Perez Martinez, C. Calvo Sainz and A. Ramos-Cormenzana

Identification of the Halocin H4 Gene in *Haloferax mediterranei* . . 426
 B. Gambin, G. Juez, F. Rodriguez-Valera, M. Betlach and H. W. Boyer

Halococcus saccharolyticus sp. nov., a New Group of Extremely Halophilic Archaebacterial Cocci 427
 A. Ventosa, C. G. Montero, F. Rodriguez-Valera, M. Kates, N. Moldoveanu and F. Ruiz-Berraquero

Restriction Enzyme Screening on Bacterial Species of Azores and Madeira Mineral Waters 428
 J. M. B. Vitor and R. V. Correia de Silva

Growth Characteristics and Salt Requirement of Two New Groups of Moderately Halophilic Microorganisms 429
 M. J. Valderrama, V. Bejar, E. Quesada and A. Ramos-Cormenzana

EUROPEAN LABORATORY WITHOUT WALLS ON EXTREMOPHILES

ALFREDO AGUILAR
Division Biotechnology, Directorate Biology,
Directorate-General for Science, Research and Development,
Commission of the European Communities,
200 Rue de la Loi, B-1049, Brussels

The biotechnological interest of extremophile microorganisms is well perceived by scientists and by the most active biotechnological industries. During the last few years, the research efforts on extremophiles have multiplied, thanks to the additional resources allocated by national and international organizations. The pioneer Community Programme on Biotechnology known as "BEP" (Biomolecular Engineering Programme: 1982-1986), reflected the importance of extremophiles and in particular supported the work of Professor M. Rossi in Naples on the potential applications of the thermophilic bacterium Sulfolobus sulfataricus to biotechnology.

The ongoing Biotechnology Action Programme, BAP, (1986-1989) of the Commission of the European Communities, is the successor of BEP. This programme is inherently pre-competitive and is orientated towards medium and long-term objectives essential for the strategic strength of Europe. The two major aims are the establishment of a supportive infrastructure for biotechnology research in Europe, and the elimination through research and training, of bottlenecks which prevent the exploitation by industry and agriculture of the materials and methods originated from molecular biology.

The research part of the programme is divided into two sections, Contextual Measures and Basic Biotechnology. Two of the major areas covered by Contextual Measures are bioinformatics, considered as the interface between biotechnology and information technology, and collections of microorganisms and animal cell lines. It must be noted, in this context, that the DSM, the German Collection of microorganisms, which participates in BAP within the framework of MINE (Microbial Information Network in Europe), is the only service collection in Europe holding a collection of extremophile microorganisms.

The other major research effort in BAP is dedicated to Basic Biotechnology. This sub-programme covers essential areas such as enzyme engineering (stability of enzymes, protein design, etc.), bioreactors, the application of genetic engineering to plants, industrial microorganisms, and to animal husbandry, as well as the assessment of risks possibly associated with modern biotechnology.

It is in the frame of this sub-programme for basic biotechnology that Community research on extremophiles is conducted by a number of laboratories working in transnational associations.

Professors W. Harder, (Groningen, NL), and D.W. Tempest (Sheffield, UK) collaborate on the environmental control of metabolic fluxes and its biotechnological importance upon the thermophile <u>Bacillus stearothermophilus</u>, and the recently isolated thermotolerant, methanol-utilizing, <u>Bacillus</u> species.

The groups of Professors A. Fontana, (Padova, I), and R. Jaenicke, (Regensburg, FRG), aim to characterize the functional, structural and folding properties of thermophilic enzymes (proteases, dehydrogenases, galactosidases), with emphasis upon those isolated from extreme thermophiles and archaebacteria.

Finally, Professor M. Rossi, (Napoli,I) is studying, within a transnational group of laboratories from France, Greece and the FRG, the biotechnological properties of a number of enzymes (alcohol dehydrogenase, malic enzyme, β-galactosidase DNA polymerase, etc.), from <u>Sulfolobus solfataricus</u>.

A detailed description of the state of the art of these research groups can either be found elsewhere in this book as well as in the 1987 and 1988 BAP Progress Reports (1,2).

As BAP progressed, it became apparent to the scientific project managers of BAP that an initiative was needed to strengthen the co-ordination of efforts in the Community. Thus, a number of meetings between BAP contractors and other European scientists not directly participating in BAP, including representatives from industries, have taken place, which led to the creation of a "European Laboratory Without Walls" (ELWW), on Extremophiles (for more information on the structure, nature and objectives of ELWW, see Ref. 3). The scientists participating in the ELWW agreed to focus their attention on those aspects of extremophile research which are considered as being of immediate importance:

a) physiology and metabolic regulation
b) cell walls, membranes and proteins
c) properties as whole cell biocatalysts, and
d) molecular biology and genetics

There is a consensus among the members of the ELWW on Extremophiles that the current level of knowledge on the field is still rather superficial, and that increased efforts into the exploitation of extreme environments for novel organisms, combined with a co-ordinated approach in the areas mentioned above, will contribute significantly to biotechnological innovation.

In many areas of biotechnology, and extremophiles constitute one of them, there is clearly a need for the Community to promote exploratory research ventures, particularly when basic knowledge is insufficient, and the scope of applying molecular methods still limited. The primary requirement in this prevalent situation is to increase the multi-disciplinarity of research by fostering temporary combinaitons of backgrounds and skills. The particular nature of this research calls for a reasonable dosage of efforts and of supporting funds, and its importance necessitates an active partnership with industries. Such an involvement of industries may take place in a diversity of ways; direct participation to the programme, co-financing of the activities implemented by one or several laboratories within the ELWW, supply of materials, access to infrastructure, expression of interest for the exploitation of research results, etc.

All these requirements can best be satisfied through the creation (or consolidation) of European Laboratories Without Walls, such as they have been successfully promoted and established in the framework of BAP. The task is to be performed in the framework of the next Community Programme for Biotechnology, BRIDGE (Biotechnology Research for Innovation, Development and Growth in Europe), which is planned for 1990-1994 with a total budget of 100 Mio ECU. It is in the framework of BRIDGE that Community R&D proposals on Extremophiles will be eligible for contract negotiations. Additional information on the objectives, orientations and implementation modalities of BRIDGE may be obtained from the author of the present note.

REFERENCES

1. Biotechnology Action Programme. Progress Report, 1987 (2 volumes). Ed.: E. Magnien. Commission of the European Communities, EUR 11138 EN, Luxembourg, 1987

2. Biotechnology Action Programme. Progress Report, 1988 (4 volumes). Ed.: E. Magnien. Commission of the European Communities, EUR 11650 EN, Luxembourg, 1988

3. van der Meer, R., E. Magnien and D.de nettancourt, 1987. European Laboratories Without Walls: focused pre-competitive research. Trends in Biotechnology, 4: 318-321

PHYSIOLOGY OF CLOSTRIDIAL HOMOACETOGENS: AUTOTROPHY ENERGY METABOLISM AND POTENTIAL FOR INDUSTRIAL PRODUCTION OF ACETATE

LARS G. LJUNGDAHL, JEROEN HUGENHOLTZ, AMARESH DAS and JUERGEN WIEGEL
Center for Biological Resource Recovery and
Departments of Biochemistry and Microbiology
University of Georgia, Athens, GA. 30602, U.S.A.

ABSTRACT

Acetate is produced by the homoacetogenic clostridia in fermentations of glucose, fructose and xylose, when they grow on methanol, formate or other 1-carbon sources, and when they grow autotrophically on CO_2/H_2 or CO using a newly discovered pathway yielding acetyl-CoA as the first 2-carbon product. The primary reactions of autotrophic CO_2-fixation are catalyzed by a tungsten-selenium-iron containing formate dehydrogenase and a nickel-zinc-iron carbon monoxide dehydrogenase. The latter enzyme catalyzes also the final step in the synthesis of acetyl-CoA. Generation of ATP during autotrophic growth is by electron transport phosphorylation coupled to the synthesis of acetate from CO_2. In $\underline{C}$. thermoaceticum and $\underline{C}$. thermoautotrophicum this involves the membrane bound enzymes methylene-H_4folate reductase, carbon monoxide dehydrogenase and hydrogenase of the autotrophic pathway and ATPase, b-type cytochromes and menaquinone. $\underline{C}$. thermoaceticum and $\underline{C}$. thermoautotrophicum may be used for the production of calcium-magnesium acetate (CMA) from hydrolyzed corn starch by fermentation and dolime. The theoretical yield of CMA is 90% and the cost U.S. $0.48/kg.

INTRODUCTION

In this presentation we will discuss results obtained in studies of the anaerobic thermophilic homoacetogenic bacteria <u>Clostridium thermoaceticum</u> discovered by Fontaine et al. [1] and <u>Clostridium thermoautotrophicum</u> isolated by Wiegel et al

[2]. The homoacetogenic bacteria are distinctive in that they convert one mol of glucose almost quantitatively to three mol of acetate. Of these two mol are formed by fermentation of the glucose. The third mol of the acetate is formed by fixation of CO_2 [1,3,4]. This is shown in reactions 1 to 3.

$$C_6H_{12}O_6 + 2H_2O \longrightarrow 2CH_3COOH + 2CO_2 + 8H^+ + 8e \qquad (1)$$
$$2CO_2 + 8H^+ + 8e \longrightarrow 3CH_3COOH + 2H_2O \qquad (2)$$

$$\text{Net reaction: } C_6H_{12}O_6 \longrightarrow 3CH_3COOH \qquad (3)$$

Reaction 2 constitutes a synthesis of acetate from CO_2. It has now been established that this synthesis occurs via a pathway of autotrophic CO_2 fixation which leads to the formation of acetyl-CoA as the first 2-carbon product [5,6,7]. We will discuss features of this newly discovered acetyl-CoA pathway.

Many homoacetogenic clostridia grow autotrophically on a gas mixture of H_2/CO_2 or carbon monoxide as the only sources of energy and carbon [2,8]. It follows that the reduction of CO_2 with synthesis of acetate occurs in conjunction with a generation of energy. This possibility was suggested some years ago for *C. thermoaceticum* on the basis of growth yields [9] and the presence of cytochromes and menaquinone in the membranes [10]. Recent studies have established that these membranes contain ATPase, several oxidoreductases and other electron carrying components. A possible electron transport and energy generating systems will be presented.

The third part of this discussion deals with the possible use of homoacetogenic clostridia for the industrial production of acetate from hydrolyzed corn starch. The homoacetogens as shown in reaction 3 above convert glucose essentially quantitatively to acetate. They also ferment xylose with acetate as the only product. The high yield and the fact that acetate is the only product and can be recovered without a separation from other products make the acetogenic fermentation a very attractive industrial process.

THE ACETYL-CoA AUTOTROPHIC PATHWAY

Reactions 1 and 2 given above show that electrons generated by fermentation of glucose are used for the reduction of CO_2 and the synthesis of acetate. Other sugars including fructose and xylose are similarly quantitatively converted to acetate. The discovery that the homoacetogenic clostridia grow autotrophically on H_2/CO_2 (Reaction 4) and CO (Reaction 5) [2,8] demonstrates that electrons for the acetate synthesis can be generated also from hydrogen, CO and other sources. These

$$2CO_2 + 4H_2 \longrightarrow CH_3COOH + 2H_2O \qquad (4)$$

$$4CO + 2H_2O \longrightarrow CH_3COOH + 2CO_2 \qquad (5)$$

sources include formate [11] methanol [12, 13] and methyl groups of methyl phenolethers [14, 15]. The fermentation of methanol may be summarized as follows (Reactions 6-9) [16].

$$CH_3OH + H_2O \longrightarrow CO_2 + 6H^+ + 6e \qquad (6)$$
$$3CO_2 + 6H^+ + 6e \longrightarrow 3CO + 3H_2O \qquad (7)$$
$$3CH_3OH + 3CO \longrightarrow 3CH_3COOH \qquad (8)$$

Net reaction: $4CH_3OH + 2CO_2 \longrightarrow 3CH_3COOH + 2H_2O \qquad (9)$

The present concept of the metabolism of glucose and of the autotrophic pathway via which the homoacetogens utilize CO_2, H_2, CO, formate and methanol for the synthesis of acetyl-CoA is shown in Fig. 1. The pathway is outlined with heavy arrows. It is detailed in recent reviews [5,6,7]. Carbon dioxide destined to form the methyl group of acetyl-CoA is reduced to formate, that in turn reacts with tetrahydrofolate (H_4F) to form formyl-H_4F. The latter is in a sequence of reactions reduced to methyl-H_4F, the methyl group of which is transferred

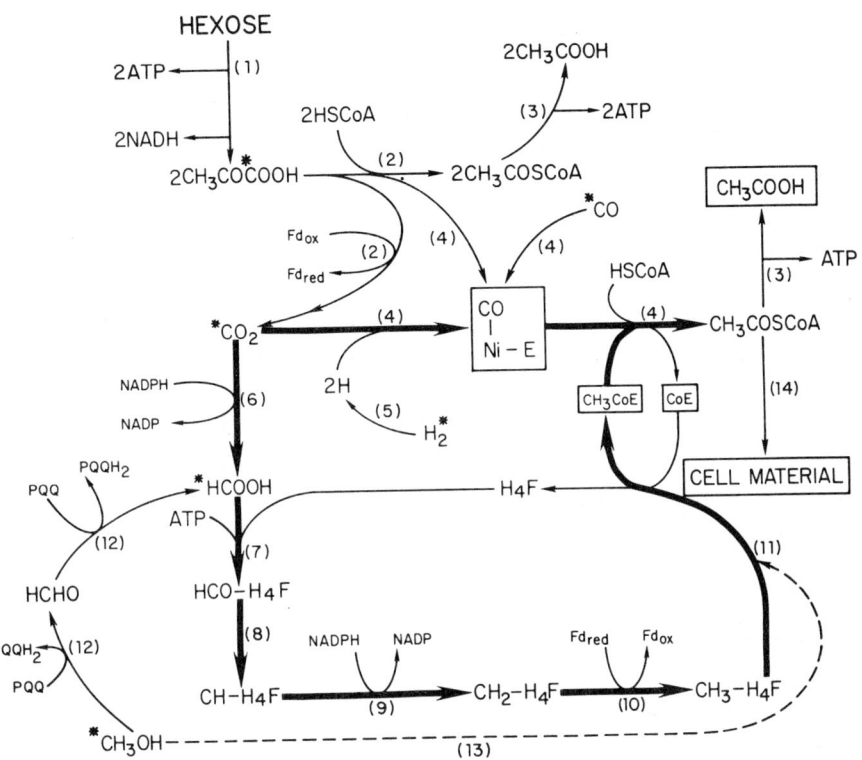

FIGURE 1. Present concept of the autotrophic acetyl-CoA pathway (heavy arrows) in homoacetogenic clostridia and the metabolism of hexoses, methanol and CO. H_4F = tetrahydrofolate, CoE = corrinoid enzyme, CO-Ni-E = carbon monoxide dehydrogenase with CO moiety bound to nickel, PQQ = pyrrolo quinoline quinone, Fd = ferredoxin. Enzymes or reaction sequences are as follows: 1 glycolysis; 2, pyruvate-ferredoxin oxidoreductase; 3, phosphotransacetylase and acetate kinase; 4 carbon monoxide dehydrogenase; 5, hydrogenase; 6, formate dehydrogenase, 7, formyl-H_4folate synthetase; 8, methenyl-H_4folate cyclohydrolase; 9, methylene-H_4folate dehydrogenase; 10, methylene-H_4folate reductase; 11, transmethylase; 12, methanol dehydrogenase; 13, methanol-cobamide methyltransferase; 14, anabolism.

to the cobalt atom of a corrinoid attached to a protein (CoE).

Carbon dioxide to become the carbonyl group of acetyl-CoA is apparently first reduced to carbon monoxide in a reaction catalyzed by CO dehydrogenase. However, it is not necessary that free CO is formed, it may instead be bound to a nickel atom of the CO dehydrogenase (CO-Ni-E in Fig. 1). Carbon monoxide dehydrogenase finally completes the synthesis of acetyl-CoA by condensing the methyl group of CH_3-CoE, the nickel bound carbonyl moiety and CoA to form acetyl-CoA. Acetyl-CoA is then either converted to acetate or is used for the synthesis of cell material.

All enzymes of the acetyl-CoA pathway have been purified [for references see review 6]. Several of these enzymes contain metals. Formate dehydrogenase has 2 each of 2 different subunits and holds 2 tungsten, 2 selenium and about 36 iron per molecule [17]. The tungsten is attached to a pterine forming a tungsten-cofactor similar to the molybdenum-pterin cofactor of molybdoenzymes [18]. Selenium is present as selenocysteine in the larger of the subunits. Iron is in the form of ferredoxin-like non-heme-iron-sulfur centers, but they have not been closely examined. The finding of tungsten in the formate dehydrogenase of C. thermoaceticum was the first demonstration of a biological role for this metal [19]. The enzyme is very sensitive to oxygen. Formate dehydrogenase from different sources differ in composition and in type of electron acceptor [20]. The enzyme in C. thermoaceticum and C. thermoautotrophicum is the only known NADP-dependent. NAD, FAD, cytochromes, ferredoxin or rubredoxin do not function.

Carbon monoxide dehydrogenase first discovered in homoacetogenic clostridia by Diekert and Thauer [21] and shown to catalyze the synthesis of acetyl-CoA by Drake et al [22] contain 6 nickel, 3 zinc and about 36 iron per molecule [23]. Recent reviews are available on the physiology and the chemistry of this enzyme [24,25]. It has already been pointed out that the CO dehydrogenase binds a carbonyl group, that is the direct precursor of the carbonyl group of acetyl-CoA. It is shown in Fig. 2 that there are three sources for this carbonyl group; CO, CO_2 and the carboxyl group of pyruvate [26,27,28].

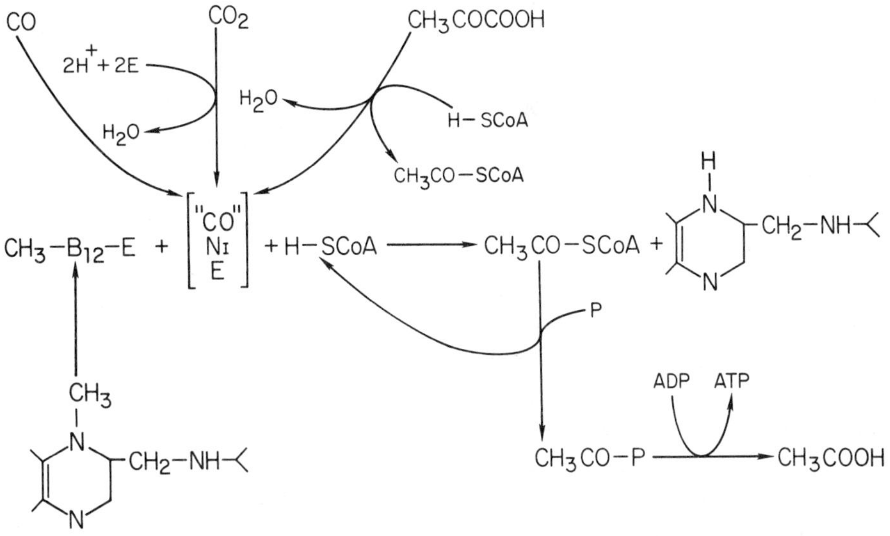

FIGURE 2. Reactions catalyzed by carbon monoxide dehydrogenase designated [NiE] and with a carbonyl moiety ["CO"NiE]. B_{12}-E is the corrinoid protein.

CO dehydrogenase in addition to catalyzing the reduction of CO_2 to CO and the synthesis of acetyl-CoA catalyzes also an exchange reaction between acetyl-CoA and CO as shown in reaction (10) [29]. This exchange requires that the methyl,

$$^{14}CO + CH_3{}^{12}CO\text{-SCoA} \longleftrightarrow CH_3{}^{14}CO\text{-SCoA} + {}^{12}CO \qquad (10)$$

carbonyl and CoA moieties separate, which in turn requires separate binding sites for each group on the enzyme. We have already mentioned that the carbonyl group binds to nickel [26], that may be associated with iron in an iron-nickel complex [27]. Shanmugasundaram et al. [30] found that tryptophan residues are involved in the CoA binding site. The methyl binding site has been identified as a SH-group of a cysteine [31].

Other metal-containing proteins of the autotrophic pathway are methylene-H$_4$folate reductase [32,33], the corrinoid protein [34] and non-heme iron sulfur proteins such as ferredoxins [35] and rubredoxins [36]. The metylene-H$_4$folate reductase contains one [Fe$_4$S$_4$]-cluster and FAD. The corrinoid protein is a dimer having two different subunits, 5-methoxybenzimidazolylcobamide and one [Fe$_4$S$_4$] cluster. The iron cluster is most likely involved in the reduction of the cobamide, which cobalt atom must be reduced to a Co$^+$-state before a transfer of the methyl group of methyl-H$_4$folate occurs. Iron sulfur clusters are most likely also present in hydrogenase. This enzyme discovered in C. thermoaceticum by Drake [37], has not been purified and characterized. However, hydrogenases are known for their complex requirements of iron, nickel and also selenium [38,39].

Of special interest is the methanol dehydrogenase, that oxidizes methanol to formate apparently with formaldehyde as free intermediate (Fig. 1). This enzyme was recently purified from C. thermoautotrophicum [16]. It contains per mol two mol each of pyrroloquinoline quinone [PQQ] and zinc. PQQ-dependent methanol dehydrogenases from methylotrophs have been studied extensively [40] whereas the purification of the enzyme from an anaerobe has previously not been reported. The oxidation of methanol to formaldehyde and formate allows the entry of methanol-carbon in the autotrophic pathway. Methanol may also be a more direct precursor of the methyl group of acetyl-CoA and enter the pathway on the level of the corrinoid enzyme (Fig. 1).

ENERGY GENERATION IN ACETOGENS

When homoacetogenic bacteria ferment sugars energy is generated by substrate level phosphorylation, and as is shown in Fig. 1 four mol of ATP is formed per mol of fermented glucose. However, when the acetogens grow autotrophically a net gain of ATP is not possible. This suggest that ATP is generated by a chemiosmotic mechanism that is coupled to the reduction of CO_2 to form acetyl-CoA. Studies in our laboratory have led to a mechanism of electron transport and energy generation as outlined in Fig. 3.

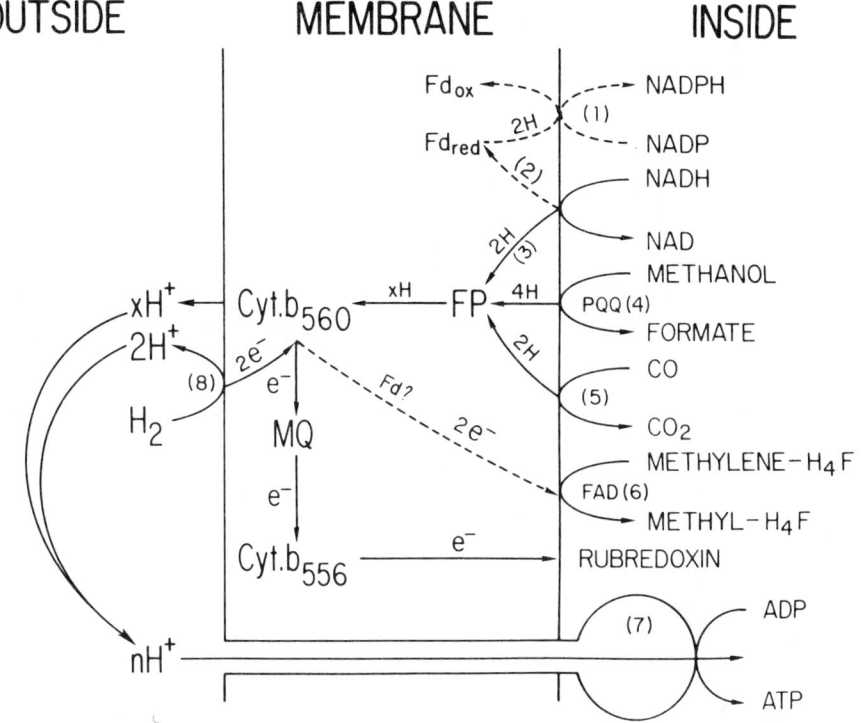

FIGURE 3. Postulated mechanism of electron transport and energy generation in the homoacetogen C. thermoautotrophicum. Enzymes and abbreviations are 1, NADPH-ferredoxin oxidoreductase; 2, NADH-ferredoxin oxidoreductase; 3, NADH dehydrogenase; 4, methanol dehydrogenase; 5, carbon monoxide dehydrogenase; 6, methylene-H_4folate reductase; 7, H^+-ATPase; 8, hydrogenase; MQ, menaquinone; Fd, ferredoxin; FP, flavoprotein; PQQ, pyrroloquinoline quinone.

C. thermoaceticum and C. thermoautotrophicum contain an ATPase [41,42,43], which is rather similar to F_1F_0 ATPases of several other bacteria [44]. The F_1 part of the homoacetogenic ATPase is, however, composed of only 4 different subunits with the stoichiometry $\alpha_3\beta_3\gamma\delta$ and is lacking the subunit found in F_1 ATPase of E. coli. The F_0 part consists of 2 different subunits and lacks the c subunit of other F_1F_o ATPases. The C. thermoautotrophicum F_1F_o ATPase has been reconstituted into proteoliposomes [43].

Several membrane associated electron carriers have been identified and isolated from the two homoacetogenic clostridia. Ivey [43] isolated cytochrome b_{560} and cytochrome b_{556} from C. thermoautotrophicum and determined their midpoint redox potentials to be -200 mV and -48 mV, respectively. Almost identical cytochromes were isolated from C. thermoaceticum. He found also a flavoprotein (FP) with E_m -220 mV that catalyzes the oxidation of NADH and proposed the reaction sequence outlined in reaction 11, in which MQ is menaquinone. The MQ has been identified as 2-methyl-3-heptaprenyl-1,4-naphthoquinone (Mk_7) and in spectrophotometric studies shown

$$NADH \longrightarrow FP \longrightarrow Cyt\ b_{560} \longrightarrow MQ \longrightarrow Cyt\ b556 \qquad (11)$$

to act between the two cytochromes (Unpublished data).

Hydrogenase has been found in the membrane fraction of C. thermoautotrophicum [45]. It catalyzes the reduction of the cytochromes with molecular hydrogen. Of special significance are the findings that carbon monoxide dehydrogenase and methylene-H_4folate reductase are membrane bound [46]. These enzymes have been postulated to be involved in energy generation in acetogenic bacteria [7,47]. Membranes with high CO dehydrogenase when exposed to CO generate a proton motive force (Δp), when the artificial electron acceptor $K_3[Fe(CN_6]$ is present. The Δp drives ATP synthesis [41], and also uptake of amino acids (alanine) [45]. Diekert et al. [48] using whole cell suspensions of the acetogen Acetobacterium woodii have also demonstrated energy generation by the oxidation of CO.

When growing on methanol and CO_2 the clostridia must rely on electrons generated from methanol for energy generation. Evidence has been obtained that the PQQ-dependent methanol dehydrogenase is membrane associated [16]. However, whether electrons from PQQ are transferred to the flavoprotein as indicated in Fig. 3 or not has not been established.

INDUSTRIAL PRODUCTION OF ACETATE USING HOMOACETOGENIC BACTERIA

There is an interest to use clostridia for the production of organic acids by fermentation of resources such as starch, lignocellulose or other biomass materials [49,50,51,52]. A lot of this interest is focused upon the production of acetic acid [53,54,55]. Firstly, because acetic acid is used in a large amount (1.3 x 10^9 kg in the U.S.A. per year, U.S. Int. Trade Com., 1986). Secondly, the U.S. Federal Highway Administration has found calcium magnesium acetate (CMA) to be a good and apparently ecologically safe deicer [56]. If CMA is going to be used as a deicer, the U.S. consumption of acetate probably will double [57]. Thirdly, the homoacetate fermentation yields an almost complete conversion of glucose (Reaction 3) and xylose to acetate as the only product. This means, that a fermentation of 100 g of sugar produces 100 g of acetate. In practice the yield is from 90 to 95% of the theoretical. The fact that acetate is the only product simplifies the recovery since it will not involve a separation step from other products.

We have evaluated acetate production by C. thermoaceticum, and C. thermoautotrophicum using hydrolyzed corn starch as substrate. Culture conditions included batch fermentations, continuous culturing with and without cell recycling, and the use of a rotary fermentor [58]. The anaerobic technique used and the set up for continuous fermentations have been described [59] and details of the rotary fermentor are given in a report to the U.S. Federal Highway Administration [58].

Table 1 shows examples of rates of acetate production per liter fermentor volume and hour and the concentration of acetate in the culture fluid after the fermentation.

TABLE 1
Production of acetate from hydrolyzed starch using
C. thermoaceticum*

Culture Method	Rate $g \cdot l^{-1} \cdot h^{-1}$	Concentration $g \cdot l^{-1}$
Batch	0.9	120
Continuous	2.5	7.3
Cont. with cell recycling	4	22
Rotary fermentor: a	10	3.7
: b	3.9	10

*Two results with the rotary fermentor are given: a) at pH 6.6 with a retention time of 0.32 h and b) with the culture effluent at pH 4.8 and a retention time of 3.9 h. Other fermentations were at pH 6.6. The data are from [58].

The batch fermentation was in reality a semicontinuous process using a CMA-tolerant strain. The fermentation medium contained at start about 65 g of CMA per liter. Hydrolyzed corn starch was added intermittently to keep the concentration of glucose between 20 to 100 mM. The pH was controlled by addition of dolime. When an acetate concentration of 120 g per liter was reached about half of the culture was harvested and an equal volume of fresh medium without CMA was added to the fermentor. The rotary fermentor contains disks to which the bacteria attach. The disks are slowly rotated. The medium is introduced at the bottom of the fermenter at about pH 7 using a medium of low buffering capacity. The glucose concentration was adjusted so that it was almost completely fermented before the culture fluid was collected at the top of the fermentor.

The results demonstrate that the highest concentration is achieved with the batch culture. It is unfortunately a relatively slow process. The fastest fermentation was obtained with the rotary fermentor, but the problem here is that the concentration of acetate reached in the culture fluid is low.

Our results are comparable to those obtained by Reed and Bogdan [53] and Wang and Wang [54]. Reed and Bogdan used cell recycling with bacterial cell adsorption on corn cob granules and activated carbon. They obtained an acetate production of 14.3 $g \cdot l^{-1} \cdot h^{-1}$ and a concentration of 7.1 $g \cdot l^{-1}$. Wang and Wang used a batch culture procedure with glucose addition during the fermentation. They reached a production rate of 0.8 $g \cdot l^{-1} \cdot h^{-1}$ and concentration of 45 $g \cdot l^{-1}$.

Economical evaluations of the acetate fermentations using the results of Table 1 indicated that CMA can be produced for U.S. $0.48 per kg [58] in an industrial plant with a capacity of 1000 ton per day. Recent cost analysis by Wise and Augenstein [57] using woody biomass residues as starting material and a plant with the capacity of 454 ton per day projected a cost for calcium acetate of U.S. $0.258/kg. Although much research remains before an industrial process for acetate fermentation can be a reality, we feel that results so far obtained are very encouraging.

REFERENCES

1. Fontaine, F. E. Peterson, W. H. McCoy, E. Johnson, M. J. and Ritter, G. J., A new type of glucose fermentation by Clostridium thermoaceticum n. sp. J. Bacteriol., 1942, 43, 701-15.

2. Wiegel, J. Braun, M. and Gottschalk, G., Clostridium thermoautotrophicum species novum, a thermophile producing acetate from molecular hydrogen and carbon dioxide. Curr. Microbiol. 1981, 5, 255-60.

3. Barker, H. A. and Kamen, M. D., Carbon dioxide utilization in the synthesis of acetic acid by Clostridium thermoaceticum. Proc. Natl. Acad. Sci. USA, 1945, 31, 219-25.

4. Wood, H. G., A study of carbon dioxide fixation by mass determination of the types of C^{13}-acetate. J. Biol. Chem., 1952, 194, 905-31.

5. Wood, H. G. Ragsdale, S. W. Pezacka, E., The acetyl-CoA pathway: a newly discovered pathway of autotrophic growth. Trends Biochem. Sci., 1986, 11, 14-8.

6. Ljungdahl, L. G., The autotrophic pathway of acetate synthesis in acetogenic bacteria. Ann. Rev. Microbiol. 1986, 40, 415-50.

7. Fuchs, G., CO_2-fixation in acetogenic bacteria: variations on a theme. FEMS Microbiol. Rev. 1986, 39, 181-213.

8. Kerby, R. and Zeikus, J. G., Growth of Clostridium thermoaceticum on H_2/CO_2 or CO as energy source. Curr. Microbiol. 1983, 8, 27-30.

9. Andreesen, J. R. Schaupp, A. Neurauter, C. Brown, A. and Ljungdahl, L. G., Fermentation of glucose, fructose, and xylose by Clostridium thermoaceticum: Effect of metals on growth yield, enzymes, and the synthesis of acetate from CO_2. J. Bacteriol. 1973, 114, 743-51.

10. Gottwald, M. Andreesen, J. R. LeGall, J. and Ljungdahl, L. G., Presence of cytochrome and menaquinone in Clostridium formicoaceticum and Clostridium thermoaceticum. J. Bacteriol., 1975, 122, 325-28.

11. Gottwald, M., Untersuchungen zum Pentose-stoffwechsel bei Homoacetatgärern. Diplomarbeit, Universität Göttingen, FRG. 1973.

12. Wiegel, J. and Garrison, R., Utilization of methanol by *Clostridium thermoaceticum*, Abstr. Ann. Meet. Am. Soc. Microbiol. 1985 I-115, p. 165.

13. Winters-Ivey, D. Ljungdahl, L. G. and Wiegel, J. Metabolism of methanol in *Clostridium thermoautotrophicum*. Abstr. Ann. Meet. Am. Soc. Microbiol. 1985, K-66, p. 182.

14. Bache, R. and Pfennig, N., Selective isolation of *Acetobacterium woodii* on methoxylated aromatic acids and determination of growth yields. Arch. Microbiol. 1981, 130, 255-61.

15. Daniel, S. L. and Drake, H. L., Acetogenesis from methoxylated aromatic acids by *Clostridium thermoaceticum*. Abstr. Ann. Meet. Am. Soc. Microbiol. 1988, I-105, p. 198.

16. Winters, D. K. and Ljungdahl, L. G., PQQ-dependent methanol dehydrogenase from *Clostridium thermoautotrophicum*. In 1st. Int. Symp. PQQ Quinoproteins, eds. J. A. Duine, and M. Ameyama, Delft, The Netherlands, 1988, In press.

17. Yamamoto, I. Saiki, T. Liu, S.-M. and Ljungdahl, L. G., Purification and properties of NADP-dependent formate dehydrogenase from *Clostridium thermoaceticum*, a tungsten-selenium-iron protein. J. Biol. Chem., 1983, 258, 1826-32.

18. Hageman, R. V. and Rajagopalan, K. V., Assay and detection of the molybdenum cofactor. Meth. Enzymol., 1986, 122, 399-412.

19. Andreesen, J. R. and Ljungdahl, L. G., Formate dehydrogenase of *Clostridium thermoaceticum*: Incorporation of selenium-75, and the effects of selenite, molybdate, and tungstate on the enzyme. J. Bacteriol., 1973, 116, 867-73.

20. Adams, M. W. W. and Mortenson, L. E., Mo reductases:Nitrate reductase and formate dehydrogenase. In Molybdenum Enzymes, ed. T. G. Spiro, John Wiley and Sons, Inc., N.Y., 1985, 519-93.

21. Diekert, G. B. and Thauer, R. K., Carbon monoxide oxidation by *Clostridium* thermoaceticum and *Clostridium* formicoaceticum, *J. Bacteriol.*, 1978, 136, 599-606.

22. Drake, H. L., Hu S.-I. and Wood, H. G., Purification of five components from *Clostridium* thermoaceticum which catalyze synthesis of acetate from pyruvate and methyltetrahydrofolate. Properties of phosphotransacetylase. *J. Biol. Chem.*, 1981, 256, 11137-44.

23. Ragsdale, S. W. Clark, J. E. Ljungdahl, L. G. Lundie, L. L. and Drake, H. L., Properties of purified carbon monoxide dehydrogenase from *Clostridium* thermoaceticum, a nickel, iron-sulfur protein. *J. Biol. Chem.*, 1983, 258, 2364-69.

24. Diekert, G., Carbon monoxide dehydrogenase of acetogens. In *The Bioinorganic Chemistry of Nickel*, ed., J. R. Lancaster, Jr., VCH Publishers Inc., N.Y., N.Y., 1988, 299-309.

25. Ragsdale, S. W. Wood, H. G. Morton, T. A. Ljungdahl, L. G. and DerVartanian, D. V., Nickel in CO dehydrogenase. In *The Bioinorganic Chemistry of Nickel*, ed. J. R. Lancaster, Jr., VCH Publishers Inc., N.Y., N.Y., 1988, 311-32.

26. Ragsdale, S. W. Ljungdahl, L. G. and DerVartanian, D. V., ^{13}C and ^{61}Ni isotope substitutions confirm the presence of a nickel(II)-carbon species in acetogenic CO dehydrogenase. *Biochem. Biophys. Res. Commun.*, 1983, 115, 658-65.

27. Ragsdale, S. W. Wood, H. G. and Antholine, W. E., Evidence that an iron-nickel-carbon complex is formed by reaction of CO with the CO dehydrogenase from *Clostridium* thermoaceticum. *Proc. Natl. Acad. Sci. USA*, 1985, 82, 6811-14.

28. Pezacka, E. and Wood, H. G., Role of carbon monoxide dehydrogenase in the autotrophic pathway used by acetogenic bacteria. *Proc. Natl. Acad. Sci. USA*, 1984, 81, 6261-65.

29. Ragsdale, S. W. and Wood, H. G., Acetate biosynthesis by acetogenic bacteria, Evidence that carbon monoxide dehydrogenase is the condensing enzyme that catalyzes the final steps in the synthesis, *J. Biol. Chem.* 1985, 260, 3970-77.

30. Shanmugasundaram, T. Kumar, G. H. and Wood, H. G., Involvement of tryptophan residues at the coenzyme A binding site of carbon monoxide dehydrogenase from Clostridium thermoaceticum, Biochemistry 1988, 27, 6499-503.

31. Pezacka, E. and Wood, H. G., Methyl-5-cysteine, the binding site for methyl in carbon monoxide dehydrogenase. J. Biol. Chem. 1988, In press.

32. Clark, J. E. and Ljungdahl, L. G., Purification and properties of 5,10-methylenetetrahydrofolate reductase, an iron-sulfur flavoprotein from Clostridium formicoaceticum. J. Biol. Chem. 1984, 259, 10845-49.

33. Han, E. Y., Purification and characterization of methylenetetrahydrofolate reductase from Clostridium thermoaceticum. Thesis, University of Georgia, Athens, GA., 1987.

34. Ragsdale, S. W. Lindahl, P. A. and Munck, E., Mossbauer, EPR, and optical studies of the corrinoid/iron sulfur protein involved in the synthesis of acetyl coenzyme A by Clostridium thermoaceticum. J. Biol. Chem. 1987, 262, 14289-97.

35. Elliott, J. I. and Ljungdahl, L. G., Isolation and characterization of an Fe_8S_8 ferredoxin (ferredoxin II) from Clostridium thermoaceticum. J. Bacteriol. 1982, 151, 328-33.

36. Yang, S.-S. Ljungdahl, L. G. DerVartanian, D. and Watt, G. D., Isolation and characterization of two rubredoxins from Clostridium thermoaceticum. Biochim. Biophys. Acta 1980, 590, 24-33.

37. Drake, H. L., Demonstration of hydrogenase in extracts of the homoacetate-fermenting bacterium Clostridium thermoaceticum. J. Bacteriol. 1982, 150, 702-9.

38. Moura, J. J. G. Teixeira, M. Moura, I. and LeGall, J., (Ni,Fe) Hydrogenases from sulfate-reducing bacteria:Nickel catalytic and regulatory roles. In The Bioinorganic Chemistry of Nickel, ed. J. R. Lancaster, Jr., VCH Publishers Inc., N.Y., N.Y., 1988, 191-226.

39. Bastian, N. R. Wink, D. A. Wackett, L. P. Livingston, D. J. Jordan, L. M. Fox, J. Orme-Johnson, W. H. and Walsh, C. T., Hydrogenases of Methanobacterium thermoautotrophicum strain H. In The Bioinorganic Chemistry of Nickel, ed. J. R. Lancaster, Jr., VCH Publishers Inc., N.Y., N.Y. 1988, 227-47.

40. Duine, J. A. Frank Izn, J. and Jongegan, J. A., PQQ and quino protein enzymes in microbial oxidations. FEMS Microbiol. Rev. 1986, 32, 165-78.

41. Ivey, D. M. and Ljungdahl, L. G., Purification and characterization of the F_1-ATPase of Clostridium thermoaceticum. J. Bacteriol. 1986, 165, 252-7.

42. Mayer, F. Ivey, D. M. and Ljungdahl, L. G., Macromolecular organization of F_1-ATPase isolated from Clostridium thermoaceticum as revealed by electron microscopy. J. Bacteriol. 1986, 166, 1128-30.

43. Ivey, D. M., Generation of energy during CO_2 fixation in acetogenic bacteria Dissertation, University of Georgia, Athens, GA. 1987.

44. Schneider, E. and Altendorf, K. Bacterial adenosine 5'-triphosphate synthase (F_1F_o):Purification and reconstitution of F_o complexes and biochemical and functional characterization of their subunits. Microbiol. Rev. 1987, 51, 477-97.

45. Hugenholtz, J. and Ljungdahl, L. G. The bioenergetics of Clostridium thermoautotrophicum. Ann. Meet. Am. Soc. 1988 Abstr. K153, p. 232.

46. Hugenholtz, J. Ivey, D. M. and Ljungdahl, L. G. Carbon monoxide-driven electron transport in Clostridium thermoautotrophicum membranes, J. Bacteriol. 1987, 169, 5845-7.

47. Thauer, R. K. Jungerman, K. and Decker, K. Energy conservation in chemotrophic anaerobic bacteria. Bacteriol. Rev. 1977, 41, 100-80.

48. Diekert, G. B. Schräder, E. and Harder, W. Energetics of CO formation and CO oxidation in cell suspensions of Acetobacterium woodii, Arch. Microbiol. 1986, 144, 386-92.

49. Zeikus, J. G. Chemical and fuel production by anaerobic bacteria. Ann. Rev. Microbiol. 1980, 34, 423-64.

50. Wise, D. L. ed. Organic Chemicals from Biomass, Benjamin/Cummings Publ. Comp. Menlo Park, CA., 1983.

51. Rogers, P., Genetics and biochemistry of *Clostridium* relevant to development of fermentation process, *Adv. Appl. Microbiol.* 1986, *31*, 1-60.

52. Wiegel, J. and Ljungdahl, L. G. The importance of thermophilic bacteria in biotechnology. *CRC Crit. Rev. Biotech.* 1986, *3*, 39-108.

53. Reed, W. M. and Bogdan, M. E., Application of cell recycling to continuous fermentative acetic acid production. *Biotech. Bioeng. Symp.* 1985, *15*, 641-7.

54. Wang, G. and Wang, D., Elucidation of growth inhibition and acetic acid production by *Clostridium thermoaceticum*, *Appl. Environ. Microbiol.* 1984, *47*, 294-8.

55. Ljungdahl, L. G. Carreira, L. H. Garrison, R. J. Rabek, N. E. and Wiegel, J., Comparison of three thermophilic acetogenic bacteria for production of calcium magnesium acetate, *Biotech. Bioeng. Symp.*, 1985, *15*, 207-23.

56. Chollar, B. H., Federal Highway Administration research on calcium magnesium acetate-an alternative deicer, *Public Roads*, 1984, *47*, 113-8.

57. Wise, D. L. and Augenstein, D., An evaluation of the bioconversion of woody biomass to calcium acetate deicing salt, *Solar Energy*, 1988, *41*, 453-63.

58. Ljungdahl, L. G. Carreira, L. H. Garrison, R. J. Rabek, N. E. Gunter, L. F. and Wiegel, J., *CMA Manufacture* [II]. Improved bacterial strain for acetate production, U. S. Dept. of Transportation, Federal Highway Administration Report No. FHWA/RD-86/117, Technical Information Service, Springfield, Virginia 22161, U.S.A.

59. Ljungdahl, L. G. and Wiegel, J., Working with anaerobic bacteria. In *Manual of Industrial Microbiology and Biotechnology*, eds. A. L. Demain and N. A. Soloman. Am. Soc. Microbiol., Washington, D.C. 1986, pp. 84-96.

ACIDOPHILIC, MINERAL-OXIDIZING BACTERIA: THE UTILIZATION OF CARBON DIOXIDE WITH PARTICULAR REFERENCE TO AUTOTROPHY IN SULFOLOBUS

PAUL NORRIS, ANDREW NIXON AND ALWYN HART
Department of Biological Sciences,
University of Warwick,
Coventry CV4 7AL, UK.

ABSTRACT

The concentration of CO_2 and the different mechanisms of its assimilation are considered as factors which can variably affect growth and therefore mineral oxidation by phylogenetically-distinct, autotrophic, iron-oxidizing bacteria. The rates of pyrite oxidation during the autotrophic growth of T. ferrooxidans and a strain of Sulfolobus were reduced by about 15% and 26% respectively when cultures were gassed with air in place of CO_2-enriched air. In contrast, the activity of moderately thermophilic, mineral-oxidizing bacteria was greatly reduced (83%) in the absence of CO_2 enrichment. A comparison was made of CO_2 uptake and fixation by Sulfolobus grown on thiosulphate in continuous culture under low and high CO_2 concentrations. Preliminary data indicated that the capacity of Sulfolobus to respond to CO_2 limitation involved the increased production of a protein which catalysed the ATP-dependent carboxylation of acetyl CoA.

INTRODUCTION

Most of the research towards developing commercial processes which would utilize acidophilic, mineral-oxidizing bacteria in reactors for the extraction of metals from mineral sulphides has involved the mesophile Thiobacillus ferrooxidans. However, the possible application of thermophilic mineral-oxidizing bacteria is also worthy of consideration and has been reviewed [1]. High temperature bacterial mineral oxidation can produce higher rates of metal extraction and, where the extent of mineral dissolution can be related to the temperature, greater yields of target metals in solution. These advantages have been illustrated with comparisons of the low and high temperature bacterial oxidation of chalcopyrite [2,3,4] and nickeliferous pyrrhotite [5]. Examples of the potential advantages of using thermophilic bacteria for the treatment of gold-bearing concentrates and ores, which is currently of particular commercial interest, have also been described [6]. High temperature bacterial coal desulphurization could also improve on the process rates obtained with T. ferrooxidans [7].

The acidophilic bacteria which oxidize mineral sulphides at temperatures above those permitting the growth of the more thoroughly studied mesophilic, mineral-oxidizing bacteria appear to fall into two groups. Firstly, there are iron- and sulphur-oxidizing eubacteria which grow at $60^{\circ}C$ and which might be considered moderately thermophilic. Most of these strains have an optimum temperature of about $50^{\circ}C$ but with some it is slightly lower and with most of them growing quite well at $30^{\circ}C$, they are perhaps at the borders of thermotolerance and thermophily. Secondly, at higher temperatures, there are strains of Sulfolobus which can rapidly oxidize mineral sulphides. Morphologically resembling Sulfolobus, other archaebacteria of the genus Acidianus and relatively new isolates which are active at temperatures of at least $85^{\circ}C$ can be included in this category. The bacteria active at the higher temperatures are known to have a higher GC content at 45 mol% [8] than Sulfolobus (37 mol%) but any differences in their behaviour in the context of mineral oxidation, apart from the growth temperature, remain to be elaborated.

The moderate thermophiles

Several isolates of the acidophilic bacteria which oxidize iron, sulphur and mineral sulphides optimally at about $50^{\circ}C$ show sufficient differences in morphology and in some of their growth characteristics to warrant the designation of several genera. Comparative nucleic acid analyses [9] have supported this contention but only one genus, Sulfobacillus with S. thermosulfidooxidans as the type strain [10], has so far been proposed. The isolation and some growth characteristics of other relatively well-studied isolates have been noted [11]. These isolates include strain TH3 (69 mol% GC) from a copper leach dump, strain LM2 (60 mol% GC) from a hot spring and strain ALV (57 mol% GC) from coal spoil. Strains TH1 and BC1 (50 mol% GC) with very close DNA:DNA homology (A. Harrison, personal communication) were isolated from a hot spring and coal spoil respectively and represent the species which is most widespread and most readily isolated. The growth characteristics and DNA base content of this type are similar to those of S. thermosulfidooxidans which suggests that they are at least closely affiliated but they have not been directly compared. The cell wall structure and sporulation of S. thermosulfidooxidans and 16S rRNA base sequence analyses of strains ALV, TH3 and BC1 (A. Harrison, personal communication) have indicated that these bacteria are Gram positive but any relationship to other Gram positive bacteria is not yet clear. Strain TH3 appears to have the least in common with the other isolates but they all show considerable nutritional versatility in comparison with the Gram negative, mesophilic, mineral-oxidizing Thiobacillus ferrooxidans and Leptospirillum ferrooxidans.

Growth characteristics: Doubling times of about 7 hours have been recorded for heterotrophic growth of strains BC1, ALV and TH3 and 12 hours for strain LM2 on yeast extract (0.25 g.l^{-1}) at pH 2 and 50°C in the absence of ferrous iron (D. Hinson and P. Norris, unpublished data). These bacteria probably grow most rapidly chemolitho-heterotrophically on ferrous iron and yeast extract [12] with the clear majority of cell carbon obtained from the organic supplement [13] but can also grow mixotrophically with the simultaneous utilization of glucose and carbon dioxide during iron oxidation [14,15]. Glucose utilization by strain TH3 has not been demonstrated [11]. Most of the thermophiles appear unable to utilize sulphate for biosynthesis and require a source of reduced sulphur [11] but otherwise grow autotrophically on ferrous iron in the simple defined salts media used for the mesophiles, a capacity which is the foundation of their potential utility in commercial mineral treatment. As exceptions to the general behaviour pattern [11], strain ALV can utilize sulphate during autotrophic growth on iron and autotrophic growth of strain TH3 has not been confirmed. Growth of strains ALV and LM2 on sulphur (flowers) both autotrophically and with concurrent glucose utilization has been described [16]. Strain TH3 grows and oxidizes sulphur in the presence of yeast extract (unpublished work) and sulphur oxidation has been noted in a review of the characteristics of S. thermosulfidooxidans [17].

Acidophilic bacteria which oxidize sulphur optimally at 45-50°C but which, in contrast to the moderate thermophiles noted above, can also oxidize thiosulphate and tetrathionate but not iron have been readily isolated from hot springs, coal spoil and metal leach dumps. These bacteria are Gram negative and show the characteristics which would be expected of a "thermotolerant Thiobacillus thiooxidans" although the 60 mol%GC content (A. Harrison, personal communication) of a typical strain, BC13, is higher than that of the mesophilic T. thiooxidans (53 mol%GC). Growth and sulphur oxidation at 45-50°C by strain BC13 and similar isolates is more rapid than that of T. thiooxidans and T. ferrooxidans at 30°C while sulphur oxidation by the moderately thermophilic iron-oxidizing bacteria is slower than that of the mesophiles [16].

Mineral oxidation: Iron-oxidizing, moderate thermophiles appear abundant in some copper leach dumps [17,18] which indicates that they could usefully catalyse some metal extraction where rising temperatures as a consequence of exothermic mineral sulphide oxidation restrict or preclude the activity of mesophiles such as T. ferrooxidans. The first laboratory demonstrations of the dissolution of finely-ground mineral sulphides by the moderate thermophiles indicated a requirement for yeast extract for rapid growth [19,20]. Later demonstrations of the growth-associated, rapid oxidation of minerals under enhanced carbon dioxide concentrations but importantly in the absence of yeast extract [21] further established the potential of such bacteria for the commercial treatment of mineral concentrates in reactors.

Sulfolobus

Various aspects of the molecular biology of Sulfolobus strains have received considerable attention during the last ten years while the relationship of phenotypic and physiological characteristics to taxonomic designations remains uncertain. The first division of the genus gave S. acidocaldarius, S. solfataricus and S. brierleyi [22] with the latter species subsequently claimed by the genus Acidianus to join A. infernus as organisms capable of anaerobic sulphur reduction as well as aerobic sulphur oxidation [23].

Growth characteristics: There are differences in the growth characteristics of the various species which are of fundamental significance to their capacity for mineral oxidation. Sulfolobus B6-2 [24], apparently not closely related to the other species on the basis of different electrophoretic protein profiles [16], oxidizes sulphur but not ferrous iron which almost certainly restricts any capacity to solubilize most mineral sulphides. The apparent inability of autotrophically-growing S. brierleyi to oxidize sulphur (flowers) when cultures are shaken has been noted [5,16] and the rapid growth of the same strain on pyrite in shaken flasks required the presence of yeast extract in contrast to the rapid autotrophic growth and pyrite dissolution by strains of Sulfolobus [25] which show the characteristics originally attributed to S. acidocaldarius [26,27]. Approximately ten years after the description of S. acidocaldarius, sulphur and mineral oxidation by the type strain and by S. solfataricus from culture collections could not be demonstrated [25] although heterotrophic growth on yeast extract was rapid. A comparison of the growth on yeast extract of the sulphur-oxidizing strains and the obligate heterotrophs has suggested the selection of heterotrophic bacteria during the maintenance of S. acidocaldarius rather than the loss of the sulphur-oxidizing capacity by a pure culture (data not shown).

Mineral oxidation: As with the development of mineral oxidation studies with the moderately thermophilic bacteria, the media used in the pioneering laboratory demonstrations of the mineral-oxidizing activity of Sulfolobus strains generally contained yeast extract [28] and involved relatively large mineral particles, column percolation experiments and consequently slow mineral leaching rates [29,30]. Subsequent reports of high rates of oxidation of finely ground mineral sulphides during autotrophic growth of Sulfolobus strains [5,25] demonstrated the commercial potential of these thermophiles for mineral concentrate treatment in reactors.

Factors affecting bacterial growth and mineral oxidation
Most of the laboratory demonstrations of efficient, high temperature mineral oxidation which have encouraged the consideration of the thermophiles for industrial application have involved ideal experimental

conditions which have avoided many of the stresses that the bacteria would face in an industrial process. For example, an apparently greater sensitivity of Sulfolobus, in comparison with mesophilic mineral-oxidizing bacteria, to agitation in the presence of minerals has necessitated the use of air-lift reactors rather than stirred reactors to demonstrate the full potential of the high temperature oxidation of pyrite [31]. Apart from the obvious influence of the temperature, different mineral-oxidizing bacteria can also be selectively affected by the acidity and the concentrations of iron and potentially toxic metals [3]. The concentration of CO_2 which might have to be maintained during the aeration of reactors in order to maximize the bacterial leaching activity could also influence the cost effectiveness of the application of certain bacteria if their efficiency of CO_2 utilization were significantly different. In general, it seems likely that differences in fundamental aspects of metabolism which are directly involved in the autotrophic growth on minerals provide the basis for there being different primary stresses or limitations restricting the mineral-oxidizing activity of particular bacteria. This paper assesses the utilization of CO_2 as one of the factors affecting growth with particular reference to Sulfolobus in which a reductive carboxylic acid pathway is believed to operate [32] rather than the reductive pentose phosphate pathway which is probably the primary route of CO_2 assimilation in the mesophile T. ferrooxidans and the moderately thermophilic mineral-oxidizing bacteria.

MATERIALS AND METHODS

Organisms: The isolation and some growth characteristics of the moderate thermophile strain BC1 [11,12,13,33] and of the Sulfolobus strains BC and LM [16,25,34] have been described. These Sulfolobus isolates were from a coal spoil and hot spring respectively but are considered to represent the same species on the basis of identical electrophoretic protein profiles [16] and growth characteristics. Thiobacillus ferrooxidans DSM 583 has been used previously in some comparisons of mesophilic and thermophilic bacterial mineral oxidation [3,21,33] and the influence of CO_2 on its growth has been described [35].

Growth conditions: Growth of bacteria on pyrite (<75 µm particle size diameter; 40% w/w Fe) was in air-lift reactors (500 ml) with gas introduction (250 ml min^{-1}) at the base of a central draught tube. The medium, initially at pH 2, also contained (g l^{-1}) $(NH_4)_2SO_4$ (0.4), $MgSO_4 \cdot 7H_2O$ (0.4) and K_2HPO_4 (0.2).

The medium for growth of Sulfolobus strain LM on potassium tetrathionate or sodium thiosulphate in chemostat cultures contained (g l^{-1}) $(NH_4)_2SO_4$ (0.2), $MgSO_4 \cdot 7H_2O$ (0.4), K_2HPO_4 (0.1), KCl (0.1) and $FeSO_4 \cdot 7H_2O$ (0.01), initially at pH 3. 10 mM tetrathionate or 20 mM thiosulphate were used as substrates fed to the chemostats with the salts (tetrathionate) or separately in water at pH 9 (200 mM thiosulphate) at one tenth of the salts solution flow rate. The cultures (725 ml) were grown at 70°C and gassed with either air or 5% (v/v) CO_2 in air (100 ml min^{-1}).

Oxygen electrode studies: 3 ml samples from the Sulfolobus chemostat cultures (pH 1.6) were placed directly in the oxygen electrode (Rank Brothers, Bottisham, U.K.) and maintained at 65°C for the characterization of tetrathionate oxidation kinetics.

Polyacrylamide gel electrophoresis: Standard techniques of SDS-PAGE (12% acrylamide gels) and non-denaturing PAGE (a 7% acrylamide gel) were used. The estimation of protein molecular weight used the method discussed by Bryan [36] with 7-13% non-denaturing gels and the procedure and marker proteins as described in the Sigma Technical Bulletin MKR-137.

Sulfolobus CO_2 assimilation assays: 1. Whole cells. 25 ml samples (pH 1.6) from chemostat cultures were stirred continuously in suba-sealed, conical flasks (50 ml) at 65°C. 1 ml samples were filtered at time intervals following the addition of potassium tetrathionate (final concentration, 0.2 mM) and 20 μl 36 mM $Na_2{}^{14}CO_3$ (55 mCi mmol^{-1}). Filters were placed in 10 ml LKB OptiPhase Safe scintillant for counting.

2. Cell lysates and sucrose gradient fractions. 300 ml of the CO_2-limited, thiosulphate-oxidizing, chemostat culture (42 mg protein l^{-1}) were centrifuged and the cell pellet was resuspended in 2 ml Tris-HCl (pH 8) to give a cell lysate. The same culture volume was centrifuged and the pellet resuspended in Tris-HCl (20 mM, pH 8), $MgCl_2 \cdot 6H_2O$ (10 mM), $NaHCO_3$ (50 mM), EDTA (10 mM) and β-mercaptoethanol (5 mM) for lysis before loading on a discontinuous sucrose gradient comprising layers (each 2 ml) of 0.8-0.2 M sucrose in 0.1 M steps on a 60% (w/v) sucrose base (2 ml). 1 ml fractions were collected from the bottom of the gradient after centrifugation in a

fixed-angle rotor at 240,000g for 2.5 h and stored at -20°C. Assays using the whole cell lysate or the sucrose gradient fractions were carried out at 65°C in the reaction mixtures (1 ml) indicated in the text. Reactions were stopped with 100 μl 20 M formic acid and the residue after drying at 90°C was taken up in 1 ml H_2O and 9 ml scintillant as above.

RESULTS

The influence of CO_2 on mineral oxidation by various bacteria and on growth on thiosulphate by Sulfolobus

Pyrite oxidation rates during the autotrophic growth of T. ferrooxidans, moderate thermophile strain BC1 and Sulfolobus strain BC were compared under conditions which allowed CO_2 to be the principal factor limiting the growth and consequently the mineral-oxidizing activity of the thermophiles (Table 1).

TABLE 1
The rates of pyrite (1% w/v) dissolution, estimated as iron solubilization, by mineral-oxidizing bacteria in air-lift reactors supplied with air or CO_2-enriched air.

Bacteria	Temp. (°C)	Iron Solubilization (mg l^{-1} h^{-1})	
		5% (v/v) CO_2 in air	Air
T. ferrooxidans	30	26	22
Strain BC1	50	48	8
Sulfolobus BC	70	70	52

The activity of T. ferrooxidans was only slightly reduced by the use of air without additional CO_2. Previous work [37] has shown that the concentration of CO_2 in air limited mineral oxidation by T. ferrooxidans only at higher solids concentrations but the effect could depend to some extent on the strain as reports have described the growth rate of ferrous iron-oxidizing T. ferrooxidans strains as independent of [35] and dependent on [38] the concentration of CO_2. The capacity of

T. ferrooxidans to respond to CO_2 limitation probably depends on an increase in the concentration of ribulose 1,5-bisphosphate carboxylase/oxygenase (RuBisCo) [see 39].

The assimilation of CO_2 by the moderate thermophiles has been little studied but RuBisCo activity has been demonstrated. It was not significantly reduced during mixotrophic growth on ferrous iron in the presence of citrate or glucose [15] but the synthesis of the enzyme was repressed during the utilization of yeast extract (data not shown). The autotrophic growth rates of strain BC1 [12] and another similar isolate [15] on ferrous iron were reduced about fourfold under air in place of a CO_2-enriched atmosphere. The inability to respond as effectively as *T. ferrooxidans* to CO_2 limitation and the consequently poor mineral oxidation in the absence of additional CO_2 (Table 1) could result from the absence of an increase in RuBisCo synthesis or from less efficient CO_2 uptake.

The rate of pyrite oxidation by *Sulfolobus* was not reduced as severely as that by the moderate thermophile in the absence of additional CO_2 (Table 1). In order to investigate further the response of *Sulfolobus* to CO_2 limitation with greater practical convenience than was possible with growth of the bacteria on minerals, chemostat cultures were established with thiosulphate as the substrate and with the growth limited by either the energy source or by CO_2. The operation of the chemostats and some characteristics of the bacterial growth will be described elsewhere (A. Nixon and P.R. Norris, in preparation). Tetrathionate was used as the initial substrate to establish the continuous cultures but was then replaced with the less expensive thiosulphate which was metabolized via an initial, extracellular conversion to tetrathionate (data not shown). Only trace levels of thiosulphate or tetrathionate were detectable during thiosulphate- or CO_2-limited steady-state growth (specific growth rate, 0.028 h^{-1}). During operation of the chemostats over a period of one year, CO_2 limitation generally resulted in a halving of the biomass yield although there appeared to be some increase in yield when the substrate concentration was increased despite the growth being principally limited by CO_2 (data not shown). On some occasions when samples were taken for examination of the characteristics of substrate oxidation (in an oxygen electrode, Fig. 1) and CO_2 assimilation (Fig. 2), the ratio of the biomass concentration in the thiosulphate and CO_2-limited cultures had varied because of changes in the separate mineral salts and substrate feed rates.

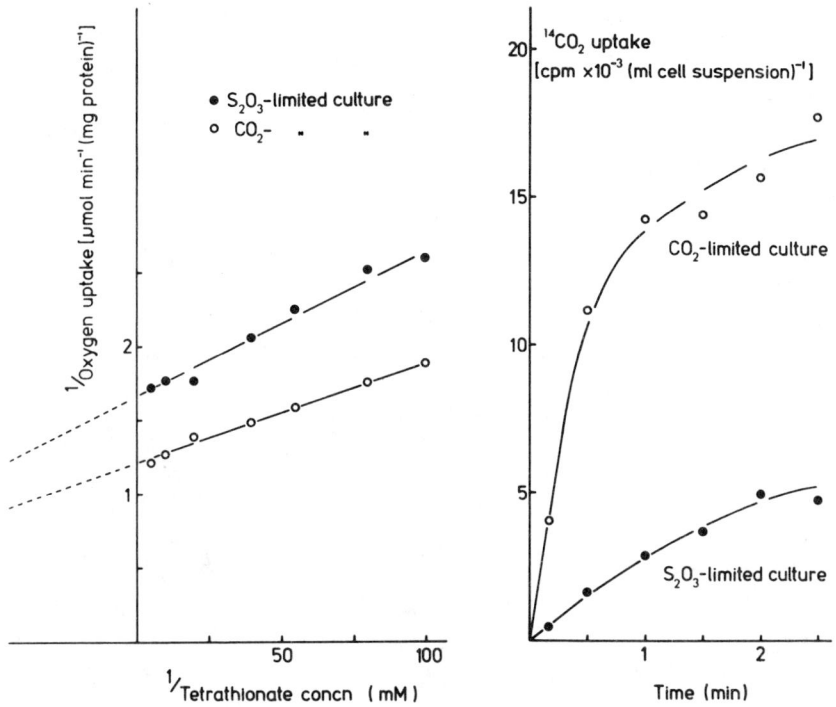

Figure 1 (left). Reciprocal plot of the tetrathionate concentration against oxygen uptake by Sulfolobus strain LM from CO_2-limited (42 mg cell protein l^{-1}) or thiosulphate-limited (58 mg cell protein l^{-1}) cultures.
Figure 2 (right). CO_2 uptake by Sulfolobus strain LM from CO_2-limited (26 mg cell protein l^{-1}) and thiosulphate-limited (61 mg cell protein^{-1}) chemostat cultures (see Materials and Methods).

The CO_2-limited culture containing about 30% less biomass than the thiosulphate-limited culture showed a 30% greater oxidation capacity (Fig. 1) reflecting the complete substrate oxidation in both cultures and some uncoupling of the oxidation from biosynthesis under CO_2 limitation. Tetrathionate was used in the oxygen electrode so that the measured rates and affinities of substrate oxidation did not reflect the initial oxygen-consuming conversion of thiosulphate to tetrathionate. The affinity for tetrathionate oxidation was the same in both CO_2- and thiosulphate-limited cultures with the K_m just below 6 µM tetrathionate.

In short term assays, the rate of CO_2 accumulation by bacteria from the CO_2-limited culture was initially over 10 times that, per unit biomass, by those grown with excess CO_2 (Fig. 2).

Irrespective of whether pyrite or thiosulphate was the substrate when growth was limited by CO_2, SDS-PAGE protein profiles revealed a marked increase in the quantity of a 59 kDa polypeptide (Fig. 3, B and H).

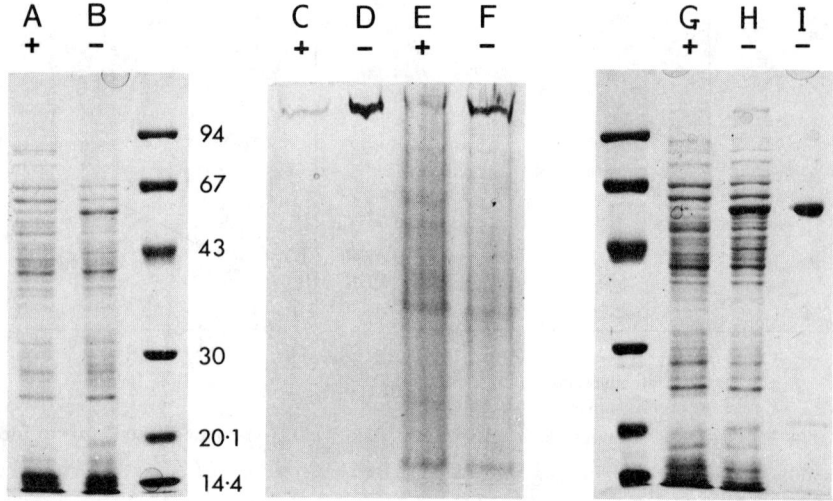

Figure 3. SDS-PAGE of cell lysates of <u>Sulfolobus</u> strain BC grown on pyrite (A and B) and strain LM on thiosulphate (G and H) in cultures gassed with 5% (v/v) CO_2 in air (+) or with air (-). Non-denaturing PAGE of bacteria grown on thiosulphate at high (+) or low (-) CO_2 concentrations (C-F): samples (40 ml) from chemostat cultures were centrifuged, the pellets lysed in Tris buffer (20 mM, pH 7.4) and about 1, 0.5, 0.5 and 0.25 mg protein was added to lanes C, D, E and F respectively. Coomassie blue staining was used throughout. The protein was eluted from the heavily-stained band in lane D and subjected to SDS-PAGE (I). Molecular weight standards (kDa) are shown in lanes of the SDS-PAGE gels.

The major protein which was produced in response to the CO_2 limitation was revealed by non-denaturing PAGE (Fig. 3, C-F) and estimated to have an apparent molecular mass of about 330 kDa (see Materials and Methods). SDS-PAGE of the major protein after elution from the non-denaturing gel showed putative subunits of about 59 and 19.5 kDa (Fig. 3, I), suggesting possibly 4 large and 4 small subunits in the native protein.

The role of the protein (330 kDa) in CO_2 assimilation?

CO_2 fixation by cell lysates and by sucrose gradient fractions of cell lysates was measured.

TABLE 2

CO_2 fixation by a whole cell lysate (left) and a lysate sucrose gradient fraction (right) of <u>Sulfolobus</u> strain LM. The whole cell lysate reaction mixture contained Tris-H_2SO_4 (70 mM, pH 8), $MgSO_4 \cdot 7H_2O$ (20 mM) and reduced glutathione (2 mM). The reaction mixture to which fraction 6 was added contained Tris-HCl (78 mM, pH 8), $MgCl_2$ (25 mM) and reduced glutathione (10 mM). Other reagents were added, as indicated, to the concentrations noted below. All reaction mixtures (1 ml) contained 40 mM $Na_2^{14}CO_3$ (200 µCi mmol^{-1}) and were incubated at 65°C for 10 min.

Additions to cell lysate	cpm $^{14}CO_2$ fixed	Additions to fraction 6	cpm $^{14}CO_2$ fixed
none	451	none	101
ATP	1,765	ATP	129
Acetyl CoA	432	Acetyl CoA	122
PEP	3,194	PEP + Acetyl CoA	129
PEP + Acetyl CoA	30,433*	ATP + Acetyl CoA	31,818
ATP + Acetyl CoA	21,344		

Acetyl CoA (0.5 mM), ATP (3 mM), PEP (92 mM), cell lysate (50 µl, see Materials and Methods).
 * Equivalent to fixation of about 25 nmol CO_2 mg protein^{-1} min^{-1}.

Acetyl CoA (1 mM), ATP (3 mM), PEP (92 mM), sucrose gradient fraction 6 (50 µl, see Materials and Methods and Fig. 4)

The activity in the gradient fractions is discussed below (see Fig. 4). The activity in the whole cell extract which appeared dependent on acetyl CoA and phosphoenolpyruvate (PEP) (Table 2) was threefold greater than

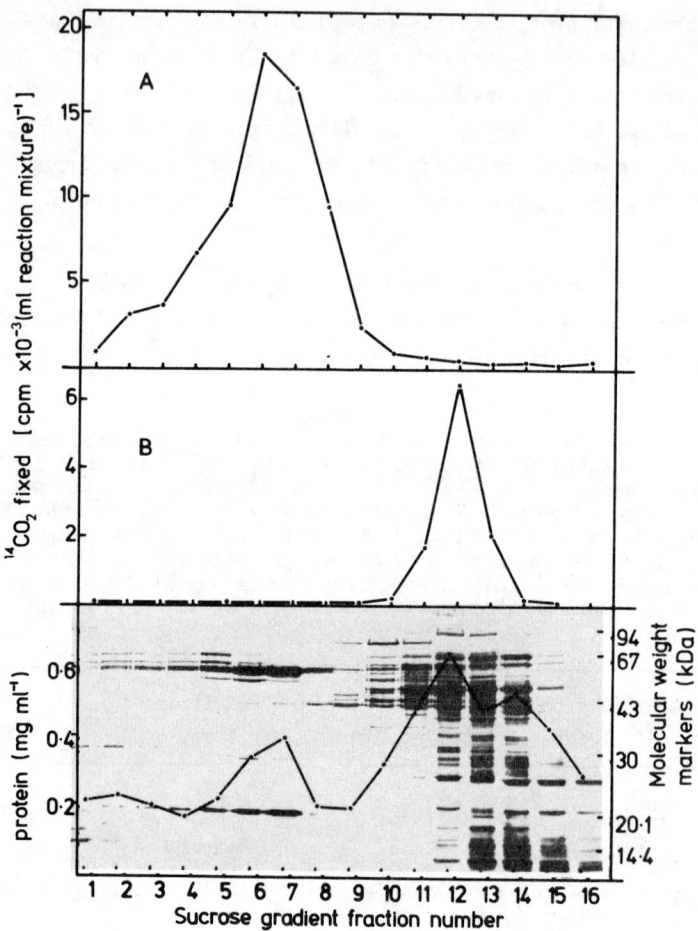

Fig. 4. CO_2 fixation in sucrose gradient fractions of a lysate of Sulfolobus strain LM (10 min incubation at 65°C). Reaction mixture A (top) containing acetyl CoA and ATP was as given in Table 2 for the assay of fraction 6 activity. Reaction mixture B (centre) was as used in PEP carboxylase assays with thiobacilli [40] and contained 70 mM Tris-H_2SO_4 (pH 8), 20 mM $MgSO_4 \cdot 7H_2O$, 2.0 mM glutathione, 0.32 mM acetyl CoA, 92 mM PEP and 40 mM $Na_2{}^{14}CO_3$ (200 µCi mmol^{-1}). The result of protein SDS-PAGE (silver-stained) of the gradient fractions is illustrated.

with an extract similarly produced from bacteria grown with excess CO_2 (data not shown), a difference of just over 4.5-times with correction to a comparable extract protein concentration. On this latter basis, the CO_2 fixation in the presence of acetyl CoA and ATP was just over threefold greater in an extract of bacteria from the CO_2-limited culture. These are comparisons of initial assays however, which require repetition after the optimization of the assay conditions. On another occasion, the CO_2 fixation which occurred following the addition of only PEP was up to 50% of that following additions of PEP and acetyl CoA.

Acetyl CoA and ATP-dependent CO_2 fixation appeared linked to the protein with putative subunits of about 59 and 19.5 kDa with fractions 6 and 7 of the sucrose gradient of a cell lysate containing most of this protein (Fig. 4) and showing the greatest activity (Fig. 4A) with fixation of about 300 nmol CO_2 mg protein^{-1} min^{-1}. The rate of CO_2 fixation was linear for at least 20 min (data not shown) and in contrast to the activity in the whole cell lysate, ATP but not PEP was able to promote the carboxylation of acetyl CoA (Table 3).

A second peak of CO_2 fixation activity in the gradient fractions was obtained in the presence of acetyl CoA and PEP (Fig. 4B). An acetyl CoA-dependent PEP carboxylase could have been present in fraction 12 but significant CO_2 fixation was also obtained in this fraction following the addition of acetyl CoA and ATP in the absence of PEP but otherwise in the mixture used for the "PEP carboxylase" assay (data not shown).

A study of the effect of the assay conditions, including substrate and effector concentrations, purification of the active proteins in the extracts and identification of the products are all required for characterization of the reactions and enzymes which have produced the CO_2 fixation demonstrated in these cell extracts of Sulfolobus.

DISCUSSION

The behaviour of Sulfolobus strain LM with respect to its energy and carbon sources during autotrophic growth on thiosulphate at least outwardly resembles the pattern observed with several strains of thiobacilli. The initial, extracellular conversion of thiosulphate to tetrathionate occurs with most thiobacilli [see 41]. The high affinity for tetrathionate compares with that of T. ferrooxidans [42]. The uncoupled substrate oxidation during CO_2-limited growth resembles that seen with thiosulphate-oxidizing, CO_2-limited cultures of thiobacilli [40,43] and the pattern of increased CO_2 uptake by Sulfolobus grown under CO_2 limitation (Fig. 2) was similar to that observed with CO_2 or thiosulphate-limited cultures of Thiobacillus neapolitanus [44]. The pathway of thiosulphate and tetrathionate oxidation by Sulfolobus has not been defined and could be similar to that in some thiobacilli. The primary route of CO_2 assimilation in the autotrophically-growing, thermophilic archaebacterium is, however, clearly different from that in the sulphur-oxidizing eubacteria. Growth under CO_2-limitation has been used to indicate potentially key reactions and to reveal potentially key proteins in CO_2 fixation by Sulfolobus. The concentration of a roughly 43 kDa polypeptide increased with CO_2 limitation (Fig. 3, B and H). The increased synthesis of a 42 kDa polypeptide, in response to CO_2 limitation, in Anacystis nidulans has also been described [45] but it is not known if there is any similarity, other than of size, between these proteins and their involvement in CO_2 fixation has not been demonstrated. In contrast, the major protein produced by Sulfolobus (Fig. 3) in response to CO_2 limitation did appear active in CO_2 fixation (Fig. 4).

There are probably some variations in the pathway of CO_2 assimilation among the bacteria, including the sulphur-dependent archaebacteria, which lack RuBisCo and have acetyl CoA as a central intermediate in the operation of a "reductive carboxylic acid cycle" [46]. The major, essential anaplerotic reaction in the pathway in Chloroflexus [47], Thermoproteus [48] and Desulfobacter [49] is the carboxylation of phosphoenolpyruvate to oxaloacetate with the reductive carboxylation of acetyl CoA apparently catalysed by pyruvate synthase or, in Desulfobacter, possibly by a pyruvate:X oxidoreductase. The early labelling of alanine in Chloroflexus during $^{14}CO_2$ assimilation [47], indicative of pyruvate formation from acetyl CoA, was not found in similar experiments with Sulfolobus (Acidianus) brierleyi [32] or with Sulfolobus strain LM in the present study (data not shown). In the latter case, malate, aspartate and glutamate were labelled before phosphoenolpyruvate and phosphoglyceric acid, consistent with the absence of RuBisCo, but the first two labelled products were not identified at the time of this presentation. Further study of the major enzyme which was produced in response to CO_2 limitation, which catalysed the acetyl CoA carboxylation and which appeared to be both stable and amenable to purification, should allow clarification of one of the important reactions concerned with the proposed operation of a yet to be defined reductive carboxylic acid cycle in Sulfolobus.

Further work is also required to compare the effect of CO_2 limitation on the growth on minerals of T. ferrooxidans and Sulfolobus; the descibed reductions in activity possibly concealed different contributions of uncoupled mineral oxidation in the nominally growth-associated solubilization of iron. Certainly, with cell numbers estimated by microscopy (data not shown), the growth of Sulfolobus on pyrite in the absence of CO_2 enrichment appeared more reduced than the 26% reduction in the iron extraction rate, indicating that, as when thiosulphate was the substrate, significant oxidation was uncoupled from growth. The rate of pyrite oxidation in the Sulfolobus culture gassed with air was twice that obtained with T. ferrooxidans even with CO_2 enrichment (Table 1). Doubling the pyrite concentration (1→2% w/v) resulted in twice the iron extraction rate by Sulfolobus strain BC (70→142 mg l^{-1} h^{-1}) under 5% (v/v) CO_2 in air but in a small increase (52→62 mg l^{-1} h^{-1}) in the absence of CO_2 enrichment (unpublished work). It remains to be confirmed, particularly with factors such as the mineral concentration more appropriate to an industrial process, that through differences in bacterial CO_2 assimilation routes or uptake characteristics, different minimum concentrations of CO_2 from culture gassing would be effective in the maintenance of the maximum rates of mineral oxidation of which particular bacteria were capable.

Acknowledgements: We are grateful to the Science and Engineering Research Council for support of this work, to Graeme MacDonald and Nick Mann for sucrose gradient work and to Professor Don Kelly for analyses and discussions of CO_2 fixation work.

REFERENCES

1. Brierley, J.A. and Brierley, C.L., Microbial mining using thermophilic microorganisms. In Thermophiles: General, Molecular and Applied Microbiology, ed., T.D. Brock, John Wiley and Sons, New York, 1986, pp. 279-305.

2. Brierley, J.A. and Brierley, C.L., Microbial leaching of copper at ambient and elevated temperatures. In Metallurgical Applications of Bacterial Leaching and Related Microbiological Phenomena, ed., L.E. Murr, A.E. Torma and J.A. Brierley, Academic Press, New York, 1978, pp. 477-90.

3. Norris, P.R., Bacterial diversity in reactor mineral leaching. In Proceedings of the 8th International Biotechnology Symposium, Paris, 1988 (in press).

4. Le Roux, N.W. and Wakerley, D.S., Leaching of chalcopyrite ($CuFeS_2$) at 70°C using Sulfolobus. In Biohydrometallurgy: International Symposium 1987, ed., P.R. Norris and D.P. Kelly, Science and Technology Letters, Kew, 1988, pp. 305-17.

5. Norris, P.R. and Parrott, L., High temperature, mineral concentrate dissolution with Sulfolobus. In Fundamental and Applied Biohydro-metallurgy, ed., R.W. Lawrence, R.M.R. Branion and H.G. Ebner, Elsevier, Amsterdam, 1986, pp. 355-65.

6. Hutchins, S.R., Brierley, J.A. and Brierley, C.L., Microbial pre-treatment of refractory sulfide and carbonaceous gold ores. In Process Mineralogy VII, ed., A.H. Vassiliou, D.M. Hausen and D.J.T. Carson, The Metallurgical Society, Inc., 1987, pp. 53-66.

7. Detz, C.M. and Barvinchak, G., Microbial desulfurization of coal. Mining Congress Journal, 1979, 65, 75-82 and 86.

8. Huber, G. and Stetter, K.O., Properties of newly isolated metal-mobilizing Sulfolobus-like organisms. In Archaebacteria '85, ed., O. Kandler and W. Zillig, Gustav Fischer Verlag, Stuttgart, 1986, p. 413.

9. Harrison, A.P., Characteristics of Thiobacillus ferrooxidans and other iron-oxidizing bacteria, with emphasis on nucleic acid analyses. Biotechnol. Appl. Biochem., 1986, 8, 249-57.

10. Golovacheva, R.S. and Karavaiko, G.I., A new genus of thermophilic spore-forming bacteria, Sulfobacillus. Microbiology, 1979, 47, 658-65.

11. Norris, P.R. and Barr, D.W., Growth and iron oxidation by acidophilic moderate thermophiles. FEMS Microbiol. Lett., , 1985, 28, 221-4.

12. Marsh, R.M. and Norris, P.R., The isolation of some thermophilic, autotrophic, iron- and sulphur-oxidizing bacteria. FEMS Microbiol. Lett., 1983, 17, 311-5.

13. Wood, A.P. and Kelly, D.P., Autotrophic and mixotrophic growth of three thermoacidophilic iron-oxidizing bacteria. FEMS Microbiol. Lett., 1983, 20, 107-12.

14. Wood, A.P. and Kelly, D.P., Growth and sugar metabolism of a thermoacidophilic iron-oxidizing mixotrophic bacterium. J. Gen. Microbiol., 1984, **130**, 1337-49.

15. Wood, A.P. and Kelly, D.P., Autotrophic and mixotrophic growth and metabolism of some moderately thermoacidophilic iron-oxidizing bacteria. In Planetary Ecology, ed., D.E. Caldwell, J.A. Brierley and C.L. Brierley, Van Nostrand Reinhold Co., New York, 1985, pp. 251-62.

16. Norris, P.R., Marsh, R.M. and Lindstrom, E.B., Growth of mesophilic and thermophilic acidophilic bacteria on sulfur and tetrathionate. Biotechnol. Appl. Biochem., 1986, **8**, 318-29.

17. Karavaiko, G.I., Golovacheva, R.S., Pivovarova, T.A., Tzaplina, I.A. and Vartanjan, N.S., Thermophilic bacteria of the genus Sulfobacillus. In Biohydrometallurgy: International Symposium 1987, ed., P.R. Norris and D.P. Kelly, Science and Technology Letters, Kew, 1988, pp. 29-41.

18. Brierley, J.A., Thermophilic iron-oxidizing bacteria found in copper leaching dumps. Appl. Environ. Microbiol., 1978, **36**, 523-25.

19. Le Roux, N.W., Wakerley, D.S. and Hunt, S.D., Thermophilic Thiobacillus-type bacteria from Icelandic thermal areas. J. Gen. Microbiol., 1977, **100**, 197-201.

20. Brierley, J.A. and Le Roux, N.W., A facultative thermophilic Thiobacillus-like bacterium: oxidation of iron and pyrite. In Conference Bacterial Leaching, ed., W. Schwartz, Verlag Chemie, Weinheim, 1977, pp. 55-66.

21. Marsh, R.M. and Norris, P.R., Mineral sulphide oxidation by moderately thermophilic acidophilic bacteria. Biotechnol. Lett., 1983, **5**, 585-90.

22. Zillig, W., Stetter, K.O., Wunderl, S., Schulz, W., Priess, H. and Scholz, I., The Sulfolobus-"Caldariella" group: taxonomy on the basis of the structure of DNA-dependent RNA polymerases. Arch. Microbiol., 1980, **125**, 259-69.

23. Segerer, A., Neuner, A., Kristjansson, J.K. and Stetter, K.O., Acidianus infernus gen. nov., sp. nov., and Acidianus brierleyi comb. nov.: facultatively aerobic, extremely acidophilic thermophilic sulfur-metabolizing archaebacteria. Int. J. Syst. Bacteriol., 1986, **36**, 559-64.

24. Konig, H., Skorko, R., Zillig, W. and Reiter, W.-D., Glycogen in thermoacidophilic archaebacteria of the genera Sulfolobus, Thermoproteus, Desulfurococcus and Thermococcus. Arch. Microbiol., 1982, **132**, 297-303.

25. Marsh, R.M., Norris, P.R. and Le Roux, N.W., Growth and mineral oxidation studies with Sulfolobus. In Recent Progress in Biohydrometallurgy, ed., G. Rossi and A.E. Torma, Associazione Mineraria Sarda, Iglesias, 1983, pp. 71-81.

26. Brock, T.D., Brock, K.M., Belly, R.T. and Weiss, R.L., Sulfolobus: a new genus of sulfur-oxidizing bacteria living at low pH and high temperature. Arch. Microbiol., 1972, **84**, 54-68.

27. Brock, T.D., Thermophilic Microorganisms and Life at High Temperatures, Springer-Verlag, New York, 1978.

28. Brierley, C.L. and Murr, L.E., Leaching: use of a thermophilic and chemoautotrophic microbe. Science, 1973, **179**, 488-90.

29. Berry, V.K. and Murr, L.E., Direct observations of bacteria and quantitative studies of their catalytic role in the leaching of low-grade, copper-bearing waste. In Metallurgical Applications of Bacterial Leaching and Related Microbiological Phenomena, ed., L.E. Murr, A.E. Torma and J.A. Brierley, Academic Press, New York, 1978, pp. 103-36.

30. Brierley, C.L., Biogenic extraction of uranium from ores of the Grants region. In Metallurgical Applications of Bacterial Leaching and Related Microbiological Phenomena, ed., L.E. Murr, A.E. Torma and J.A. Brierley, Academic Press, New York, 1978, pp. 345-63.

31. Norris, P.R. and Barr, D.W., Bacterial oxidation of pyrite in high temperature reactors. In Biohydrometallurgy: International Symposium 1987, ed., P.R. Norris and D.P. Kelly, Science and Technology Letters, Kew, 1988, pp. 532-5.

32. Kandler, O. and Stetter, K.O., Evidence for autotrophic CO_2 assimilation in Sulfolobus brierleyi via a reductive carboxylic acid pathway. Zbl. Bakt. Hyg., I. Abt. Orig. C., 1981, **2**, 111-21.

33. Norris, P.R., Parrott, L. and Marsh, R.M., Moderately thermophilic mineral-oxidizing bacteria. In Workshop on Biotechnology for the Mining, Metal-Refining and Fossil Fuel Processing Industries, ed., H.L. Ehrlich and D.S. Holmes, John Wiley and Sons, New York, 1986, pp. 253-62.

34. Wood, A.P., Kelly, D.P. and Norris, P.R., Autotrophic growth of four Sulfolobus strains on tetrathionate and the effect of organic nutrients. Arch. Microbiol., 1987, **146**, 382-9.

35. Kelly, D.P. and Jones, C.A., Factors affecting metabolism and ferrous iron oxidation in suspensions and batch cultures of Thiobacillus ferrooxidans: relevance to ferric iron leach solution regeneration. In Metallurgical Applications of Bacterial Leaching and Related Microbiological Phenomena, ed., L. E. Murr, A. E. Torma and J.A. Brierley, Academic Press, New York, 1978, pp. 19-44.

36. Bryan, J.K., Molecular weights of protein multimers from poly-acrylamide gel electrophoresis. Anal. Biochem., 1977, **78**, 513-9.

37. Torma, A.E., Walden, C.C., Duncan, D.W., and Branion, R.M.R., The effect of carbon dioxide and particle surface area on the

microbiological leaching of a zinc sulfide concentrate. <u>Biotechnol. Bioeng.</u>, 1972, **14**, 777-86.

38. Holuigue, L., Herrera, L., Phillips, O.M., Young, M. and Allende, J.E., CO_2 fixation by mineral-leaching bacteria: characteristics of the ribulose bisphosphate carboxylase - oxygenase of <u>Thiobacillus ferrooxidans.</u> <u>Biotechnol. Appl. Biochem.</u>, 1987, **9**, 497-505.

39. Codd, G.A. and Kuenen, J.G., Physiology and biochemistry of autotrophic bacteria. <u>Antonie van Leeuwenhoek,</u> 1987, **53**, 3-14.

40. Smith, A.L., Kelly, D.P. and Wood, A.P., Metabolism of <u>Thiobacillus</u> A2 grown under autotrophic, mixotrophic and heterotrophic conditions in chemostat culture. <u>J. Gen. Microbiol.</u>, 1980, **121**, 127-38.

41. Kelly, D.P., Oxidation of sulphur compounds. In <u>The Nitrogen and Sulphur Cycles,</u> ed., J.A. Cole and S.J. Ferguson, Cambridge University Press, Cambridge, 1988, pp. 65-98.

42. Hazeu, W., Bijleveld, W., Grotenhuis, J.T.C., Kakes, E. and Kuenen, J.G., Kinetics and energetics of reduced sulfur oxidation by chemostat cultures of <u>Thiobacillus ferrooxidans.</u> <u>Antonie van Leeuwenhoek,</u> 1986, **52**, 507-18.

43. Beudeker, R.F., Cannon, G.C., Kuenen, J.G. and Shively, J.M., Relations between D-ribulose-1,5-Bisphosphate carboxylase, carboxysomes and CO_2 fixing capacity in the obligate chemolithotroph <u>Thiobacillus</u> neapolitanus grown under different limitations in the chemostat. <u>Arch. Microbiol.,</u> 1980, **124**, 185-9.

44. Holthuijzen, Y.A., van Dissel-Emiliani, F.F.M., Kuenen, J.G. and Konings, W.N., Energetic aspects of CO_2 uptake in <u>Thiobacillus neapolitanus.</u> <u>Arch. Microbiol.</u>, 1987, **147**, 285-90.

45. Omata, T. and Ogawa, T., Changes in the polypeptide composition of the cytoplasmic membrane in the cyanobacterium <u>Anacystis</u> nidulans during adaptation to low CO_2 concentrations. <u>Plant Cell Physiol.,</u> 1985, **26**, 1075-81.

46. Fuchs, G. and Stupperich, E., Carbon assimilation pathways in archaebacteria. In <u>Archaebacteria</u> '85, ed., O. Kandler and W. Zillig, Gustav Fischer Verlag, Stuttgart, 1986, pp. 364-9.

47. Holo, H. and Sirevag, R., Autotrophic growth and CO_2 fixation of <u>Chloroflexus aurantiacus.</u> <u>Arch. Microbiol.</u>, 1986, **145**, 173-80.

48. Schafer, S., Barkowski, C. and Fuchs, G., Carbon assimilation by the autotrophic thermophilic archaebacterium <u>Thermoproteus neutrophilus.</u> <u>Arch. Microbiol.,</u> **146**, 1986, 301-8.

49. Schauder, R., Widdel, F. and Fuchs, G., Carbon assimilation pathways in sulfate-reducing bacteria. II. Enzymes of a reductive citric acid cycle in the autotrophic <u>Desulfobacter hydrogenophilus.</u> <u>Arch. Microbiol.,</u> **148**, 1987, 218-25.

ISOLATION AND CHARACTERISATION OF PLASMIDS PRESENT IN THERMUS STRAINS
FROM YELLOWSTONE NATIONAL PARK

N.D.H. RAVEN AND R.A.D. WILLIAMS
Biochemistry Department,
The London Hospital Medical College,
Turner Street, London E1 2AD, U.K.

ABSTRACT

The plasmid profiles of forty-eight Thermus strains from Yellowstone National Park have been determined. Two plasmids, pTYS14-1 (2.0Kb) and pTYS45-1 (5.9Kb) have been characterised by restriction endonuclease mapping to facilitate their use as cloning vectors in Thermus. Plasmid pTYS45-2 (12.1Kb) has been shown to be cured at high frequency from Thermus strain YS45 by freeze-drying. Phenotypic comparisons between the cured derivative and the wild type strain revealed three significant differences. None of these differences proved usable as a selectable marker. Chromosomal DNA from streptomycin-resistant mutants was used to investigate transformation in these isolates. A plasmid minus strain YS44 has been shown to be transformable by both chromosomal and plasmid DNA.

INTRODUCTION

Plasmid purification procedures are an obvious prerequisite for plasmid cloning and are used both for the isolation of the initial vector DNA and in the characterisation of recombinant clones. The presence of plasmid DNA in the genus Thermus has been demonstrated [1,2,3,4] and restriction endonuclease maps of a number of these plasmids have been produced [5,6,7].

This report describes the development of rapid reproducible, small and large scale plasmid isolation procedures optimised for the genus Thermus as assessed by their application to a group of strains isolated from Yellowstone National Park. These procedures should also prove applicable to the rapid screening of recombinant clones in Thermus that would be required with the development of an efficient host-vector cloning system.

MATERIALS AND METHODS
Organisms and growth conditions
Forty-eight Thermus strains isolated from Yellowstone National Park, USA [3,8] were kindly provided by Dr. R.J. Sharp. Eight reference strains were also screened: Thermus aquaticus YT1 and YVII-51B [9], Thermus sp. strain X1 [10], Thermus thermophilus HB8 [11], Thermus flavus AT62 [12], Thermus sp. strain T2 [13], Thermus sp. strain B [14] and Thermus ruber [15]. All of the strains were cultured with shaking at 65°C in medium D [16] supplemented with 0.3% Difco Bacto-Tryptone and 0.1% Difco Bacto yeast extract (pH 7.6 at 25°C).

Isolation of plasmid DNA - small scale
The described procedure is for 1.5 ml aliquots of an overnight culture of Thermus or fresh single colonies from plates. All manipulations are carried out at room temperature on the bench. Eppendorf-type tubes (1.5 ml) are used throughout and all centrifugations are carried out at maximum speed in a microcentrifuge capable of generating 8-10,000 x g.

1. Centrifuge the 1.5 ml culture aliquots for 2 minutes, decant the supernatant and resuspend each pellet in 100 µl of STE buffer (25% sucrose - 10 mM Tris HCl, 0.1 mM EDTA, pH 8.0).
2. Add 200 µl NSE buffer (0.3 M NaOH, 4% SDS, 100 mM EDTA), mix and incubate for 5 minutes.
3. Add 200 µl PEB III (3 M sodium acetate, pH 6.0), vortex for 5 seconds then centrifuge for 5 minutes.
4. Transfer 450 µl of the supernatant, avoiding the pellet and pellicle, to a new tube and add 2 volumes of ethanol. Mix by inversion then centrifuge for 10 minutes. Discard the supernatant.
5. Dissolve the pellet in 50 µl of PEB I (50 mM glucose, 25 mM Tris HCl, 10 mM EDTA, pH 8.0), add 100 µl of PEB II (0.1 M NaOH, 1% SDS), mix by inversion and incubate for 5 minutes [17,18].
6. Add 75 µl of PEB III, vortex for 5 seconds than centrifuge for 5 minutes.
7. Transfer the supernatant to a new tube, add 2 volumes of ethanol, mix by inversion, then centrifuge for 10 minutes. Discard the supernatant.
8. Dissolve the pellet in 50 µl T.E. buffer (10 mM Tris HCl, 0.1 mM EDTA, pH 8.0). Add 50 µl 5 M ammonium acetate, incubate the mixture for 10 minutes, then centrifuge for 5 minutes.
9. Transfer the supernatant to a new tube, add 2 volumes of ethanol and centrifuge for 10 minutes. Drain the tube of liquid, wash the pellet with a 70% ethanol solution then dry under vacuum.
10. Dissolve the pellet in 20 µl T.E. buffer, add 2 µl 25% ficoll, 0.1% bromophenol blue and electrophorese on a 0.8% agarose gel.

Isolation of plasmid DNA – large scale

The small scale plasmid isolation procedure could be scaled up readily from 1.5 ml volumes to 1 litre or greater with only a number of minor modifications.

1. In order to reduce the total quantities of reagents used, harvested cells were resuspended in proportionately smaller volumes (e.g. 40 ml of STE buffer per litre of original culture). Other reagent volumes were adjusted pro rata.
2. To ensure sufficiently alkaline conditions in the first alkaline denaturation, the pH was monitored using narrow range pH papers (Merck, Spezialindikator pH 11.0-13.0). Dropwise addition with mixing of a 0.5 M sodium hydroxide solution was used where necessary to bring the final pH to 12.1-12.5
3. Residual protein, present after the ammonium acetate precipitation step, was removed by extraction with chloroform/isoamylalcohol (24:1). The supernatant was vortexed for 30 seconds with an equal volume of the organic phase before being centrifuged at 10,000 x g for 5 minutes. The upper phase was then recovered. Extractions were repeated until no material was observed at the interface (normally 3-4 times). Ethanol precipitation of the supernatant was then performed as in the small scale isolation procedure except that the pellets were redissolved in 10 mM Tris HCl, pH 7.5 rather than T.E. buffer.

Separation of plasmid DNA from low molecular weight RNA

At this stage of the purification procedure the only major remaining contaminant of the plasmid is low molecular weight RNA. A rapid method for the removal of this material was devised based upon the observation that HA-Ultrogel (IBF) has a high capacity for low molecular weight nucleic acids such as transfer RNA (approximately 750 µg per ml of gel) but a negligible capacity for nucleic acid molecules of a size above its exclusion limit such as plasmid DNA.

A 1.6 x 10 cm column (total bed volume approximately 20 ml) was sufficient to totally remove low molecular weight RNA from the products of approximately 500 ml of *Thermus* culture. The column was first washed with 2-3 column volumes of 0.5 M potassium phosphate, pH 6.8, at 2 ml per minute, followed by 2-3 column volumes of 10 mM Tris HCl, pH 7.5. The low molecular weight RNA/plasmid mixture also in 10 mM Tris HCl, pH 7.5, was then applied to the column. The plasmid DNA was recovered in the void volume of the column and detected as a sharp peak of high $A_{260/280}$ ratio (>1.8). The DNA was concentrated by butan-2-ol extraction or by the addition of 0.5 volumes of 7.5 M ammonium acetate followed by ethanol precipitation.

The DNA isolated at this stage was normally sufficiently pure that restriction endonuclease digestions could be performed directly. Where residual contaminating materials were suspected caesium chloride/ ethidium bromide equilibrium density gradient centrifugation was performed. The recovered DNA was extracted with water-saturated butan-2-ol to remove ethidium bromide and the caesium chloride removed by dialysis against T.E. buffer [19].

Many strains were found to bear more than one plasmid. Preparative isolation of individual plasmids from these strains was therefore performed by electrophoresing DNA samples in wide slots across an agarose gel. Individual bands were excised and the DNA electroeluted within dialysis bags [20]. The volume of the recovered eluate was reduced by extraction with butan-2-ol. Ether extraction was used to remove residual butan-2-ol followed by the passage of nitrogen gas over the sample to remove traces of ether. The DNA solution obtained was then subjected to caesium chloride/ethidium bromide equilibrium density gradient centrifugation to separate the DNA from soluble impurities, mainly acidic polysaccharides co-eluted from the gel slice. Plasmid DNA was then freed of ethidium bromide and caesium chloride as above.

Gel electrophoresis

Horizontal agarose gels (BRL Ultrapure agarose) were run at 5-8 V/cm in 40 mM Tris acetate, 1 mM EDTA, pH 8.0, containing 0.5 µg/ml ethidium bromide. To calibrate gels, bacteriophage lambda DNA digested with HindIII and a 123bp ladder (BRL) were used as molecular weight standards. Gels were transilluminated with u.v. light (302 nm) and photographed on Ilford FP4 film (5 or 10 second exposure) through a Kodak 23A Wratten gelatin filter.

Restriction enzymes

Most were obtained from Northumbria Biologicals or BRL and used under the conditions described by the manufacturers.

Plasmid curing

Freeze-dried cultures of <u>Thermus</u> strain YS45 were revived by resuspension in 10 ml of growth media followed by overnight incubation at 65°C. Dilutions of the culture were plated out on solidified growth medium (25 g agar per litre) to isolate single colonies. Twelve cultures of the strain were then grown, each derived from 10 pooled colonies. A plasmid preparation was then made from each of the cultures and dilutions of the one showing the least amount of pTYS45-2 were plated out. Single colonies from these plates were then screened for "curing" of pTYS45-2.

Plasmid curing was also attempted using novobiocin by the method of Vasquez et al [21] as modified by Becker et al [4].

Antibiotic and chemical sensitivity
Plates containing solidified growth medium were flooded with cultures of wild-type strain YS45 and its cured derivative then air-dried until no free fluid was observed on the agar surface. Oxoid "Multodiscs" or single test discs were applied to the plates for antibiotic sensitivity testing. For chemical sensitivity 10 µl aliquots of 0.1 M stock solutions of various salts were dried onto sterile filter paper discs. These discs were then applied to the plates in the same manner. The zone sizes, from the edge of the discs to the border of the bacterial culture, were recorded after 2 days incubation at $65^{\circ}C$.

Bacterial transformation
Transformation was carried out essentially as described by Koyama et al [22]. Plasmid transformants were detected by colony hybridisation [23] using photobiotin-labelled [24], RsaI digested, pTYS14-1 DNA as a probe. High backgrounds in the procedure were avoided by use of more effective washing steps performed in a temperature-controlled chamber on a roller bottle apparatus [25].

RESULTS AND DISCUSSION
Thermus cell lysis
Probably the two most widely used rapid plasmid isolation procedures are the boiling method of Holmes and Quigley [26] and the alkaline lysis method of Birnboim and Doly [18]. The boiling method is entirely unsuitable for thermophiles as it relies upon rapid thermal inactivation and denaturation of host proteins. The alkaline lysis method was therefore chosen as the method most likely to give positive results with Thermus. Plasmid bands were detected in some strains using this method, however, many strains were found either not to be lysed by this procedure or to be lysed unacceptably slowly. More vigorous lysis procedures were therefore sought.

Witholt et al [27] observed that stationary phase cells of many E.coli strains appeared to be insensitive to lysozyme as assessed by their lack of conversion to spheroplasts. Exposure to plasmolysing buffers followed by a mild osmotic shock (2-fold dilution) in the presence of EDTA and lysozyme, however, rapidly caused conversion. This procedure was tested with Thermus YS45 which was insensitive to the Birnboim and Doly procedure. Cells were resuspended in 25% sucrose in T.E. buffer, pH 8.0, with 1 mg/ml lysozyme (L.STE buffer) at $0^{\circ}C$, room temperature and 42°C. Samples were then diluted with an equal volume of 0.1 M EDTA containing alkaline SDS. Lysis was immediately apparent in samples held at room temperature and at 42°C; the mixtures became clear and their viscosities increased considerably.

The high sucrose/mild osmotic shock procedure which lysed Thermus YS45 was tested with a selection of other Thermus strains. Rapid and complete lysis was observed in all cases indicating the method's broad applicability as a rapid lysis procedure for Thermus. Subsequent studies revealed that lysozyme was not required for lysis and that room temperature was more satisfactory than 42°C as endogenous nucleases were less active at the lower temperature.

Alkaline extraction of Thermus lysates

To investigate the conditions appropriate for alkaline extraction of Thermus lysates both the SDS and sodium hydroxide concentrations were varied. Cell pellets were resuspended in 100 µl STE buffer and 50 µl of 0.25 M EDTA, pH 7.0 was added, followed rapidly by from 5-50 µl of 20% SDS and 60 µl of 1 M sodium hydroxide solution. It was found that 50-60 µl of 1 M sodium hydroxide was sufficient to bring the lysate to pH 12.0-12.5. Rapid neutralization was achieved by addition with mixing of 200 µl of 3 M sodium acetate, pH 6.0 (PEB III). The nucleic acids were then recovered as in the Birnboim and Doly procedure. Agarose gel electrophoresis resulted in the detection of three DNA bands in addition to low molecular weight RNA. The use of 20-40 µl of 20% SDS gave similar results under these conditions. With less than 20 µl, more background was observed on the gels, as well as fluorescent material in the gel wells due to residual DNA-binding proteins. Above 40 µl an additional weakly absorbing band due to residual SDS was observed on the gels. To investigate the optimum alkaline conditions for the extraction 30 µl of 20% SDS was therefore chosen. HindIII restriction enzyme digestion of the DNA contained in the upper two bands observed on the gels gave discrete (but different) banding patterns, suggesting that they were plasmids. The highest molecular weight band gave a smear indicating that this band was residual chromosomal DNA.

The pH at which strands of DNA separate is dependent not only upon conformation but also upon the G+C content of the DNA [28]. Thermus strains so far tested have been found to have a G+C content 12-15% greater than Escherichia coli, the organism upon which the Birnboim and Doly procedure was developed. The pH at which the onset of denaturation occurs in Thermus DNA would be approximately 0.3 pH units higher than for E.coli by comparison with the examples used by Zimmer. It was, therefore, thought that a higher pH might be necessary to completely denature the Thermus chromosomal DNA. To test this 30 samples of Thermus YS45 were resuspended in STE buffer as before. Sixty µl of

0.25 M EDTA and 30 µl of 20% SDS were added, followed by from 5-150 µl of 1 M sodium hydroxide in 5 µl intervals. Only the highest alkali additions (140-150 µl) removed all of the chromosomal DNA, but this treatment also removed all of the plasmid DNA (Fig. 1). It was therefore apparent that at no pH could the chromosomal DNA be separated from the plasmid DNA by a single alkaline denaturation. The optimum separation appeared to be given at around 60 µl of alkali as in the original procedure. Under more alkaline conditions increasing amounts of irreversibly denatured supercoiled DNA were observed, at lower pH values increasing amounts of background fluorescence and inadequate lysis were observed.

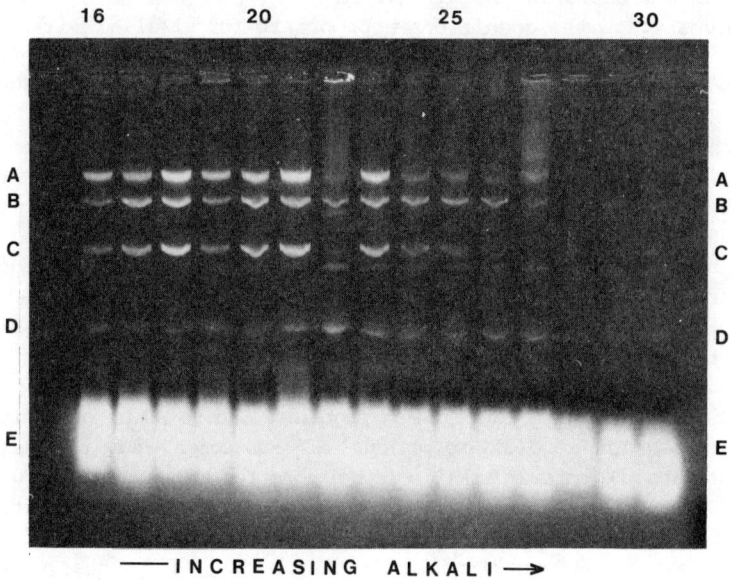

Figure 1. Effect of increasing alkali upon plasmid isolation and the elimination of chromosomal DNA. Samples 16-30 are additions of 80-150 µl 0.5 M NaOH in 5 µl intervals.
A, chromosomal DNA; B, pTYS45-2; C, pTYS45-1; D, collapsed supercoiled DNA; E, low molecular weight RNA.

In larger scale versions of the Birnboim and Doly procedure [17,29], residual chromosomal DNA was also frequently observed and a second alkaline denaturation step was required to remove this material. Pellets obtained by the high sucrose/alkaline denaturation step for Thermus YS45 were, therefore, subjected to a second alkaline denaturation, identical to that of Birnboim and Doly except that the pH of the neutralizing buffer was 6.0 not 4.8. Agarose gel electrophoresis of the samples obtained gave the two plasmid bands alone in over 80% of cases. Where four replicas were used the presence of two plasmids could be reliably deduced in over 95% of cases (Fig. 2).

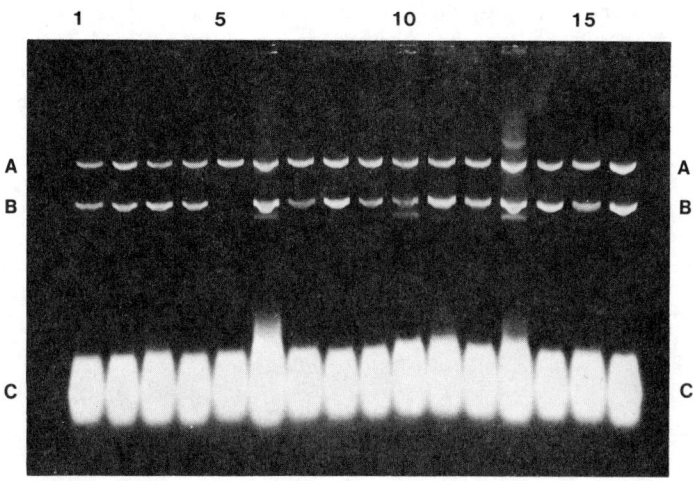

Figure 2. Four x four replica (1-4, 5-8, 9-12, 13-16) plasmid preparations from Thermus YS45. The presence of two plasmids can be observed in each case.
    A, pTYS45-2; B, pTYS45-1; C, low molecular weight RNA.

High salt extraction
Trials of the plasmid isolation procedure were extended to additional Thermus strains. Generally reasonable plasmid preparations were obtained using this method, but certain strains still exhibited relatively high backgrounds on agarose gels, with evidence of residual material in the wells. In the procedure of Marko et al [17] 5 M lithium chloride was used to reduce the background attributed to residual ribosomal RNA and single-stranded DNA. Cornelis et al [30] used 2 M ammonium sulphate to achieve the same result. It was found by experiment, however, that an equivalent separation was given with 2.5 M ammonium acetate. This had the additional advantage that proteins are generally soluble in alcoholic solutions of 2.5 M ammonium acetate [31] resulting in ethanol precipitates with a higher proportion of nucleic acid. Agarose gel electrophoresis of these precipitates, washed with 70% ethanol to remove residual ammonium acetate, and dissolved in T.E. buffer, pH 8.0, gave sufficiently clear and reproducible results to suggest the overall method's general applicability as a rapid plasmid isolation procedure for Thermus.

Plasmid isolation

The 56 Thermus strains represented in this study are those that were screened for plasmid content by Munster et al [3], using an in-situ lysis procedure [32]. These were chosen as they had been shown by Munster et al to be a rich source of plasmids. Also, apart from the reference strains, they had all been isolated at the same time and handled in a similar manner. Any differences in plasmid content could not, therefore, be ascribed to differences in treatment or storage. It was felt that this could improve the prospect of correlating bacterial phenotypes with the presence of particular plasmids.

TABLE 1
Plasmid bands identified by agarose gel electrophoresis

Strain	No.of bands	Strain	No.of bands
YT1	4	AT62	2
YVII-51B	2	T2	1
HB8	2	B	3
X1	6	T.ruber	1
YS01	2	YS26	3
YS02	1	YS27	2
YS03	1	YS28	7
YS04	4	YS29	6
YS05	2	YS30	2
YS07	2	YS31	2
YS08	2	YS32	3
YS09	1	YS33	7
YS10	1	YS34	8
YS11	1	YS35	1
YS12	3	YS36	2
YS13	4	YS37	2
YS14	2	YS38	2
YS15	1	YS39	1
YS16	3	YS40	1
YS17	2	YS41	3
YS18	1	YS44	0
YS19	1	YS45	2
YS20	2	YS47	2
YS21	3	YS48	2
YS22	2	YS49	2
YS23	5	YS50	5
YS24	3	YS51	1
YS25	1	YS52	2

Only four of the strains examined exhibited less DNA bands than were observed in the above study. Providing that each band can be confirmed to be a plasmid (as opposed to being chromosomal DNA or conformational variants of a single plasmid), it would appear that this method is able to detect more plasmids than the in-situ lysis procedure. Thermus YS45 was confirmed as carrying two plasmids during the development of the alkaline extraction procedure. The detection of only one of these plasmids in the in-situ lysis procedure appears to be due to co-migration, under the conditions used, of one of the plasmids with the chromosomal DNA band. Both methods clearly show, however, that many different plasmids are present in the Thermus strains, and that a large number of strains bear multiple plasmids (Table 1).

Restriction endonuclease mapping

A restriction map of the Thermus plasmid pTYS45-1 has already been reported [7]. Subsequent to the mapping of this plasmid, it was discovered that the host strain Thermus YS45 does not bear the Taq I (TCGA) restriction/modification system. In order to investigate transformation in other strains, many of which do possess the Taq I system, it was felt that the isolation and characterisation of a plasmid bearing this methylation pattern would be desirable. Thermus strain YS14 was observed to possess a small Taq I-insensitive plasmid which could be isolated in quantity, without recourse to an agarose gel electrophoresis step (as the second, much larger, low copy number plasmid detected in the strain was isolated in very low amounts). This plasmid, designated pTYS14-1 was then digested with 36 different restriction enzymes, of which six were found to possess single sites. Double-digests were performed with these six enzymes and a coherent circular map of the sites was produced (Fig. 3).

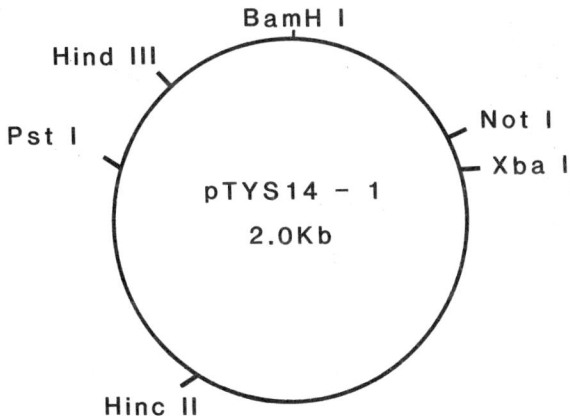

Figure 3. Restriction map of pTYS14-1.

Plasmid curing

Previous large scale isolations of Thermus YS45 plasmid DNA indicated that the yield of pTYS45-2 varied considerably when isolated from cultures which had been grown from freeze-dried ampoules. This effect was not observed for pTYS45-1 suggesting that freeze-drying had "cured" pTYS45-2 from a proportion of the cells. Plasmid preparations of cultures derived from a small number of founder cells exhibited even greater variability in the amount of pTYS45-2 (Fig. 4).

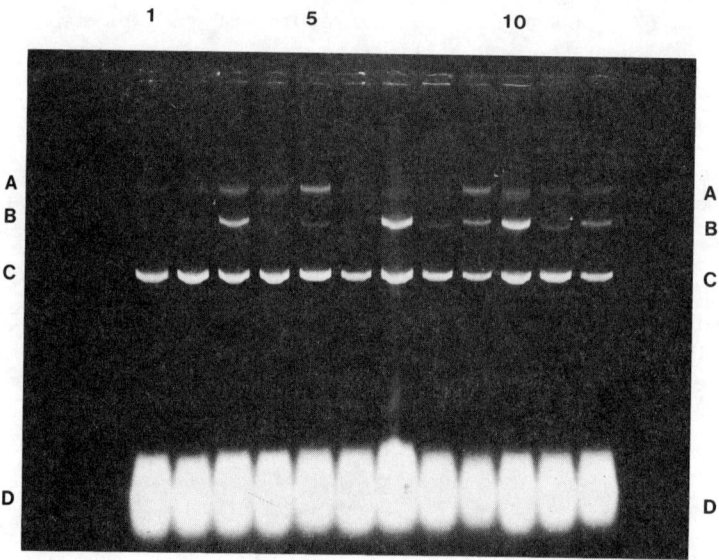

Figure 4. Single alkaline denaturation plasmid preparations from 12 Thermus YS45 cultures, each derived from 10 colonies. Initial individual colonies were isolated by plating out dilutions of a 10 ml overnight culture of Thermus YS45 derived from a freeze-dried ampoule of the organism.
A,chromosomal DNA; B,pTYS45-2; C,pTYS45-1; D,low molecular weight RNA.

Plasmid screening of single colonies derived from the culture exhibiting the least amount of pTYS45-2 (Fig. 5) revealed cured isolates at high frequency (>50%).

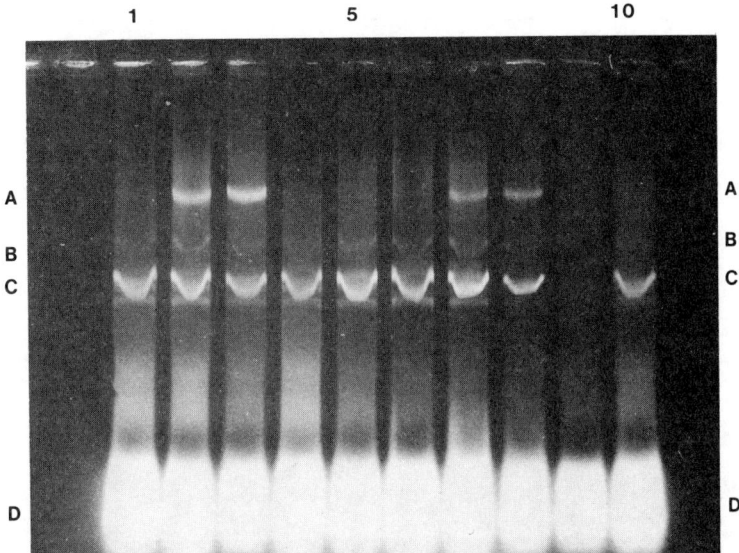

Figure 5. Plasmid preparations from 10 Thermus YS45 cultures, each grown from individual colonies isolated by plating out dilutions of Thermus YS45-6 (Fig. 4, lane 6). The plasmid preparation in lane 9 was later shown to be that of a contaminating B.stearothermophilus-like organism. Five of the remaining nine cultures were confirmed to be cured of pTYS45-2.
A, pTYS45-2; B, o.c. pTYS45-1; C, c.c.c, pTYS45-1; D, low molecular weight RNA.

No curing effect was observed using freeze-drying with either pTYS45-1 or pTYS14-1, therefore the use of the DNA gyrase inhibitor, novobiocin was tested.

No curing was observed for pTYS45-1 or pTYS14-1, however, very little covalently closed circular pTYS45-2 was observed suggesting that, like the freeze-drying procedure, novobiocin has an effect on its stability (Fig. 6).

Antibiotic and chemical sensitivity

The pTYS45-2 cured YS45 isolate was compared with the wild isolate for any differences in sensitivity to a wide range of antimicrobial and chemical agents. Inhibition zone sizes were sufficiently large with particular antibiotics, e.g. beta-lactams, that multodiscs could not be used in the normal manner. Larger diameter plates (14 cm) were used along with multodiscs missing alternate spokes (i.e. four discs per plate). The two isolates exhibited no significant differences in relative sensitivity to the following antibiotics: ampicillin,

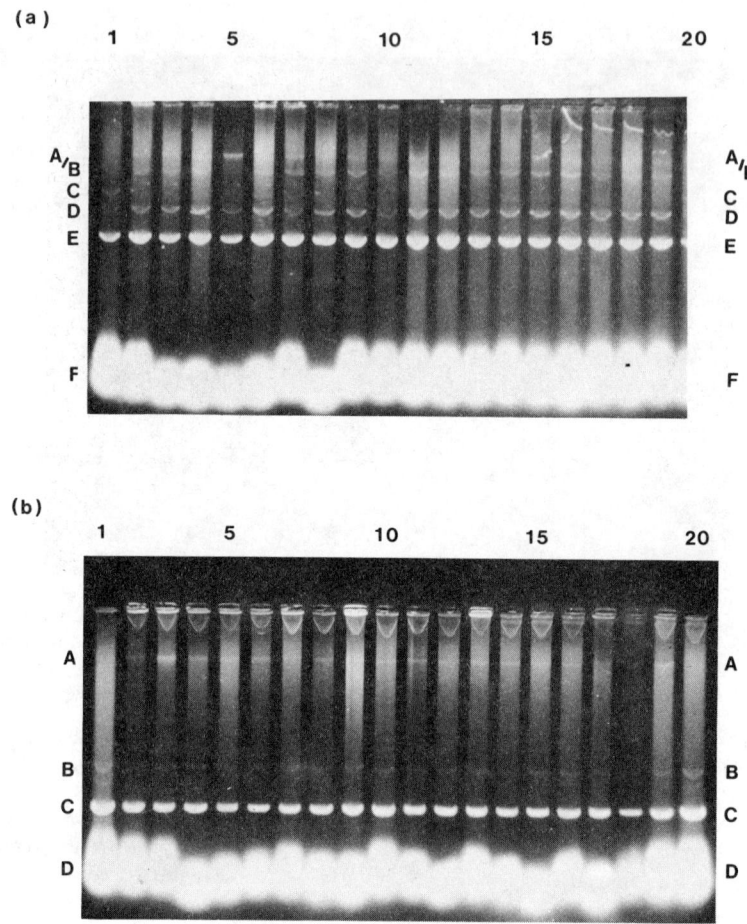

Figure 6(a). Plasmid preparation from untreated Thermus YS45 cultures (lanes 1 and 20) and novobiocin treated Thermus YS45 cultures (lanes 2-19). No covalently closed circular pTYS45-2 is observed in the treated cultures.
A/B, linear/open circular forms of pTYS45-2; C, covalently closed circular pTYS45-2; D, open circular pTYS45-1; E, c.c.c. pTYS45-1; F, low molecular weight RNA.

Figure 6(b). Plasmid preparations from untreated Thermus YS14 cultures (lanes 1 and 20) and novobiocin treated Thermus YS14 cultures (lanes 2-19).
A, pTYS14-2; B, open circular pTYS14-1; C, c.c.c. pTYS14-1; D, low molecular weight RNA.

cephaloridine, chloramphenicol, chlortetracycline, cloxacillin, colistin sulphate, co-trimoxazole, erythromycin, fusidic acid, gentamicin, kanamycin, lincomycin, nalidixic acid, neomycin, novobiocin, oleandomycin, oxytetracycline, penicillin G, rifampicin, spectinomycin, streptomycin, sulphafurazole, tetracycline and vancomycin. Two significant differences were observed: furazolidone gave the same size inhibition zone for both the wild isolate and its cured derivative, however, resistant colonies within this zone were found only with the wild isolate. Secondly the cured isolate had an inhibition zone four times larger than the wild isolate with nitrofurantoin. In neither case, however, could a concentration of antibiotic be found where only the pTYS45-2 bearing wild isolate was able to grow.

No significant differences were observed between the two isolates with the following cations: antimony, bismuth, cadmium, cobalt, lead, mercury, nickel, silver, thallium and zinc, nor the anions: arsenate, arsenite, borate and chromate. The cured isolate was, however, found to be more sensitive to tellurite than the wild-type. No concentration of tellurite was found where the wild isolate was able to grow but the cured derivative was suppressed. At 10^{-4} M tellurite approximately 80 times as many colonies were produced by the wild isolate as the cured isolate. Due to the frequency with which chromosomal mutations arose this difference in susceptibility could not, however, be amplified to make the altered tellurite resistance a selectable marker for the presence of pTYS45-2.

Bacterial transformation
With the demonstration of transformation in Thermus by Koyama et al [22] a more direct approach to the detection of selectable markers carried by Thermus plasmids could be attempted. In order to assess which of the Yellowstone National Park isolates might be most suitable as a host for plasmid transformation experiments their ability to be transformed by exogenous DNA was tested. Chromosomal DNA from streptomycin-resistant mutants of strains YS14 and YS45, bearing their differing methylation patterns, was used to investigate both the ability of the strains to take up DNA and also to indicate in which strains appropriate methylation of the transforming DNA might be required.

Twenty-two of the strains could not be transformed to streptomycin resistance with DNA from either source (Table 2). One of these strains was YS14 itself, suggesting that inappropriate methylation of the DNA was not responsible for the absence of transformation in at least some of these strains. In the 26 strains where transformants were produced it was possible to examine the effect of this methylation (Table 3).

TABLE 2
Transformation (chromosomal DNA) per 5×10^7 cells
1 µg/ml Str^R DNA YS14 and YS45
Streptomycin resistants (50 µg/ml)

	Range	Mean	No.of strains
Controls (no DNA)	0-35	12 $\pm$ 11	48
No transformants	4-31	15 $\pm$ 8	22
Low frequency transformation	550-2880	1540 $\pm$ 745	10
High frequency transformation	4640-16800	9560 $\pm$ 4300	16

TABLE 3
Strain dependence on transformation (26 strains)
Ratios YS14:YS45

	Range	Mean	No.of strains
No bias	0.78- 1.55	1.13 $\pm$ 0.25	10
YS45>YS14	0.31- 0.56	0.41 $\pm$ 0.11	5
YS45>>YS14	0.05	-	1
YS14>YS45	2.01- 5.71	4.31 $\pm$ 1.41	5
YS14>>YS45	9.17-27.90	16.53 $\pm$ 7.23	5

Of the 48 strains, only six were found that gave transformants at a high level with DNA from both strains. Fortuitously one of these strains was YS44, the only strain which had not been observed to bear a plasmid during screening. This strain was selected for plasmid transformation experiments as, although it did not have the highest transformation frequency found, it could not generate falsely-positive transformants due to the presence of (or by recombination with) an endogenous homologous plasmid.

Plasmid pTYS14-1 was used to transform _Thermus_ strain YS44 as described by Koyama et al [22]. Seventy positive clones were detected out of 10^5 screened. Sub-culturing of the replica colonies encompassing a positive signal was used to isolate a single positive clone. Plasmid DNA isolated from this clone was of the same size and had the same restriction enzyme digest pattern as pTYS14-1.

The minimum inhibitory concentrations of a wide range of chemical and antibacterial agents for _Thermus_ YS44 have been determined. Additionally, other biochemical properties of this strain were determined by Munster et al during a numerical phenetic survey of all of the Yellowstone National Park isolates [8]. Any change in the phenotypes of Thermus strain YS44 generated by transformation with a native Thermus plasmid (or by transformation with plasmid DNA ligated to a fragment encoding a selectable marker) can now be readily determined.

CONCLUSIONS

The Yellowstone National Park Thermus isolates of Munster et al [3] are an abundant source of plasmid DNA. The demonstration of transformation in these strains allows the rapid screening of these plasmids for selectable markers of potential use in a Thermus cloning system in a more direct manner than by plasmid curing.

REFERENCES

1 Hishinuma, F., Tanaka, T. and Sakaguchi, K., Isolation of extrachromosomal deoxyribonucleic acids from extremely thermophilic bacteria. J. Gen. Microbiol., 1978, **104**, 193-9.

2. Koh, C-K., Detection and purification of plasmids present in Thermus strains from Iceland hot springs. MIRCEN Journal, 1985, **1**, 77-81.

3. Munster, M.J., Munster, A.P. and Sharp, R.J., Incidence of plasmids in Thermus sp. isolated in Yellowstone National Park. Appl. Environ. Microbiol., 1985, **50**, 1325-7.

4. Becker, D.A., Glass, K.A. and Starzyk, M.J., Isolation and curing of extrachromosomal DNA in Thermus aquaticus. Microbios, 1986, **48**, 71-9.

5. Eberhard, M.D., Vasquez, C., Valenzuela, P., Vicuna, R. and Yudelevich, A., Physical characterization of a plasmid (pTT1) isolated from Thermus thermophilus. Plasmid. 1981, **6**, 1-6.

6. Vasquez, C., Venegas, A. and Vicuna, R., Characterization and cloning of a plasmid isolated from the extreme thermophile Thermus flavus AT-62. Biochem. Int., 1981, 3, 291-9.

7. Raven, N.D.H. and Williams, R.A.D., Isolation and partial characterization of two cryptic plasmids from an extreme thermophile. Biochem. Soc. Trans., 1985, 13, 214.

8. Munster, M.J., Munster, A.P., Woodrow, J.R. and Sharp, R.J., Isolation and preliminary taxonomic studies of Thermus strains. J. Gen. Microbiol., 1986, **132**, 1677-83.

9. Brock, T.D. and Freeze, H., Thermus aquaticus gen. n. and sp. n., a non-sporulating extreme thermophile. J. Bacteriol., 1969, 98, 289-97.

10. Ramaley, R.J. and Hixson, J., Isolation of a non-pigmented thermophilic bacterium similar to Thermus aquaticus. J. Bacteriol., 1970, 103, 527-8.

11. Oshima, T., Studies on an extreme thermophile Flavobacterium thermophilum HB8. In Molecular Evolution: Pre-Biological and Biological, ed. D.L. Rohlfing and A.I. Oparin, Plenum Press, New York and London, 1972, pp.399-423.

12. Saiki, T., Kimura, R. and Arima, K., Isolation and characterization of extremely thermophilic bacteria from hot springs. Agr. Biol. Chem., 1973, 36, 2357-66.

13. Ulrich, J.T., McFeters, G.A. and Temple, K.L., Induction and characterization of beta-galactosidase in an extreme thermophile. J. Bacteriol., 1973, 110, 691-8.

14. Williams, R.A.D., Caldoactive and thermophilic bacteria and their thermostable proteins. Sci. Prog., Oxf., 1975, 62, 373-93.

15. Loginova, L.G. and Egorova, L.A., An obligately thermophilic bacterium Thermus ruber from hot springs in Kamchatka. Mikrobiologiya, 1975, 44, 661-5.

16. Castenholz, R.W., Aggregation in a thermophilic Oscillatoria. Nature, 1967, 215, 1285-6.

17. Marko, M.A., Chipperfield, R. and Birnboim, H.C., A procedure for the large-scale isolation of highly purified plasmid DNA using alkaline extraction and binding to glass powder. Anal. Biochem., 1982, 121, 382-7.

18. Birnboim, H.C. and Doly, J., A rapid alkaline extraction procedure for screening recombinant plasmid DNA. Nucl. Acids Res., 1979, 7, 1513-23.

19. Maniatis, T., Fritsch, E.F. and Sambrook, J., Purification of closed-circular DNA by centrifugation to equilibrium in cesium chloride-ethidium bromide gradients. In Molecular Cloning: A Laboratory Manual, Cold Spring Harbor Laboratory, New York, 1982, pp.93-94.

20. Maniatis, T., Fritsch, E.F. and Sambrook, J., Recovery of DNA from agarose gels. In Molecular Cloning: A Laboratory Manual, Cold Spring Harbor Laboratory, New York, 1983, pp.164-65.

21. Vasquez, C., Villanueva, J. and Vicuna, R., Plasmid curing in Thermus thermophilus and Thermus flavus. FEBS Letters, 1983, 158, 339-42.

22. Koyama, Y., Hoshino, T., Tomizuka, N. and Furukawa, K., Genetic transformation of the extreme thermophile Thermus thermophilus and other Thermus spp. J. Bacteriol., 1986, 166, 338-40.

23. Mason, P.J. and Williams, J.G., Hybridisation in the analysis of recombinant DNA. In Nucleic Acid Hybridisation - A Practical Approach, ed., Hames, B.D. and Higgins, S.J., IRL Press, Oxford, 1985, pp.113-37.

24. Forster, A.C., McInnes, J.L., Skingle, D.C. and Symons, R.H., Non-radioactive hybridization probes prepared by the chemical labelling of DNA and RNA with a novel reagent, photobiotin. Nucl. Acids Res., 1985, 13, 745-61.

25. Howell, M.D., Austin, R.K. and Kagnoff, M.F., An improved method for nucleic acid blot hybridization. Focus (BRL), 1987, 9:2, 10.

26. Holmes, D.S. and Quigley, M., A rapid boiling method for the preparation of bacterial plasmids. Anal. Biochem., 1981, 114, 193-7.

27. Witholt, B., Van Heerikhuizen, H. and De Leij, L., How does lysozyme penetrate through the bacterial outer membrane? Biochim. Biophys. Acta, 1976, 443, 534-44.

28. Zimmer, Ch., Alkaline denaturation of DNAs from various sources. Biochim. Biophys. Acta, 1968, 161, 584-6.

29. Birnboim, H.C., A rapid alkaline extraction method for the isolation of plasmid DNA. Methods in Enzymology, Recombinant DNA (Part B), 1983, 100, 243-55.

30. Cornelis, P., Digneffe, C., Willemot and Colson, C., Purification of Escherichia coli amplifiable plasmids by high-salt sepharose chromatography. Plasmid, 1981, 5, 221-3.

31. Crouse, J. and Amorese, D., Ethanol precipitation: ammonium acetate as an alternative to sodium acetate. Focus (BRL), 1987, 9:2, 3-5.

32. Eckhardt, T., A rapid method for the identification of plasmid DNA in bacteria. Plasmid, 1978, 1, 584-8.

TAXONOMIC AND GENETIC STUDIES OF BACILLUS THERMOPHILUS

RICHARD SHARP, MICHAEL MUNSTER, ALAN VIVIAN*, SHAMIN AHMAD[+]
AND TONY ATKINSON
Division of Biotechnology, PHLS Centre for Applied Microbiology & Research, Porton Down, Salisbury; * Bristol Polytechnic, Bristol; [+] Trent Polytechnic, Nottingham, United Kingdom.

ABSTRACT

Until recently thermophilic Bacilli were classified as either B. coagulans or B. stearothermophilus, but it is becoming increasingly evident that they comprise a diverse group of organisms. In addition to these two, there are at least eight other groups which appear to merit the rank of species. Genetic studies of thermophilic Bacilli have also emphasised these differences. Genetic techniques which have been developed for B. stearothermophilus are not readily applicable to other thermophilic species of Bacillus.

TAXONOMIC STUDIES OF THERMOPHILIC BACILLI

It is 100 years since Miquel [1] isolated a thermophilic aerobic spore-former from the river Seine. The organism was able to grow at 73°C and was named Bacillus thermophilus.

In the first edition of Bergeys manual [2], 15 species of thermophilic Bacillus had been described. In the sixth edition [3] some eighteen thermophilic species able to grow above 60°C were described. Many other isolates were described, although not listed as separate species due to insufficient distinguishing characteristics.

The first systematic study of aerobic thermophilic sporeformers was made by Ruth Gordon and Nathan Smith in which they examine 216 cultures from a variety of sources. Their extensive investigations [4], [5] resulted in the 7th Edition of Bergeys Manual (Breed et al, 1957) listing only two species with optimum growth at 55°C or above. These two species were B. stearothermophilus [6] and B. coagulans [7]. Similar taxonomic studies carried out by Gyllenberg [8], Stark and Tetrault [9] and Allen [10] appeared on the whole to confirm the conclusions of Gordon and Smith although Stark and Tetrault considered their strains more closely resembled the description of B. kaustophilus [11] than B. stearothermophilus [6]. All three papers however, reported the presence of a few strains unable to hydrolyse starch, a characterstic previously considered positive for B. stearothermopilus. Daron [12] and Epstein and Grossowicz [13] described the isolation of amylase negative strains and subsequent examination by Gordon et al [14] confirmed these isolates as amylase negative. The description of B. stearothermophilus in the eighth edition of Bergeys Manual [15] described the distinguishing features as; maximum growth 65-75°C, a minimum growth temperature of 30-45°C, inability to grow on sabauraud dextrose agar and in 0.02% (w/v) sodium azide, starch hydrolysis was considered to be variable.

Klaushofer and Hollaus [16] reported a taxonomic study of 84 thermophilic strains of Bacillus isolated during sugar beet extraction. In addition, for reference purposes, they include B. stearothermophilus NCA 26 (the original strain of Donk [6]), NCA 1503 and ATCC 8005, (B. kaustophilus, [11]). The cultures were examined using the tests of Smith et al [5]. Sixty-eight characters were examined by numerical taxonomy using the simple matching coefficient [17]. Six taxonomic groups were apparent, two were typified by their inability to hydrolyse starch, others represented B. stearothermophilus, B. thermodenitrificans, B. kaustophilus and B. coagulans.

Walker and Wolf [18] reported the result of an extensive study of the physiological, biochemical and serological properties of 230 strains of B. stearothermophilus including, the 75 strains of Smith et al, [5], 16 strains from Galesloot and Labots [19] previously described as B. calidolactis or B. thermoliquefaciens, five strains from Grinstead and Clegg [20], and a number of strains isolated from soil and milk. Using essentially the methods of Smith et al [5] they were able to divide the strains into three distinct major groups with minor sub-groups. These included one major group typified by the inability to hydrolyse starch, and other groups representing B. thermodenitrificans, B. stearothermophilus and B. kaustophilus.

Recent new isolations

Since these studies several other new isolates have been described. Heinen [21] reported the isolation of three thermophilic Bacilli from Yellowstone National Park in the USA which were named B. caldotenax, B. caldovelox and B. caldolyticus. Heinen and Heinen [22] differentiated these strains from B. stearothermophilus on the basis of their temperature optima, fatty acid pattern and sub-microscopical structure. The three strains were distinguished by their readiness to spore, their different temperature optima, and the morphological differences in their cell walls and membranes [22]. Darland and Brock, [23], reported the isolation of B. acidocaldarius, an acidophilic, thermophilic Bacillus able to grow at pH 2.0 with a maximum growth temperature of 70°C.

Golovacheva et al [25] described a strain initially considered to be a thermophilic strain of B. megaterium, which had some unusual characteristics. They considered the organisms to be a new species, and name it Bacillus thermocatenulatus. The isolate produced yellowish colonies, reduced nitrate to gas, had a %G+C content of 69% and maximum growth temperature of 78°C.

Schenk and Aragno [25] described an obligate thermophile which was strictly aerobic, oxidised hydrogen in the presence of O_2 and CO_2 and grew heterotrophically producing spherical spores within a swollen sporangium. The organism which was named B. schlegelii, had an optimum growth temperature of 70°C and a % G+C content of 67-68%.

Suzuki et al [26] reported the isolation of six obligate thermophilic aerobic sporeformers from soil in Japan, which they named Bacillus thermoglucosidasius. Their main characteristics were sensitivity to azide and production of ex-oligo-1,6-glucosidase. Growth occurred from 42-69°C with an optimum of 61-63°C. The % G+C was estimated by thermal denaturation to be 45-46%.

Scholz et al [27] described nine thermophilic Bacillus strains isolated from thermally treated waste water which were considered to differ from previous isolates. The group of strains named B.pallidus had a growth temperature range of 30-70°C with optimum of 60-65°C. The % G+C of the DNA ranged from 39-41 mol%.

Zarilla and Perry [28] report the isolation of 10 obligately thermophilic Bacilli able to utilise n-alkanes as growth substrate. The organisms were isolated following enrichment of mud and water samples on a medium with n-heptadecane added as substrate. Growth occurred between 45-70°C with optimum between 55-65°C. The strains were named B. thermoleovorans and had a G+C content ranging from 52 to 58%.

A NUMERICAL TAXONOMIC STUDY OF THERMOPHILIC BACILLI

Over the past ten years we have carried out isolations of over 1000 thermophilic Bacilli from source material collected from a wide range of geographical locations including Yellowstone National Park, USA, Geothermal areas in New Zealand and soil, water and compost from Zaire, Turkey, Greece, Bulgaria, Hong Kong, Singapore, France India and United Kingdom.

These isolates show a wide variation in colony morphology from flat, rough, erose, irregular colonies (typified by B. thermodenitrificans DSM 465) to raised, smooth, circular, entire colonies (typified by B. stearothermophilus NCA 1503 and B. caldotenax). Several are motile isolates spreading on agar plates which may vary from just a few colonies producing small spreading areas of growth, to strains which covered the entire surface of the plate.

Biochemical and physiological tests showed considerable heterogeneity amongst the strains, ranging from those hydrolysing starch, casein, and gelatin, reducing nitrate and nitrite, and fermenting a wide range of carbohydrates, to those with no apparent amylolytic or proteolytic activity and fermenting very few carbohydrates.

A study of 102 of these strains analysing 96 phenotypic characters using the similarity coefficient of Gower [29] and average linkage analysis, resulted in their allocation to five major clusters at 60% similarity (Fig. 1). At 80% similarity, nine taxonomically distinct groups were evident. The average probability (P) of an erroneous result calculated on the basis of results for the 96 characters on 10 duplicate strains was 2.37%. This was well within the range of error generally considered acceptable [30].

Cluster (1): This was the largest cluster comprising 32 strains and was divided into two sub-groups (1a) and (1b). The major distinguishing characteristics of the strains in sub-group (1a) were; lack of amylolytic activity, little fermentative activity and spreading growth on TSBA (Table 1). This group appear to correspond with group IA and IB of Klaushofer and Hollaus [16]. The majority (72%) of the strains in sub-group (1b) were amylase negative, but they fermented a range of sugars, including glucose (100%), sucrose (100%), trehalose (100%), fructose (89%), glycerol (78%), inositol (89%), maltose (95%), mannose (70%), mannitol (89%), salicin (84%) and sorbitol (95%). These reactions closely resemble the description given by Walker and Wolf to the organisms in their group 2. Strain LUDA T141 from Walker and Wolf's group 2 clustered with this group, as did the starch negative strain EP136, isolated by Epstein and Grossowicz [13]. Citrate was not utilised by strains in sub-group 1a. In addition to their inability to hydrolyse starch these two groups are typified by growth in 5% saline and inability to reduce nitrate and nitrite and lack of proteolytic activity. These groups have some similarity with Bacillus pallidus [27], B. pallidus differs in weakly hydrolysing starch. Its inability to utilise citrate suggests it resembles group 1a, however, group 1a strains do not ferment glucose, fructose or sucrose.

Group 1a has limited degradative capability, and with the total lack of amylolytic acitivity the name Bacillus thermonondiasticus (derived from the original name Bacillus thermonondiasticus given to a thermophilic amylase deficient isolate by Bergey [31] is proposed to describe these strains. The typical reactions of strains within group 1a are, lack of amylolytic activity, ability to tolerate 5% (w/v) saline and growth at 37 and 65°C, no growth at 25°C. Groups 1a and 1b can be differentiated by the ability to utilise citrate and the enhanced fermentative capability of group 1b. It is proposed that strains in group 1b which were unable to liquefy gelatin be designated B. thermononliquefaciens, based on the original description of Bergey et al [31], although differing slightly to the original description.

Cluster (2): This was the second largest cluster comprising 28 strains allocated to four homogeneous sub-groups (2a), (2b), (2c) and (2d) together with a heterogeneous group of nine strains which failed to cluster at the 80% similarity level. Group (2a) was composed of eight

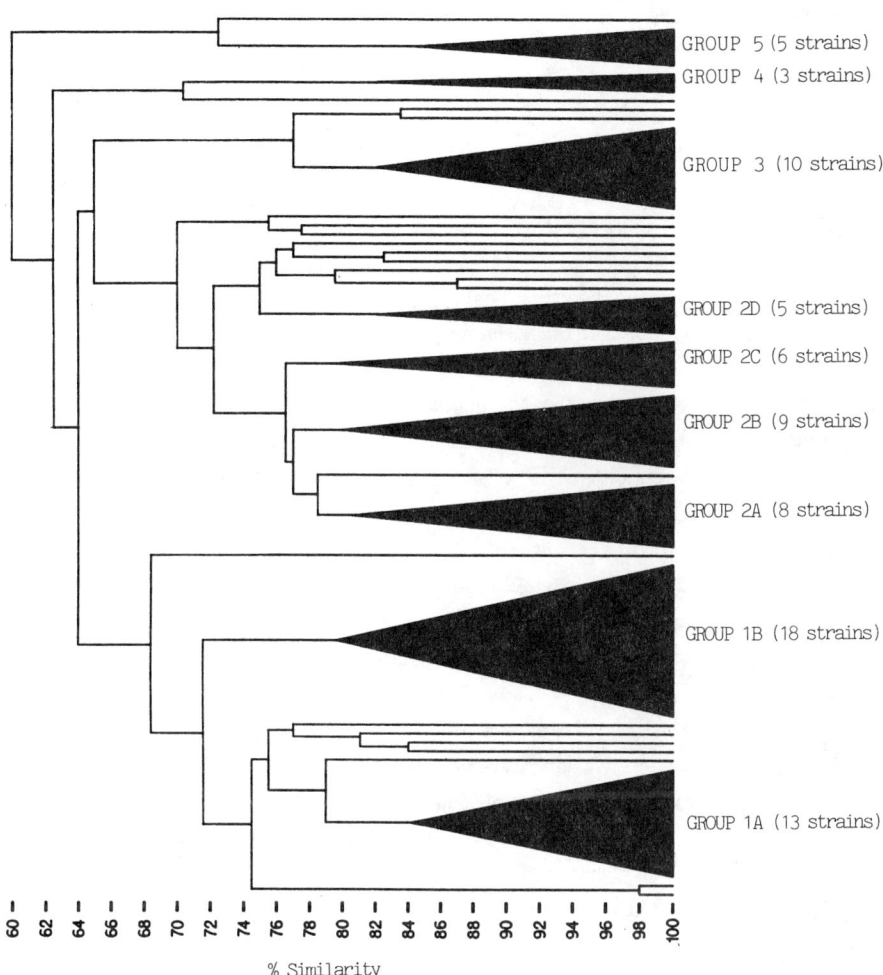

Figure 1. Dendrogram depicting relationship between strains of thermophilic
Bacillus using the SG coefficient and UPGMA

Proposed Nomenclature:
1A B. thermonondiastaticus;   1B B. thermononliquefaciens;
2A B. kaustophilus;   2B B. stearothermophilus;
2D B. thermotyrovorans;   3   B. thermodenitrificans;
4   B. coagulans   5   B. licheniformis

strains: B. kaustophilus (ATCC 8005) [11]; B. caldotenax, B. caldovelox and B. caldolyticus [22]; and strains RS 7, RS 15, RS 16 and RS 248. B. thermocatanulatus also has close similarity with this group. Two of the major characteristics differentiating sub-groups (2a) and (2b) are hippurate hydrolysis and citrate utilisation by sub-group (2a).

Group 2a demonstrated the close similarity of the three strains described by Heinen and Heinen [22] with B. kaustophilus originally described by Prickett [11]. B. caldotenax, B. caldovelox and B. caldolyticus do not appear to merit the rank of separate species and should be considered to be strains of B. kaustophilus. It is proposed that B. kaustophilus becomes the type strain for this group.

Group (2b), containing nine strains including six marker strains, B. stearothermophilus NCA 26 [6] (LUDA T210), NCA 1503 [32], NCIB 8919, NCA 1518 (LUDA T214), ATCC 10149 and NCTC 10003 [33]. All members of this group fit the classical definition of B. stearothermophilus [14, 15].

Group (2c) comprised six strains including LUDA T42 and LUDA T60 (from group 1b1 of Walker and Wolf). This group is typified in weak restricted hydrolysis of starch, all strains hydrolysed hippurate. Most strains (5/6) did not hydrolyse casein, were catalase positive and reduced nitrate to nitrite. Unlike groups 2a and 2b, most strains were resistant to sodium azide, all six strains grew in the presence of 3% saline.

All five strains in Cluster 2d grew at 70 °C, but did not grow at 37 °C. The five strains had several distinguishing characters: all five strains hydrolysed tyrosine producing a brown pigment throughout the medium, and four strains failed to grow, and one grew poorly in NB at pH 6.5. All strains denitrified nitrite to gas, four strains hydrolysed hippurate and utilised citrate. The five strains of cluster (2d) were soil isolates from: Calcutta (three); Les Eaux, France (one) and Bracknel UK (one).

The evidence indicates that cluster (2d) represents a distinct taxonomic group of Bacillus thermophiles. It is proposed that strains with these characteristics be designated Bacillus thermotyrovorans sp. nov.

Cluster (3): This comprised 10 strains typified by flat, rough, irregular, erose colonies. All strains reduced nitrate and nitrite to gas under aerobic conditions, and most (90%) utilised citrate. The majority of strains (80%) also reduced nitrate to gas under anaerobic conditions. Two strains, RS1 and RS14 which did not cluster with this group at 80% similarity do however show similar colony morphology and reduce nitrate and nitrite to gas. These two strains indicate 77% similarity with group (3).

Cluster (3) included two strains of B. thermodenitrificans (DSM 465 and DSM 466), isolated by Klaushofer and Hollaus [16], and LUDA T22 isolated by Walker and Wolf [18]. The strains in cluster (3) appeared similar to the description of Denitrobacterium thermophilum [34] (Rods 3.5-7 m by 1-1.8 m, terminal oval spores with slight swelling of the sporangium, colonies on agar resembling those of B. mycoides, nitrate reduced to gas, no hydrolysis of gelatin or starch, temperature range 37 °C to 65-70 °C). Organisms of this description were not examined by Smith et al [5] or Gordon et al [14]. Mishustin [35] included this organism in his identification key, but re-named it B. thermodenitrificans to fit the accepted nomenclature.

While cluster (3) did not represent all of the denitrifying strains, it did comprise a relatively homogeneous group of strains with little overall similarity with any other cluster. This cluster was considered to represent the species of Denitrobacterium thermophilum initially described by Ambroz [34] and later renamed B. thermodenitrificans by Mishustin [35]. It is proposed that strains with these characteristics be designated as B. thermodenitrificans.

Cluster (4): This group comprised three strains, two of which were B. coagulans marker strains.

Cluster (5): The five strains in this group had strong amylolytic and proteolytic activity and were able to grow at 25 °C. The colonies were raised, erose, shiny and circular following overnight incubation on TSBA and became mucoid after several days incubation at room temperature. The five isolates which all produced acetoin but did not utilise citrate or propionate at 55 °C. Their morphology, low growth temperature and production of acetoin suggested they were strains of B. licheniformis. Gordon et al [14] reported that acetoin production and utilisation of citrate and propionate were characteristic of B. licheniformis. On re-testing these five isolates for citrate and propionate utilisation at 37 °C all the strains were found to be positive. Studies at higher growth temperatures indicated that citrate utilisation occurred at 2-5 °C below the maximum growth temperature. Strain TB 124, [36] had relatively high similarity with members of cluster (6) and has since been independently identified as a strain of B. licheniformis.

GENETIC STUDIES OF B. CALDOTENAX

A number of bacteriophages have been isolated with the aim of developing a transfection system which could be developed to take up plasmid DNA. Fifteen bacteriophages isolated from various soil and compost samples were able to infect B. caldotenax [37]. Fourteen produced clear plaques, but one, designated JS017 produced hazy or turbid plaques. This phage had an unusual morphology with a cylindrical head, 80-90 nm long and 40-50 nm in diameter, the tail was approximately 120 nm in length. A minority of phage particles were observed to have the tail double the normal length, up to 250 nm, and the heads up to 150 nm in length (Fig. 2).

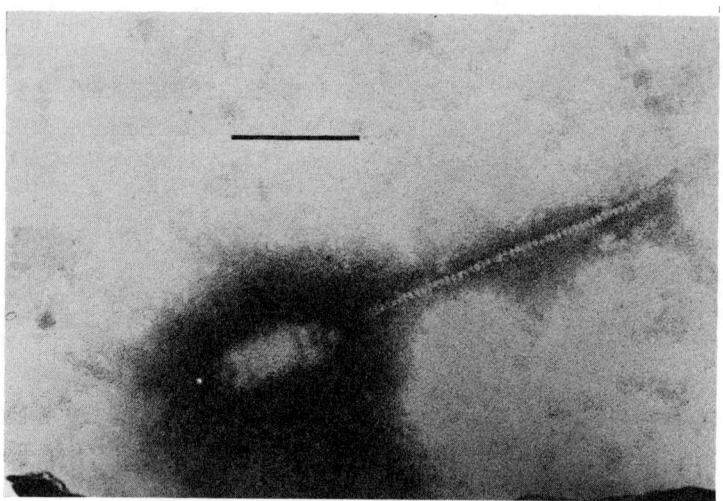

Figure 2. Electron micrograph of thermophilic bacteriophage JS017 showing an unusual tail length (Bars = 100 nm)

Transduction of B. caldotenax

Phage JS017 was able to carry out low level transduction of thymine auxotrophs of B. caldotenax at a frequency of 5×10^{-7}. A number of phages including ϕ3T are known to carry the gene for thymidylate synthetase within their genome. [38]. To determine if this was the case with JS017 it was put through 3 cycles of infection in a thymine auxotroph and a prototroph. Both phage stocks were then examined for the ability to transduce thymine auxotrophs. The phage stock produced from the auxotrophic strain was unable to transduce indicating that the phage did not carry an inherent thymidylate synthetase gene within its genome and successful transduction required it to 'pick up' the gene during a previous infection cycle.

The transduced colonies were found to be lysogenic for phage JS017 although up to 50% appeared unstable with phage detected in culture supernatant following five successive subcultures on plates. Phage were induced from the stable lysogens following exposure to UV.

After isolating stable amino acid auxotrophs, we examined transduction of histidine, methionine, isoleucine/valine and adenine auxotrophs. Methionine transductants occurred at a similar level to thymine auxotrophs and after obtaining double mutants they were found to cotransduce with an efficiency of 90% indicating close linkage of these two markers. No other markers examined showed any evidence of transduction.

Since as yet there are no genetic map of any of the thermophilic Bacilli, we looked at what is taxonomically the nearest organism. The genome of B. subtilis has the loci for thy B and met B mapping adjacent to each other at position 200 and adjacent to an attachment site for phage SPβ. In B. subtilis, SPβ transduced Kau A and Cit K genes which map at the other side of the insertion site. An analogy with B. subtilis may prove misleading but at present, it provides a working hypothesis. From our present information we consider that JS017 is a specialised transducing phage with an insertion site in the B. caldotenax genome adjacent to the met and thy loci.

The transducing frequency is very low and is highest using cells taken during early logarithmic growth (Fig. 3). The dilution of the cells with fresh medium also increases the transduction frequency. A 1/10 dilution of the cells has little or no effect on the transduction frequency. It appears that inhibitors of transduction or infection maybe diluted out or that high cell densities inhibit phage adsorption. Alternatively infection and penetration may require adsorption or penetration by more than one phage, reducing the number of cells present would therefore increase the opportunity for multiple adsorptions or infections.

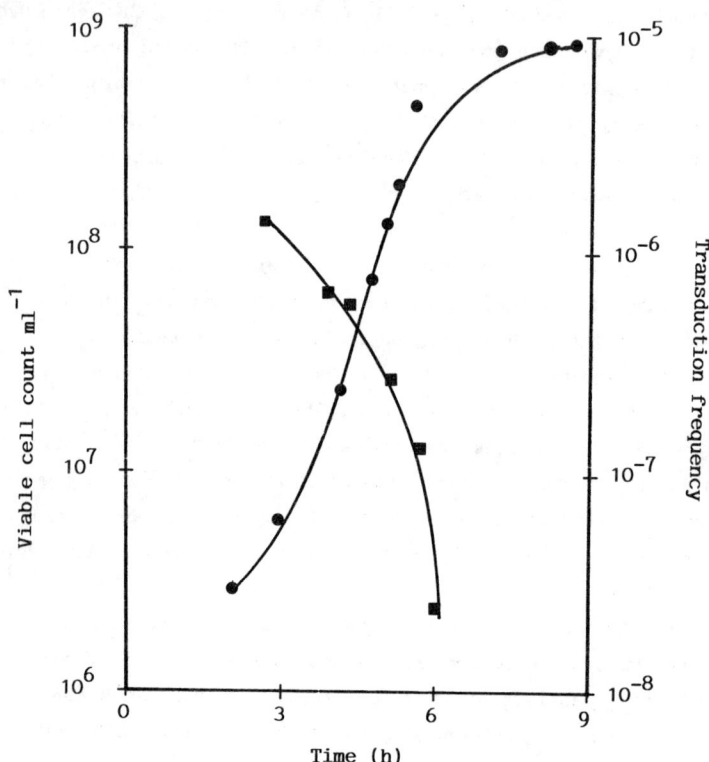

Figure 3. The effect of cell density on transduction frequency
Samples of a thymine auxotrophic strain of B. caldotenax grown in minimal medium supplemented with 50 μg.ml^{-1} of thymine was transduced at intervals throughout the growth cycle. The culture was monitored by viable cell count. The transduction frequency was calculated from the number of viable cells and transductant colonies from each sample.
● = viable cell count; ■ = transducing frequency

Transformation of B. caldotenax

Our attempts to develop a competance transformation regime in liquid medium based on that used for mesophilic Bacillus have been unsuccessful. The most reproducible transformation regime we have developed is based on a method originally devised for Thermoactinomycetes. The method relies on using agar as a support for the transformation process [39]. The culture was grown in double strength L broth to an OD of 1.6 at 540 nm washed and resuspended in pre-warmed minimal medium. 0.1 ml aliquots were mixed with 10-20 μg of DNA and spread over the surface of supplemented minimal agar plate. After overnight incubation at 60°C. The lawn of cells was washed twice and resuspended in 5 ml of sterile saline. Serial dilutions were then plated onto complex medium and selective defined medium. Colonies were visible on complex medium after 16 hours although transformants on the defined medium appeared after 48 hours. Presumptive transformants were then either replicated or streaked to selective medium to confirm their phenotype. The transformation frequency was then calculated from the transformed colonies per ml/viable count per ml. Increasing the DNA concentration up to 4 μg per plate enhanced the transformation frequency, above this level increasing concentration had little effect.

Transformation was not dependent upon the growth medium in the liquid phase, however, transformation frequencies were approximately 7-fold higher using x2 L broth compared with Tryptone Soya Broth. We suspect this may be due to the higher phosphate levels in TSB which have been shown to be inhibitory to some thermophiles [40]. The use of x2 L broth or supplemented minimal medium showed little difference in transformation frequencies although L broth became the medium of choice since the lag phase was only 2 hours compared with 5 hours for the defined medium.

Transformation did not occur when cells were plated on complex medium and it appeared that a nutritional 'step down' was required for competence similar to that required for B. subtilis.

The development of competence in the liquid medium stage was assayed by transforming samples throughout the growth cycle (Fig. 4). The results indicate two peaks in transformation frequency, the first in early exponential growth and the second in late exponential growth. No transformants were obtained using cells in mid exponential growth. Bott and Wilson [41] demonstrated maximum transformation frequencies in B. subtilis during spore germination and at the end of exponential growth. Having found two optimum growth phases for transformation in which the cells were presumed to be competent, the time required for DNA uptake to occur on the plates was examined (Fig. 5).

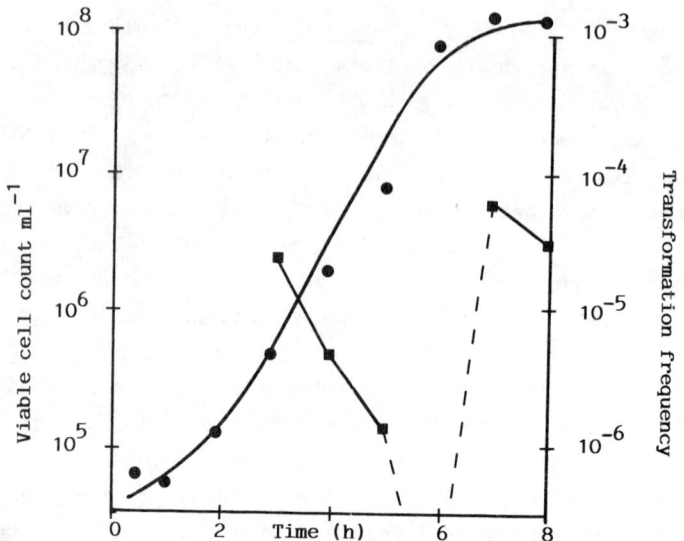

Figure 4. The relationship between growth phase and transformation frequency in an adenine auxotrophic strain of B. caldotenax
Transformations were made throughout the growth cycle.
---- = transformation frequency fell to undetectable levels.

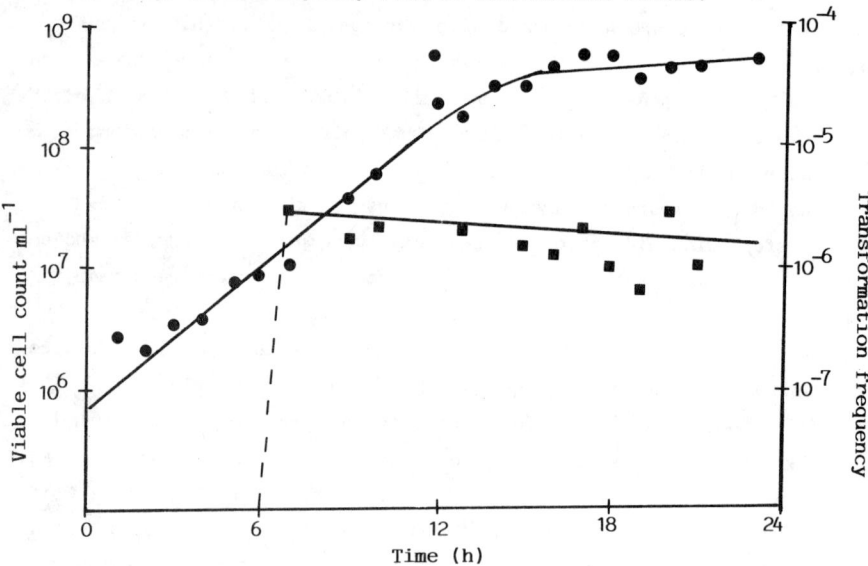

Figure 5. The occurrence of transformants with respect to the time of exposure to transforming DNA
Twenty-three transformations were carried out using an adenine auxotroph of B. caldotenax. At 1 hourly intervals, DNAase was added to one of the plates. ● = viable count; ■ = transformation frequency

TABLE 1

Transformation of B. caldotenax ade-5 mutant using DNA isolated from other thermophilic Bacillus species

Donor DNA strain	Transformation frequency	Taxonomic Group
RS 57	0	1b
RS 93 (NCIB 11400)	5×10^{-9}	1b
LUDA T141	0	1b
B. caldotenax	1×10^{-7} to 2×10^{-4}	2a
B. caldovelox	2×10^{-5}	2a
B. caldolyticus	2×10^{-5}	2a
B. stearothermophilus ATCC 8005	7×10^{-5}	2a
RS 222	0	2b
RS 240	0	2b
B. stearothermophilus NCA 1503	5×10^{-8}	2b
B. stearothermophilus NCTC 10003	1×10^{-6}	2b
LUDA T42	0	2c
LUDA T60	0	2c
LUDA T22	0	3
B. thermodenitrificans DSM 465	2×10^{-8}	3
B. thermodenitrificans DSM 466	0	3
B. coagulans ATCC 8038	0	4
B. licheniformis TB 124	0	5
RS 1	0	UA
RS 6	0	UA
RS 85	0	UA
RS 88	0	UA
B. thermocatenulatus DSM 730	4×10^{-7}	UA
B. stearothermophilus ATCC 12016	2×10^{-7}	UA
B. subtilis NCTC 3610	0	UA
B. subtilis IG 20	0	UA

UA = strain unassigned in taxonomic study

Twenty-three identical transformations were carried out and at 1 hourly intervals DNAase was added to one plate in turn. The cells were washed from the agar and plated onto selective media to select transformants, a viable cell count was also carried out.

The results indicated that the addition of DNAase to the plates totally inhibits transformation up to an incubation time of 7 hours. After 7 hours transformants were generated at a fairly constant frequency for the following 14 hours. In contrast to the competence peaks observed during the preliminary cultivation of the cells in liquid medium, a stage of competence appears to occur in cells growing as colonies on the agar during the mid log stage of growth.

Attempts to identify the production of a competence protein in the L broth stage using polyacrylamide gel electrophoresis was unsuccessful. We considered that mid log phase cells growing in L broth might produce nucleases which would inhibit transformation. Our attempts to detect nuclease activity however, were unsuccessful.

Transformation using DNA isolated from a range of thermophilic and mesophilic Bacillus species indicated considerable homogeneity (Table 1). The highest transformation frequencies were obtained with DNA from B. caldovelox, B. caldolyticus and B. stearothermophilus ATCC 8005, all of which have been shown to be taxonomically similar organisms.

Some preliminary mapping studies have been carried out but were restricted due to the availability of mutants and particularly double and multiply marked strains.

Studies have been carried out to examine the uptake of plasmid DNA using 3 plasmids, pAB124 a 2.9 megadalton plasmid isolated from a thermophilic strain of B. licheniformis and carrying a tetracycline resistance marker, pAB128 a 1.95 md deletion of pAB 124 [35] and pUB110. Using all three plasmids as linear DNA and covalently closed circular DNA attempts to transform B. caldotenax were unsuccessful. Transformation studies using DNA from phage JS017 and TP84 also proved unsuccessful.

Protoplasting studies of B. caldotenax

With the inability of the plate transformation system to take up either phage or plasmid DNA we have looked at protoplasts as a means of transforming B. caldotenax. Imanaka et al [42] found that B. stearothermophilus CU21 did not grow on succinate-based media such as DM3

TABLE 2
The effect of gelling agent on regeneration of B. caldotenax protoplasts

Gelling agent	Colony counts from regeneration plates
Noble agar	1.1×10^{8a}
Gelrite	6.5×10^{5b}
Pluronic polyol F-127	8.0×10^{7b}
Bacto agar	1.0×10^{8a}
Agarose	0

Initial variable count prior to protoplasting = 1.6×10^8 ml^{-1}

[a] Regeneration was only apparent, colonies contained protoplasts and L-froms, no bacilli

[b] Colonies contained bacilli indicating true regeneration

TABLE 3
Protoplast regeneration of thermophilic bacilli

Strain	Regeneration frequency (RF)%		
	MCL	DM3T	Pluronic/ glycerol medium
Bacillus caldotenax DSM 406	1.2	85	41[a]
Bacillus caldovelox DSM 411	1.3	85	48[a]
Bacillus caldolyticus DSM 405	2.1	95	5.2[a]
Bacillus stearothermophilus ATCC 12980	1.0	16[a]	32[a]
Bacillus stearothermophilus NCA 1503	1.05	12[a]	0.9[a]
Bacillus stearothermophilus (00136) DSM 2334	1.0	3.9	9.3[a]
Bacillus thermodenitrificans DSM 466	1.2	13.9	35.9[a]

[a] Regeneration to bacilli

MCL = Mean chain length, measured as the average number of Bacillus cells remaining attached to each other in short chains after cell division, based on counts of 300 chains.

%RF = 100 × $\dfrac{\text{Number of regenerants} - \text{number of survivors}}{\text{initial viable count} \times \text{mean chain length}}$

which was the regeneration medium used by Chang and Cohen [43] for B. subtilis. However, by replacing the succinate with sucrose as the osmotic support they were able to regenerate protoplasts of B. stearothermophilus.

This modified regeneration media however was unsuitable for the regeneration of B. caldotenax. After various modifications to the medium it was found that excessive phosphate in the medium inhibited regeneration and phosphate buffer was replaced with Eppes buffer or 0.02 M TES [44].

On close examination of the regenerant colonies on modified DM3 regeneration agar, the colonies were found to be very mucoid and on further investigation they appeared to be L forms which did not subculture or return to the bacillary form. We began to explore a range of alternative gelling agents including Noble agar, Bacto agar, Gelrite and Pluronic polyol F-127.

Colonies were evident on all of the plates except those containing agarose (Table 2). On the two agars however, the colonies consisted only of protoplasts and L-forms; Pluroronic polyol and gelrite, to a lesser extent, both supported the regeneration of protoplasts.

Pluoronic polyol is an unusual gelling agent and is a copolymer of polypropylene oxide and ethylene oxide. When incorporated into media at 25% (w/v) it forms a sol at low temperatures and sets to form a firm paste, rather than a solid gel at temperatures between 20 and 80°C.

In addition to the regeneration of protoplasts of B. caldotenax, protoplasts of B. caldolyticus, B. caldovelox, B. stearothermophilus ATCC 12980, B. stearothermophilus NCA 1503, B. stearothermophilus 00136 and B. thermodenitrificans were also regenerated (Table 3, [44]). It is interesting that on modified DM3 regeneration agar only B. stearothermophilus ATCC 12980, the strain used by Imanaka [42] and his colleagues and the closely related strain B. stearothermophilus produced true regenerants to the bacilli.

Transformation of the protoplasts with B. caldotenax chromosomal DNA encoding Novobiocin resistance has been demonstrated and protoplasts have been transformed with DNA from phage JS017 and TPIC.

Attempts to carry out plasmid tranformation of B. caldotenax, B. caldovelox and B. caldolyticus protoplasts have to date been unsuccessful.

REFERENCES

1. Miquel, P., Monographie d'un bacille vivant au-dela de 70°C. *Annales Micrographic*, 1818, 1, 3-10.

2. Bergey, D.H., Harrison, F.C., Breed, R.S., Hammer, B.W., Huntoon, F.M. *Manual of Determinative Bacteriology*. 1st edition. Baltimore: The Williams and Wilkins Company, 1923.

3. Breed, R.S., Murray, E.G.D. and Parker-Hitchens, A., *Bergey's Manual of Determinative Bacteriology*, 6th edition. London: Bailliere, Tindall and Cox, 1948.

4. Gordon, R.E. and Smith, N.R., Aerobic spore forming bacteria capable of growth at high temperatures. *J. Bact.*, 1949, 58, 327-341.

5. Smith, N.R., Gordon, R.E. and Clark, F.E. Aerobic sporeforming bacteria. Agricultural Monograph No. 16. United States Department of Agriculture, 1952.

6. Donk, R.J., A highly resistant thermophilic organism. *Journal of Bacteriology*, 1920, 5, 373-374.

7. Hammer, B.W., Bacteriological studies on the coagulation of evaporated milk. *Iowa Agricultural Experiment station Research Bulletin*, 1915, 19, 119-131.

8. Gyllenberg, H., Studies on thermophilic bacteria of the genus *Bacillus* Cohn. *Acta Agraria Fennica*, Helsinki, 1953, 73, 1-88.

9. Stark, E. and Tetrault, P., A determinative study of amylolytic, stenothermophilic bacteria isolated from soil. *Scientific Agriculture* (Canada), 1952, 32, 81-92.

10. Allen, M.B., The thermophilic aerobic spore forming bacteria. *Bact. Rev.*, 1953, 17, 125-173.

11. Prickett, P.S. Thermophilic and thermoduric microorganisms with special reference to species isolated from milk. Description of spore forming types. *New York State Agricultural Experimental Station Technical Bulletin 147*, 1928.

12. Daron, H.H. Occurrence of isocitrate lyase in a thermophilic *Bacillus* species. *J. Bact.*, 1967, 101, 145-151.

13. Epstein, I. and Grossowicz, N., Prototrophic Thermophilic *Bacillus*: Isolation, Properties and Kinetics of Growth. *J. Bact.*, 1969, 99, 414-417.

14. Gordon, R.E., Haynes, W.C. & Pang, C.H., The Genus *Bacillus*. United States Department of Agriculture, Washington D.C. 1973.

15. Gibson, T. and Gordon, R.E., Bacillus. In *Bergey's Manual of Determinative Bacteriology* ed., R.E. Buchanan and N.E. Gibbons, The Williams and Wilkins Company, Baltimore. 1974, pp. 529-555.

16. Klaushofer, H. and Hollaus, F., Zun Taxonomie der hochthermophilen, in Zuckerfabrikssaften verkommenden aeroben sporenbildner. Z. Zuckerindustr., 1970, 20, (a) 465-470.

17. Sokal, R. and Sneath, P.H.A., Principles of Numerical Taxonomy', W.H. Freeman and Co., London, 1963.

18. Walker, P.D. and Wolf, J., The taxonomy of Bacillus stearothermophilus. In Spore Research, ed., A.N. Barker, G.W. Gould and J. Wolf), London, Academic Press., 1971, 247-262.

19. Galesloot, Th. E. and Labots, H., Thermophilic bacilli in milk. Milk and Dairy Journal, Netherlands, 1959, 13, 155-179.

20. Grinsted, E. and Clegg, L.F.L., Spore-forming organisms in commercial sterilised milk. J. Dairy Research, 1955, 22, 178-190.

21. Heinen, W., Growth conditions and temperature-dependent substrate specificity of two extremely thermophilic bacteria. Arch., Microbiol., 1971, 76, 199-202.

22. Heinen, U.J. and Heinen, W., Characteristics and properties of a caldoactive bacterium producing extracellular enzymes and two related strains. Arch. Microbiol., 1972, 82, 1-23.

23. Darland, G. and Brock, T.D., Bacillus acidocaldarius sp. nov., an acidophilic thermophilic spore-forming bacterium. J. Gen. Micro., 1971, 67, 9-15.

24. Golovacheva, R.S., Loginova, L.G., Slikhov, T.A., Kolesnikov, A.A. and Zaitzeva, G.N., A new thermophilic species Bacillus thermocatenulatus nov. sp. Mikrobiologya, 1975, 44, 265-268.

25. Schenk, A. and Aragno, M., Bacillus schlegelii, a new species of thermophilic facultative chemolithotrophic bacterium oxidising molecular oxygen. J. Gen. Micro., 1979, 115, 333-341.

26. Suzuki, Y., Takashi, K., Inoue, K., Mizoguchi, Y., Eto, N., Takagi, M. and Abe, S., Bacillus thermoglucosidasius sp. nov., A new species of obligately thermophilic Bacilli. System. Appl. Microbiol. 1983, 4, 487-495.

27. Scholz, T., Demharter, W., Hensel, R., Kandler, O., Bacillus pallidus sp. nov., a new thermophilic species form sewage. System. Appl. Microbiol. 1987, 9, 91-96.

28. Zarilla, K. A. and Perry, J.J., Bacillus thermoleovorans sp. nov., a species of obligately thermophilic hydrocarbon utilising endospore forming bacteria. System. Appl. Microbiol., 1987, 9, 258-264.

29. Gower, J.C., A general coefficient of simlarity and some of its properties. Biometrics, 1971, 27, 857-874.

30. Sneath, P.H.A. & Johnson, R. The influence on numerical txonomic similarities of errors in microbiological tests. J. Gen. Microbiol., 1972, 72, 377-392.

31. Bergey, D.H., Thermophilic Bacteria, J. Bact., 1919, 4, 301.

32. Cameron, E.J. and Esty, J.R., The examination of spoiled canned foods, classification of flat sour organisms from non-acid foods, J. Infect. Dis., 1926, 39, 89-105.

33. Kelsey, J.C., The testing of sterilizers, J. Clin. Path., 1961, 14, 313-319.

34. Ambroz, A. Denitrobacterium thermophilus spec. nova. Ein Beitrag zur biologie der themophilen Bakterien. Zentralblatt fur Bakeriologie und Parasitenkunde, 1913, II, 3-16.

35. Mishustin, E.N., (1950). Quoted by Golovacheva et al. (1965).

36. Bingham, A.H.A., Bruton, C.J. and Atkinson, T., Characterisation of Bacillus stearothermophilus plasmid pAB 124 and construction of deletion variants. J. of Gen. Micro. 1980, 119. 109-115.

37. Sharp, R.J., Ahmad, S.I., Munster, A., Dowsett, B. and Atkinson, T., The isolation and characterisation of bacteriophages infecting obligately thermophilic strains of Bacillus. J. Gen. Micro. 1986, 132, 1709-1722.

38. Barner, H.D. and Cohen, S.S., Virus-induced acquisition of metabolic function. J. Biol. Chem., 1959, 234, 2987-2991.

39. Munster, M.J., Sharp, R.J., Ahmad, S., Vivian, A., Atkinson, T., Transformation and Transduction in a Bacillus thermophile. Abstracts of XIII International Congress of Microbiology, Boston, USA, 1982.

40. Rowe, J.J., Goldberg, I.D. and Amelunxen, R.E., Isolation of mutants of Bacillus stearothermophilus, J. Bact. 1975, 124, 279-284.

41. Bott, K.F. and Wilson, G.A., Developments of competence in the Bacillus subtilis transformation system. J. Mol. Biol., 1967, 93, 562-570.

42. Imanaka, T., Fuju, M., Avamori, I., and Aiba, S., Transformation of Bacillus stearothermophilus with plasmid DNA and characterisation of shuttle vector plasmids between Bacillus stearothermophilus and Bacillus subtilis. J. Bact., 1982, 149, 824-830.

43. Chang, S. and Cohen, S.N. High frequency transformation of Bacillus subtilis protoplasts by plasmid DNA, Mol. Gen. Genet. 1979, 168, 111-115.

44. Dunn, R.M., Munster, M.J., Sharp, R.J. and Dancer, B.N. A novel method for regenerating the protoplasts of a thermophilic bacilli. Arch. Microbiol. 1987, 147, 323-326.

BIOCHEMICAL TAXONOMY OF THE GENUS THERMUS

R.A.D. WILLIAMS
Biochemistry Department,
London Hospital Medical College,
Turner Street, London E1 2AD, U.K.

ABSTRACT

Strains of aerobic thermophilic bacteria allocated to the genus Thermus have been tested by techniques that have proved useful in biochemical taxonomy with other genera. Cells of the strains were tested for the composition of peptidoglycan, the occurrence of respiratory quinones, the electrophoretic patterns of cellular polypeptides, the mean base composition of DNA and the DNA:DNA homology tested by the filter hybridisation and spectrophotometric reassociation rate methods. Thermus ruber strains show great homogeneity and have little homology with other Thermus isolates. Thermus thermophilus is confirmed as a distinct genospecies with low homology with Thermus aquaticus. Amongst yellow and colourless Thermus isolates several new genospecies are detected.

INTRODUCTION

The properties of the genus Thermus and its type species Thermus aquaticus were described in 1969 (1). Strains are described as aerobic, nonsporulating heterotrophic rods with optimum pH between 7.0 and 8.0, and optimum temperature at 70°C. The Gram-stain reaction is negative, and the cell wall has an outer layer that balloons out from the inner peptidoglycan at regular intervals to give a corrugated or "annelid" appearance. Thermus aquaticus is yellow-pigmented due to a carotenoid, but colourless strains have been isolated from unilluminated sources (2,3), and Thermus ruber, which contains a red carotenoid and grows optimally at 60°C, and has been validly described(4). Strains isolated in Japan have been named Thermus flavus(5), Thermus thermophilus(6) and Thermus caldophilus(7), none of which are considered to be validly published. Patent strains, Thermus rubens and Thermus lacteus are available from the ATCC. Large collections of strains of this genus have been isolated in Iceland (8,9,10), Belgium(11) Yellowstone National Park(12), New Zealand (13) and Portugal (14) and considerable investigations carried out into their biochemical and physiological properties, as well as their use as a source of stable enzymes. Despite this information the vast majority of the isolates remain described only as strains of the genus Thermus. The only validly described species beside the type species remain two, compromising isolates with striking properties, the red pigmented Thermus ruber and the filamentous Thermus filiformis(15), of which only one strain is available. A valid taxonomy can only be constructed on the basis of detailed knowledge of large collections of isolates. Systematic investigation of biotechnologically important groups of microorganisms generally is desirable, not only as an end in itself, but because it provides a base of valuable knowledge concerning the substrates utilised, optimum growth conditions and rates, the compositions of cellular constituents and the formation of enzymes and other products. This systematic knowledge may be valuable in choosing which of many strains is best for a particular biotechnological purpose. The genus Thermus is one group of microorganisms from extreme environments whose applications are actual rather than potential, although undoubtedly many uses remain to be developed for these widespread, heterogenous but easily grown bacteria. Thermus strains have been recommended as sources of proteases, but commercially restriction endonucleases (especially Taq I) and DNA polymerase for the amplification of DNA by the polymerase chain reaction (and possibly for DNA sequencing) have a large market world-wide.

This paper reports the application of a combined biochemical approach to taxonomy which has proved useful in speciating other heterogenous groups of bacteria amongst which the physiological properties vary from strain to strain (16,17).

MATERIALS AND METHODS

Strains and their culture. Strains of Thermus used and their sources are listed in Table 1. Cells were grown at 65°C in shake flask cultures in medium containing 0.3% tryptone, 0.1% yeast extract (Difco) and Castenholtz mineral salts as described by Ramaley and Hixson(2). Cell paste was harvested by centrifugation at 6000 xg and stored at -20°C.

Respiratory quinone. Cell paste was freeze-dried and extracted with a 2:1 mixture of chloroform and methanol in the dark overnight(18). The extract was filtered and concentrated under reduced pressure, applied as a band to a preparative Keiselgel 60F thin layer chromatography plate, then developed with 15% diethyl ether in light petroleum (60-80). The band absorbing UV light (254 nm) at Rf 0.8, presumptively corresponding to menaquinones, was scraped from the plate and washed with chloroform which was then concentrated under a stream of nitrogen and applied to a reverse phase plate (RP18F254, Merck) and developed with acetone. The Rf values of extracts of Thermus strains were observed under UV light (254 nm) and compared with the menaquinones of Bacillus stearothermophilus and Thermus aquaticus YT1, and also of strains of Bacteroides and Capnacytophaga kindly supplied by Dr H.N. Shah. Ultraviolet absorption spectra of the chloroform extracts of the preparative plates were recorded between 230 and 350 nm.

Peptidoglycan preparation and analysis. Cells were extracted by the rapid peptidoglycan preparation method(19) and hydrolysed with 4M HCl(20), then the amino acids and aminosugars analysed with an LKB 4000 amino acid analyser.

TABLE 1
Sources of Thermus isolates used

Strain	Source
T. aquaticus YT1, ATCC25104, NClB11243	ATCC
T. flavus AT62, DSM674, ATCC33932	Saiki [5]
T. caldophilus GK24	Ohta [7]
T. thermophilus HB8, ATCC27624, NCIB11244	Oshima [6]
T. ruber BKMB1258, NCIB11269	Loginova [4]
T. ruber icelandic strains	Sharp & Williams [21]
Vizela and Sao Pedro do Sul isolates	da Costa, M. [14]
New Zealand isolates	Morgan, H. [13]
Thermus strain NH, NCIB11245	Pask-Hughes & Williams [3]
Thermus strain DI, NCIB11246	Pask-Hughes & Williams [3]
Thermus strain B NCIB11247	Williams [24]

Sodium dodecyl sulphate-polyacrylamide gel electrophoresis:
Frozen cells were suspended in 2% sodium dodecyl sulphate, boiled for 2 minutes, cooled and centrifuged at 20,000 x 9 for 30 mins. The supernatant was diluted to about 5 mg. protein per ml. and mixed with an equal volume of sample buffer containing 0.06M tris pH 6.8, 0.12% glycerol, 1.2% sodium dodecyl sulphate, 1.2% mercaptoethanol and 0.012% bromophenol blue. Samples were stored at -20°C and boiled for 2 mins before electrophoresis as described (16,17).

DNA purification and mean base composition: Cells were lysed and DNA purified as described for Thermus ruber (21) except that the caesium chloride density-gradient centrifugation was run for 16 hours in a Sorvall vertical rotor. The mean base composition was determined by thermal denaturation by reference to E.coli DSM2840 (21).

DNA:DNA homology: Hybridisation of ^{3}HDNA against single-stranded DNA on 5mm nitrocellulose discs was carried out as described for Thermus ruber (21). Sheared single-stranded DNA was allowed to reassociate under optimum conditions and the homology determined spectrophotometrically (22,23) as a second method.

RESULTS AND DISCUSSION

Respiratory quinones: The predominant quinone detected amongst isolates of Thermus including Thermus aquaticus YTI and many strains from Iceland and New Zealand is menaquinone 8 (18 and M.D. Collins, personal communication). The same quinone has also been reported in Thermus ruber (25) and in Dienococcus radiodurans (26), to which Thermus isolates have been reported to have a distinct, if distant, relationship (25) We have presumptively identified menaquinone 8 by ultraviolet spectrum and chromatography (Fig.1) as the major quinone of all 14 strains tested including the non-pigmented isolates NH and DI, as well as isolates described here as Thermus brockii although this was not confirmed by mass-spectroscopy.

Cell-wall analysis: All isolates of Thermus described to date have an ornithine peptidoglycan (20,25,27), which is a relatively uncommon structural type that is not found in Gram-negative bacteria. Although there is some variation between laboratories in the molar ratios of amino substances reported, the peptidoglycans of Thermus aquaticus (20,25,27), Thermus thermophilus (20,25), Thermus ruber (25) and Dienococcus radiodurans (26) are similar to those Thermus brockii YS38 and to Portuguese isolates from Vizela and Sao Pedro do Sul (Table 2). Lipopolysaccharides prepared from the cell walls of Thermus aquaticus, non-pigmented strains NH and DI and strain B shown here to belong to Thermus thermophilus were similar, and contained no heptose or ketodeoxyoctulosonate (20). More recently a major constituent of the cell wall of Thermus thermophilus

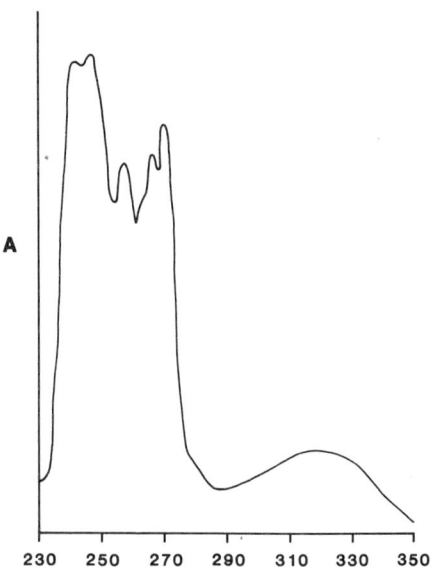

Figure 1. Ultraviolet spectrum of menaquinone of strain Vi2a in light petroleum showing peaks at 241, 246, 258, 266, 270 and 320 nm.

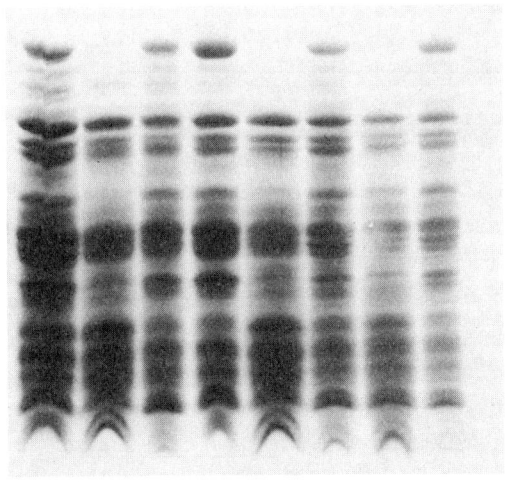

Figure 2. SDS-PAGE patterns of strains of *Thermus brockii* strains YS11, 19, 21, 30, 38, 40, 44 and 45.

has been shown to be a calcium-protein complex (28).

TABLE 2
Molar ratios of peptidoglycan components of <u>Thermus</u> strains

Strain	Muramate	Glucosamine	Glutamate	Glycine	Alanine	Orthinine	Lysine
YT1	1.0	1.2	2.0	2.4	2.9	1.0	0.2
YS38	0.8	0.7	2.0	2.8	2.6	1.1	0.4
HB8	1.4	0.9	2.0	3.3	3.8	1.7	0.3
Vi3a	0.9	0.7	2.0	3.7	3.0	1.7	0.2
Vi4a	1.1	1.2	2.0	3.8	3.5	1.5	0.2
Vi7b	0.9	0.9	2.0	3.3	2.9	1.3	0.4
Vi13	1.0	0.7	2.0	3.8	3.1	1.6	0.3
SPS17	1.0	0.7	2.0	3.6	3.9	1.7	0.4
Results for <u>Thermus</u> strains, Hensel et al. (25)							
YT1	1.4	1.6	2.0	4.0	3.6	1.5	-
HB8	1.0	n.d.	2.0	3.4	4.4	1.2	0.3
T.ruber	2.0	2.8	2.0	4.0	4.4	2.0	-
H3	1.6	3.4	2.0	3.6	3.4	2.2	-
Results for <u>Dienococcus</u> strains, Embley et al (26)							
D9	-	-	2.0	3.6	3.0	2.0	0.2
D10	-	-	2.0	3.4	2.8	2.0	0.2
D11	-	-	2.0	3.6	3.0	2.0	0.2
D12	-	-	2.0	3.6	3.6	2.0	0.2
D13	-	-	2.0	3.4	3.4	2.0	-

<u>SDS-PAGE of cellular proteins:</u> The electrophoresis of cellular proteins has been used in several ways to assist the taxonomy of microorganisms. Computer-assisted comparisons of electrophoretic patterns produced under carefully standardised conditions has provided a database for taxonomy and identification (29), but a more qualitative method has proved a good indicator of group of strains that form DNA:DNA homology clusters (16). The results with <u>Thermus</u> isolates of this type of approach (Fig.2) showed very heterogenous patterns of polypeptides amongst which some clusters of strains emerged. These clusters with similar patterns agreed with the clusters obtained by DNA:DNA homology. Thus, the Sao Pedro do Sul (Tables

5A, B) and Vizela (Table 6) DNA:DNA homology groups have distinct patterns of polypeptides. The colourless strains (NH and DI) from hot tap water (3) are unique electrophoretically. Some variation is found amongst Thermus brockii (Fig.2), patterns of which most closely resemble Thermus aquaticus.

TABLE 3
Mean DNA base composition of Thermus isolates

strain	Tm	% G+C	strain	Tm	% G+C
T aquaticus			T brockii		
YT1	80.4	64	YS7	80.2	63
YS4	78.3	60	YS11	79.2	61
YS8	79.9	63	YS19	80.1	60
YS9	80.3	63	YS36	79.3	62
YS10	80.6	64	YS37	78.7	60
YS16	78.1	59	YS38	78.5	60
YS28	80.4	64	YS44	79.3	62
YS31	80.2	63	YS45	78.4	60
YS32	78.7	60			
YS41	78.2	59			
YS48	79.7	62			
T thermophilus			Vizela		
HB8	80.4	64	Vi1a	77.3	57
B	80.1	63	Vi2a	77.5	58
GK24	78.9	61	Vi2b	78.4	60
AT62	80.6	64	Vi3a	78.6	60
			Vi4a	78.7	60
			Vi5b	79.1	61
Sao Pedro do Sul			Vi7a	78.5	60
			Vi7b	78.4	60
SPS2	78.5	60	Vi8	77.4	58
SPS7	79.4	62	Vi13	77.2	57
SPS8	80.1	63			
SPS10	79.9	63	New Zealand		
SPS11	80.0	63			
SPS12	80.1	63	T41A	81.0	64
SPS14	80.2	63	Rt358	81.0	64
SPS17	79.3	61	Tok22	80.7	63
SPS18	80.3	64	T351	80.2	62

DNA base composition and homology: The DNA mean base composition determined by thermal denaturation in dilute saline citrate (DSC) (Table 3) shows that isolates from the U.S.A., Japan, Iceland, New Zealand and Portugal have 57-64% G + C. The strains in Table 2 have been clustered according to the DNA:DNA homology results described below.

The results of filter homology determination on Yellowstone National Park strains (YS strains) showed that a substantial fraction had high homology amongst themselves and with Thermus aquaticus YT1 (Table 4A). Six of these strains belong to Cluster 1a and three to Cluster 1b of Munster et al(12). A second DNA:DNA homology group of YS strains which have been tentatively named Thermus brockii all belong to the numerical taxonomy cluster 2 of Munster et al. (12). The distinction between Thermus aquaticus YS31 and five strains of Thermus brockii has been confirmed by reassociation followed spectrophotometrically (Table 4B). A reexamination of the phenotypic properties of the YS strains by courtesy of Dr R J Sharp has revealed six tests that distinguish these two genospecies. Thermus brockii strains show a pale spreading morphology on tryptone-yeast extract plates and grow in liquid media on fructose and galactose, whereas Thermus aquaticus strains do not. Furthermore, Thermus aquaticus strains degrade casein, gelatin and starch in plates, whereas Thermus brockii strains do not.

In preliminary experiments many other strains were tested for homology against one another. Amongst these a small cluster was detected including Thermus thermophilus HB8 (6), Thermus flavus AT62 (5), Thermus caldophilus GK24 (7) and an isolate from Iceland, strain B (24). This cluster, and its distinctness from Thermus aquaticus is sufficient to re-establish Thermus thermophilus as a valid genospecies. Strains of this species grow at a slightly higher temperature and also faster than strains of Thermus aquaticus.

TABLE 4A

Filter DNA:DNA homology of Thermus aquaticus,
Thermus brockii and Thermus thermophilus

Strains tested on filters	Thermus aquaticus (5 strains*)	Thermus brockii (5 strains+)	Thermus thermophilus (4 strains)
Thermus aquaticus			
YT1^T	70-100	35-47	42-52
YS4	68-100	21-36	23-43
YS8	77- 98	20-45	22-37
YS9	84-100	36-50	45-51
YS10	84-100	27-49	26-36
YS16	77- 88	27-34	35-43
YS28	69- 89	22-52	34-50
YS31	76-100	31-43	37-57
YS32	68- 90	27-35	31-40
YS41	69- 84	28-40	31-43
YS48	82-100	24-35	N.D.
Thermus brockii			
YS7	44- 51	93-106	25-49
YS11	43- 59	78- 86	25-37
YS19	20- 49	69- 83	33-52
YS36	35- 46	70-100	25-40
YS37	30- 40	100	28-31
YS38^T	34- 49	90-100	25-31
YS44	34- 54	92-100	35-40
YS45	32- 48	73- 88	21-49
Thermus thermophilus			
HB8^T	35- 43	23- 40	82- 88
GK24	29- 46	16- 36	81-100
AT62	33- 50	20- 36	78-100
B.	20- 32	23- 27	84-100

* T. aquaticus strains YT1, YS4, YS10, YS31, YS48

+ T. brockii strains YS7, YS36, YS37, YS38^T, YS44

A group of strains have been isolated by M. da Costa from a thermal spa at Sao Pedro do Sul in central Portugal. Two of these, SPS11 and SPS17, have been used to study the effect of temperature on the lipid composition of the cell membranes [14] with results similar to those obtained with Icelandic and Japanese isolates of Thermus. The SPS isolates

TABLE 4B
Spectrophotometric DNA:DNA homology of *T. brockii*

		YS31	YS11	YS19	YS36	YS44	YS45
T.aquaticus	YS31	100					
	YS11	32	100				
	YS19	45	100	100			
T.brockii	YS36	25	91	-	100		
	YS44	37	106	75	-	100	
	YS45	47	87	82	100	75	100

were shown by filter homology to have a high DNA homology with each other, (Table 5A) and to be quite distinct from *T. aquaticus*, *T. brockii* and *T. thermophilus*. Some of the Icelandic strains of Jacob Kristjansson [9], isolated from the Hveragerthi-Hengill area of Iceland belonged to the SPS homology group, although most did not. Preliminary results indicate that some strains isolated from sites in the Azores resemble the SPS group of strains phenotypically (M. da Costa, personal communication). Spectrophotometric determination of homology confirms the relationship between SPS and JK strains (Table 5B). Therefore, this genospecies is found in two, and possibly three hot spring sites separated by as much as 1,800 miles of Atlantic Ocean. At a second site in mainland Portugal in the town of Vizela, all the isolates tested form yet another DNA:DNA homology group (Table 6), which is unrelated to any other strains tested so far. Thermal environments of moderately alkaline reaction are commonly colonised by bacteria that have been allocated to the genus *Thermus*. The similarity of environment and of physiological properties should not obscure the basic heterogeneity amongst these isolates indicated by differences in DNA:DNA homology.

TABLE 5A

Filter DNA:DNA homology of Sao Pedro do Sul and Icelandic Strains

Strains tested on filters	T.aquaticus YT1	T.brockii YS38	T.thermophilus B	SPS strains (3)*	JK strains (3)‡
T.aquaticus					
YT1	100	37	53	31- 48	37- 50
T.brockii					
YS38	-	100	43	39- 45	28
T.thermophilus					
HB8	38	32	84	31- 33	36
B	30	30	100	27- 32	32
Sao Pedro do Sul					
SPS7	-	57	58	92- 99	95
SPS8	39	38	48	94-100	82-106
SPS10	49	59	54	91-103	88-103
SPS11	44	-	-	81- 97	90-103
SPS12	-	52	-	97-103	95
SPS14	43	39	44	81-100	85-100
SPS17	32	41	-	80-100	69- 96
SPS18	50	53	51	86-102	82-104
Icelandic strains					
JK51	35	-	-	79	100
JK66	40	-	-	80	83-100
JK90	32	-	-	92	90-100
JK91	42	-	-	89	100-102

* SPS8, SPS14, SPS17   ‡ JK51, JK66, JK91

TABLE 5B
Spectrophotometric DNA:DNA homology of SPS group

	SPS11	SPS14	SPS17	JK66	JK90
SPS11	100				
SPS14	80	100			
SPS17	80	82	100		
JK66	82	69	69	100	
JK90	-	78	81	-	100

TABLE 6
Filter DNA:DNA homology of Vizela strains

	YT1	YS38	B	SPS strains [3]*	Vizela strains [3]‡
T. aquaticus					
YT1	100	37	53	31- 48	28- 36
T. brockii					
YS38		100	43	39- 45	29- 36
T. thermophilus					
HB8	38	32	84	31- 31	21- 24
B	30	30	100	27- 32	24- 28
Vizela					
Vi1a	-	52	44	46- 49	83- 91
Vi2a	39	54	50	43- 49	88- 90
Vi2b	49	63	56	36- 53	85
Vi3a	44	53	53	44- 54	87-104
Vi4a	-	-	39	45	80-100
Vi5b	-	53	49	52	100
Vi7a	-	58	40	44- 52	86- 90
Vi7b	-	54	52	34- 54	96-102
Vi8	-	50	42	38- 49	78- 87

* SPS8, SPS14, SPS17   ‡ Vi2b, Vi4a, Vi5b

REFERENCES

1. Brock, T.D. and Freeze, H. Thermus aquaticus gen.n. and sp.n. a non-sporulating extreme thermophile. Journal of Bacteriology, 1969, 98, 289-297

2. Ramaley, R.F. and Hixson, J. Isolation of a non-pigmented thermophilic bacterium similar to Thermus aquaticus. Journal of Bacteriology, 1970, 103, 527-528

3. Pask-Hughes, R.A. and Williams, R.A.D. Extremely gram-negative bacteria from hot tap water. Journal of General Microbiology, 1975, 88, 321-328

4. Loginova, L.G., Egorova, L.A., Golevacheva, R.S. and Seregina, L.M. Thermus ruber, sp.nov., hom.rev. International Journal of Systematic Bacteriology, 1984, 34, 498-499

5. Saiki, T., Kimura, R. and Arima, K. Isolation and characterisation of extremely thermophilic bacteria from hot springs. Agricultural and Biological Chemistry, 1972, 36, 2357-2366

6. Oshima, T. and Imahori, K. Description of Thermus thermophilus (Yoshida and Imahori) comb.nov. a non-sporulating thermophilic bacterium from a Japanese thermal spa. International Journal of Systematic Bacteriology, 1974, 24, 102-112

7. Taguchi, H., Yamashita, M., Matsuzawa, H. and Ohta, T. Heat-stable and fructose 1,6-bisphosphate-activated L-lactate dehydrogenase from an extremely thermophilic bacterium. Journal of Biochemistry (Japan), 1982, 91, 1343-1348

8. Pask-Hughes, R.A. and Williams, R.A.D. Yellow-pigmented strains of Thermus spp. from Icelandic hot springs. Journal of General Microbiology, 1977, 102, 375-383

9. Alfredsson, G.A., Baldursson, S. and Kristjansson, J.K. Nutritional diversity among Thermus spp. isolated from Icelandic hot springs. Systematic and Applied Microbiology, 1985, 6, 308-311

10. Cometta, S., Sonnleitner, B. and Fiechter, A. A comparative analysis of extreme thermophilic bacteria belonging to the genus Thermus. Archives of Microbiology, 1982, 117, 189-196

11. DeGryse, E., Glansdorff, N. and Pierard, A. A comparative analysis of extreme thermophilic bacteria belonging to the genus Thermus. Archives of Microbiology, 1978, 117, 189-196

12. Munster, M.J., Munster, A.P., Woodrow, J.R. and Sharp, R.J. Isolation and preliminary taxonomic studies of Thermus strains isolated from Yellowstone National Park, U.S.A. Journal of General Microbiology, 1986, 132, 1677-1683

13. Hudson, J.A., Morgan, H.W. and Daniel, R.M. A numerical classification of some Thermus isolates. Journal of General Microbiology, 1986, 132, 531-540

14. Prado, A., da Costa, M.S. and Madeira, V.M.C. Effect of growth temperature on the lipid composition of two strains of Thermus sp. Journal of General Microbiology, 1988, 134, 1653-1660

15. Hudson, J.A., Morgan, H.W. and Daniel, R.M. Thermus filiformis sp. nov. a filamentous caldoactive bacterium. International Journal of Systematic Bacteriology, 1987, 37, 431-436

16. Dent, V.E. and Williams, R.A.D. A combined biochemical approach to the taxonomy of Gram-positive rods. In Chemical Methods in Bacterial Systematics. ed., M. Goodfellow and D.E. Minnikin, Academic Press, London, 1985, pp. 341-357

17. Dent, V.E. and Williams, R.A.D. Actinomyces howellii, a new species from the dental plaque of dairy cattle. International Journal of Systematic Bacteriology, 1984, 34, 316-320

18. Collins, M.D. and Jones, D. Distribution of isoprene quinone structural types in bacteria and their taxonomic implications. Microbiological Reviews, 1981, 45, 316-354

19. Schleifer, K.H. and Kandler, O. Peptidoglycan types of bacterial cell walls and their taxonomic implications. Bacteriological Reviews, 1972, 36, 245-292

20. Pask-Hughes, R.A. and Williams, R.A.D. Cell envelope components of strains belonging to the genus Thermus. Journal of General Microbiology, 1978, 107, 65-72

21. Sharp, R.J. and Williams, R.A.D. Properties of strains of Thermus ruber isolated from Icelandic hot springs and DNA:DNA homology of Thermus ruber and Thermus aquaticus. Applied and Environmental Microbiology, 1988, 54, 2049

22. Gillis, M., Dehey, J. and DeCleene, M. The determination of molecular weight of bacterial genome DNA from renaturation rates. European Journal of Biochemistry, 1970, 12, 143-153

23. Huss, V.A.R., Festl, H. and Schleifer, K.H. Studies on

the spectrophotometric determination of DNA hybridisation from renaturation rates. Systematic and Applied Microbiology, 1983, 4, 184-192

24. Williams, R.A.D. Caldoactive and thermophilic bacteria and their thermostable proteins. Science Progress, Oxford, 1975, 62, 373-393

25. Hensel, R., Demharter, W., Kandler, O., Kroppenstedt, R.M. and Stackebrandt, E. Chemotaxonomic and moleculargenetic studies of the genus Thermus. Evidence for a phylogenic relationship of Thermus aquaticus and Thermus ruber to the genus Dienococcus. International Journal of Systematic Bacteriology, 1986, 36, 444-453

26. Embley, T.M., O'Donnell, A.G., Wait, R. and Rostron, J. Lipid and cell wall amino acid composition in the classification of members of the genus Dienococcus. Systematic and Applied Microbiology, 1987, 10, 20-27

27. Merkel, G.J., Stapleton, S.S. and Perry, J.J. Isolation and peptidoglycan of gram-negative hydrocarbon-utilising thermophilic bacteria. Journal of General Microbiology, 1978, 109, 141-148

28. Berenguer, J., Faraldo, M.L.M. and DePedro, M.A. Ca^{2+} stabilised oligomeric protein complexes are major components of the cell envelope of Thermus thermophilus. HB8 Journal of Bacteriology, 1988, 170, 2441-2447.

29. Jackman, P.J.H. Bacterial taxonomy based on electrophoretic whole-cell protein patterns. In Clinical Methods in Bacterial Systematics, ed., M. Goodfellow and D.E. Minnikin, Academic Press, London, 1985, pp. 115-129

INDUCIBLE 'SOS' REPAIR IN THE EXTREME THERMOPHILE BACILLUS CALDOTENAX

PAUL RILEY, TONY ATKINSON, ALAN VIVIAN* AND RICHARD SHARP
Division of Biotechnology, PHLS Centre for Applied Microbiology & Research
Porton Down, Salisbury; and * Bristol Polytechnic, Bristol, United Kingdom

INTRODUCTION

In a number of bacterial species, DNA damage results in the coordinated induction of a variety of genes which cause pleiotropic effects upon the cell. These effects are collectively known as the SOS response [1]. In Escherichia coli, 17 genes have so far been identified which either control or are induced during the SOS response. Effects on E. coli as a result of activating the SOS response include an increase in activity of several different DNA repair pathways as well as the induction of the error-prone repair pathway [2], the blockage of cell division leading to filamentation and induction of prophage.

Bacillus subtilis also exhibits SOS repair [3]. SOS phenomena so far identified include error-prone repair, filamentation and prophage induction. The molecular control and the number and identity of genes involved in B. subtilis SOS repair are not known.

This paper presents evidence indicating the presence of an inducible SOS-repair system in the thermophile B. caldotenax.

Filamentation

Growth of B. caldotenax in 1 g.ml^{-1} mitomycin-C resulted in the induction of filaments after approximately 120 minutes. Measurement of the filaments indicated a size range of between 16 and 25 m in length and 0.4 to 0.5 m in diameter. This compared with a size range for untreated cells of 2 to 3 m in length and 1 m in diameter.

Hence, the ability of a DNA damaging agent, Mitomycin-C, to induce filamentation was established.

De novo synthesis of proteins

Proteins from cell supernatants of B. caldotenax cells treated with mitomycin-C were compared with untreated cells using two-dimensional gel electrophoresis. The first dimension involved separation by isoelectrofocusing in the presence of urea and a neutral detergent with a pH gradient of pH 3-10. Proteins were then separated further by using an SDS polyacrylamide slab gel in the second dimension [4]. This method allowed the identification of two proteins which appeared to be induced by the presence of mitomycin-C in the growth medium. One protein had a molecular weight of approximately 43,000, the other a molecular weight of approximately 50,000. A third protein with a molecular weight of approximately 55,000 was absent from the supernatant of cells treated with mitomycin-C.

Prophage induction

A temperate bacteriophage, JS017, has been isolated which infects B. caldotenax [5]. Isolates of B. caldotenax resistant to re-infection by JS017 and therefore thought to be lysogens were purified until free of contaminating phage. UV-irradiation of these presumptive lysogens followed by over laying with soft agar containing wild-type B. caldotenax produced plaques from 11 out of 13 isolates.

Weigle reactivation

Weigle reactivation [6] is a measure of the increased ability of the host cell to repair UV-damaged phage DNA, when the host cell has itself been UV-irradiated just prior to phage infection. Any increase in survival of phage following infection in UV-irradiated cells against unirradiated cells therefore suggests the presence of an inducible 'SOS' repair system.

Two phages, JS017 - a temperate phage and an uncharacterised lytic phage, ϕ323, were UV-irradiated and used to infect a wild-type strain of B. caldotenax. As the data suggests (Tables 1 and 2) both phage gave higher phage titres (indicating repair of their DNA) when infected into UV-irradiated host cells. Non-irradiated phage, infected into irradiated and non-irradiated host cells, was used as a control.

This procedure also allowed investigation of the nature of the SOS response. Weigle reactivation did not occur when the cells were held for 20 minutes after UV-irradiation before infecting with phage. This suggested the SOS response was 'switched off' within 50 minutes of being activated. Also, using host cells from the late log phase of growth (OD_{540} > 1.7) gave no Weigle reactivation suggesting that in late log the SOS response is already activated (Table 4).

TABLE 1
Weigle reactivation of phage JS017

		B. caldotenax BT1 (host)		Weigle reactivation factor
		no UV	+ UV	
JS017	no UV	3.89×10^{10}	3.96×10^{10}	1.018
	+ UV	1.108×10^{7}	1.001×10^{8}	9.034

TABLE 2
Weigle reactivation of phage o323

		B. caldotenax BT1 (host)		Weigle reactivation factor
		no UV	+ UV	
ø323	no UV	3.72×10^{7}	3.99×10^{7}	1.07
	+ UV	4.02×10^{3}	6.93×10^{4}	17.24

TABLE 3
Effect of postirradiation incubation of B. caldotenax BT1 cells on Weigle reactivation, 20 min holding, 30 min absorption

		B. caldotenax BT1 (host)		Weigle reaction factor
		No UV	+UV	
ø323	no UV	3.86×10^{7}	3.96×10^{7}	1.025
	+ Uv	2.31×10^{3}	4.36×10^{3}	1.88

TABLE 4
Weigle reactivation of phage JS017 using a late-log culture of BT1

		B. caldotenax BT1 (host)		Weigle reaction factor
		No UV	+ UV	
ø323	no UV	4.86×10^{10}	4.79×10^{10}	0.986
	+ Uv	9.80×10^{7}	9.60×10^{7}	0.98

CONCLUSION

From the data obtained, it can be concluded that B. caldotenax has an inducible 'SOS' repair system. To date, filamentation, de novo synthesis of proteins, prophage induction and Weigle reactivation have been identified as SOS related functions. These and other phenomena have already been identified as associated with the inducible SOS response in E. coli and B. subtilis (Table 5).

TABLE 5
The pleiotropic effects of SOS repair in different bacterial species

SOS in E. coli	SOS in B. subtilis	SOS in B. caldotenax
Weigle reactivation	Weigle reactivation	Weigle reactivation
Weigle mutagenesis	? ?	? ?
RecA induction	Synthesis of new proteins	Synthesis of new proteins
Error-prone repair	Error-prone repair	? ?
Filamentation	Filamentation	Filamentation
Prophage induction	Prophage induction	Prophage induction

It is surprising that B. caldotenax maintains an inducible DNA repair capacity when it has been estimated that at 70°C, a thermophilic bacterium would lose 50 purines per generation [7]. Hence, it might be expected that any DNA repair capacity would be permanently activated. This inducible DNA repair mechanism may however, be specific for only certain types of damage such as that caused by UV and mitomycin-C, whereas depurination may be repaired by a totally different repair mechanism such as an 'insertase' which would be permanently activated [8].

REFERENCES

1. Radman, M., SOS repair hypothesis: phenomenology of an inducible repair which is accompanied by mutagensis. In Molecular Mechanisms for Repair of DNA, ed., P.C. Hanawalt and R.B. Setlow, Plenum, New York, 1975, p. 355.

2. Kato, J. and Shinoura, Y., Isolation and characterisation of mutants of Escherichia coli deficient in induction of mutation by ultraviolet light. Mol. Gen. Genet., 1977, 156, 121-131.

3. Yasbin, R. E., DNA repair in Bacillus subtilis: I. The presence of an inducible system. Mol Gen. Genet. 1977, 153, 211-218.

4. O'Farrell, P.H., High resolution two-dimensional electrophoresis of proteins. J. Biol. Chem. 1975, 250, 4007-4021.

5. Sharp, R.J., Taxonomic and genetic studies of Bacillus thermophiles. PhD Thesis, CNAA, 1982.

6. Weigle, J.J., Induction of mutations in a bacterial virus. P.N.A.S. USA, 1953, 39, 628-636.

7. Kaboev, O.K., Luchkina, L.A. and Kuziakina, T.I., Apurinic and apyrimidinic DNA endonuclease of extremely thermophilic Thermothrix thiopora. J. Bact., 1985, 164, 878-881.

8. Friedberg, E.C., Base excision repair of DNA. In Chromosome damage and repair, ed. E. Seeburg an K. Kleppe, Plenum, New York, 1981. pp. 71-83.

DEVELOPMENT OF HOST-VECTOR SYSTEMS IN THE EXTREME THERMOPHILE THERMUS THERMOPHILUS

YOSHINORI KOYAMA, SHIGERU OKAMOTO and KENSUKE FURUKAWA
Fermentation Research Institute, AIST,
Tsukuba Science City, Ibaraki 305, JAPAN

ABSTRACT

Thermus-E.coli shuttle vector plasmid pYK105 was constructed. pYK105 consists of tryptophan synthetase genes of Thermus T2, E.coli pUC13 vector and a cryptic plasmid pTT8 of T.thermophilus HB8. It transforms T.thermophilus HB27 trpB to trp$^+$ and E.coli to ampicillin registant.

INTRODUCTION

Host-vector systems are available in many bacteria. However, there has been no host-vector system for extreme thermophile. We have reported genetic transformation in extremely thermophilic bacteria Thermus species [1]. Auxotrophic strains of T.thermophilus HB27 were transformed to prototrophy at high frequecies of 1-10% when proliferating cell populations were exposed to wild type chromosome DNA. They did not require chemical treatment to induce competence and the competence was maitained throughout the growth phase. A cryptic plasmid from T.thermophilus HB8 was introduced into T.thermophilus HB27 Pro$^-$ at the frequency of 10^{-2}.

In Thermus no antibiotic registance plasmid has been found and trials to express antibiotic resistance genes from other bacteria in Thermus were unsuccesfull.

Here we describe the construction of a Thermus-E.coli

shuttle vector plasmid which contains Thermus T2 tryptophan synthetase genes as selection markers in Thermus.

RESULTS and DISCUSSION

First, we cloned tryptophan synthetase genes (trpBA) of T.thermophilus HB27. We used a new cloning system of direct plasmid transfer from replica-plated E.coli colonies to Thermus cells, developed originally for B.subtilis by J. van Randen and G. Venema [2]. Four to ten Kb fragments of MboI partially digested T.thermophilus HB27 chromosome DNA were ligated to BamHI digested pUC13. E.coli MC1009 was transformed with this DNA and was spread on plates containing ampicillin. The original E.coli clone, carrying the recombinant plasmid (pKA2), was selected directly from a mixture of plated E.coli clones by replicating these clones onto minimal agar plates without tryptophan, spread just before with T.thermophilus HB27 trpB mutant. After 48hr incubation at 70C, a cluster of small transformed Thermus colonies had developed exclusively in a E.coli/pKA2 colony print. Plasmid pKA2 was shown to contain T.thermophilus tryptophan synthetase genes on 5.5Kb insert DNA by DNA sequence analysis.

However, T.thermophilus HB27 trp genes were not suitable for selection marker genes in T.thermophilus HB27 trpB hosts, because DNA fragments of T.thermophilus HB27 trp genes, despite their small sizes, recombine with chromosomal couterpart at high frequency. Thus, we chose tryptophan synthetase genes of Thermus T2 strain as selection marker genes because T.thermophilus HB27 trpB5 was transformed to trp$^+$ at a very low frequency (10^{-5}) with chromosome DNA of wild type Thermus T2.

Thermus T2 chromosomal DNA was shot-gun cloned on E.coli pUC13 plasmid and the resulting recombinant plasmid pKA207 containing the Thermus T2 trp genes was selected by colony-hybridization with T.thermophilus trp gene as a probe. The plasmid pKA207 contains the Thermus T2 tryptophan synthetase genes on 6.8Kb insert DNA with 3 Bgl II sites outside of trp genes.

pKA207 was then digested with Bgl II and ligated to Bgl II digested Thermus cryptic plasmid pTT8. T.thermophilus HB27 trpB5 was transformed with the ligation mixture and trp$^+$ transformants were obtained. The plasmid pYK105 (Fig. 1) was isolated from a trp$^+$ transformant. pYK105 can replicate both in T.thermophilus and E.coli. It transforms T.thermophilus HB27 trpB5 to trp$^+$ and E.coli to ampicillin registant.

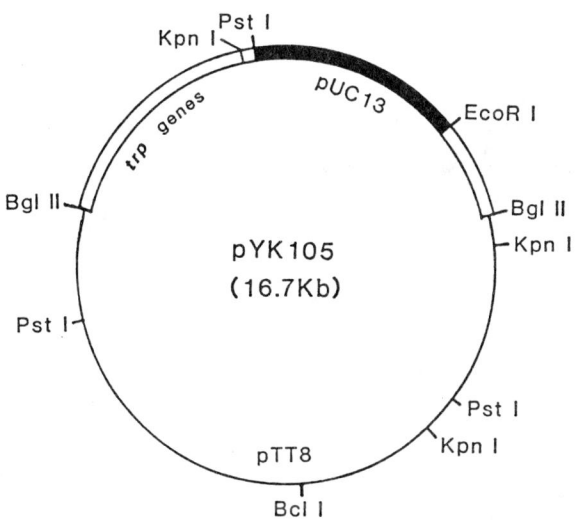

Figure 1. Plasmid pYK105.

REFERENCES

1. Koyama, Y., Hoshino, T., Tomizuka, N. and Furukawa, K., Genetic transformation of the extreme thermophile Thermus thermophilus and of other Thermus spp., J. Bacteriol., 1986, **166**, 338-340.

2. Randen, J., and Venema, G., Diirect plasmid transfer from replica-plated E.coli colonies to competent B.subtilis cells., Mol. Gen. Genet., 1984, **195**, 57-61.

BIOCATALYSIS ACTING UNDER EXTREME PHYSICO-CHEMICAL CONDITIONS

BRIGITTE HEINRITZ, MATTHIAS THIEM, CHRISTINE GWENNER,
JOACHIM SAWISTOWSKY, ROLF HEDLICH, MANFRED RINGPFEIL
Institute of Biotechnology
Academy of Sciences of the G.D.R.,
Permoserstrasse 15, Leipzig 7050, G.D.R.

ABSTRACT

Extremophilic microorganisms are ubiquitary. Thus, thermophilic Bacilli have been isolated from natural hot springs as well as from artificial hot water systems.
They are a source of thermostable enzymes, such as an extracellular protease from Bacillus stearothermophilus TP 26 and an intracellular beta-galactosidase from Bacillus stearothermophilus TP 32. Both enzymes mentioned were synthesized in a continuous and only partially protected process.

INTRODUCTION

A disadvantage of the most industrially used biocatalysts is their relatively low long-term stability. Microorganisms which need extreme physico-chemical conditions for their growth and proliferation and enzymes derived from them bear a powerful potential for overcoming this situation.

Especially thermostable enzymes derived from thermophilic microorganisms show frequently an increased long-term stability. The source of such long-term stable enzymes chosen in the most cases are thermophilic eubacteria because they can be handled easily (1 - 4).

RESULTS

Thermophilic Bacillus strains have been isolated from arti-

ficial hot water reservoirs as well as from hot springs of Vietnam.
Two strains have been selected among them which produce a protease and a beta-galactosidase, respectively.
Bacillus stearothermophilus TP 26 forms growth-coupled an extracellular thermostable protease at an optimum temperature of 65°C and an optimum pH value of 6.5. High protease secretion into the culture medium has been obtained during continuous and only partially protected fermentation of glucose or sucrose at dilution rates of 0.25 up to 0.4 h^{-1} under carbon substrate limitation.
The molecular weight of the protease is 12 000. The enzyme possesses a temperature optimum of 75°C and a pH value optimum of 7.0. The thermostability of the enzyme is reduced at temperatures greater than 60°C. The half life time is 8 minutes at 90°C.
Bacillus stearothermophilus TP 32 forms an intracellular beta-galactosidase at specific growth rates smaller than the maximum one. Beta-galactosidase activities up to 15 000 U per gram bacteria dry weight have been obtained in continuous and only partially protected fermentation of lactose at specific growth rates of 0.17 up to 0.28 h^{-1}, temperatures of 68 up to 70°C and pH values of 6.8 up to 7.5. The beta-galactosidase possesses a temperature optimum of 75°C and a pH value optimum of 6.5. The residual activity is 80 % after incubation of 60 minutes at 70°C in presence of EDTA and lactose.

CONCLUSIONS

Thermophilic Bacilli are suitable for effective synthesis of thermostable enzymes, such as a protease and a beta-galactosidase.
Whereas the protease synthesis is growth-coupled the beta-galactosidase is formed at specific growth rates smaller than the maximum one.
The described examples show the possibility of continuous and only partially protected synthesis of thermostable enzymes.

REFERENCES

1. Sonnleitner, B., Biotechnology of Thermophilic Bacteria. In Microbial Activities, ed., A. Fiechter, Akademie-Verlag, Berlin, 1984, pp. 70 - 138

2. Ng, T.K., Kenealy, W.F., Industrial Applications of Thermostable Enzymes. In Thermophiles, ed., T.D. Brock, J. Wiley and Sons, New York, Chichester, Brisbane, Toronto, Singapure, 1986, pp. 197 - 216

3. Ringpfeil, M., Heinritz, B., Biotechnological Use of Thermophilic Microorganisms. In Proc. IUPAC Congress, Sofia 1987

4. Heinritz, B., Biocatalysts Acting under Extreme Physico-Chemical Conditions. In Proc. 10. Jahrestagung der Biochemischen Gesellschaft der DDR zusammen mit der GATM in der Biologischen Gesellschaft der DDR, Neubrandenburg 1988

THERMOPHILIC ORGANISMS IN SUBMARINE FRESHWATER HOT SPRINGS IN ICELAND

SIGRIDUR HJÖRLEIFSDOTTIR[1,*], JAKOB K. KRISTJANSSON[1,2]
and GUDNI A. ALFREDSSON[1]

[1] Institute of Biology, University of Iceland, IS-108 Reykjavik, Iceland.
[2] Biotechnology Div., Technological Institute of Iceland, IS-112 Reykjavik, Iceland.
* Present address: Department of Biotechnology, Chemical Center, University of Lund P.O.Box 124, S-221 00 Lund, Sweden.

ABSTRACT

Submarine freshwater hot springs in shallow water were studied as a part of a research on microbial life in Icelandic hot springs. The study area was the Breidafjördur bay where the hot springs are located on skerries that are only accessible during few hours when the tide is very low. The effects of the higher temperatures are very localized. Thermophilic cyanobacteria form a narrow zone on the rocks closest to the hot spring openings. Further away are zones of barnacles and then the larger algae. Isolations of bacteria were done in the laboratory on nutrient agar media at 72°C and with 0 - 3% NaCl concentrations. The bacteria were mostly of the genus *Thermus* but unlike those round in terrestrial hot springs, these were salt tolerant. More than 90% of the marine strains could grow on 4% NaCl medium, whereas only 20% of terrestrial *Thermus* strains can grow in 1% NaCl. This shows the adaptation of these thermophilic bacteria to the marine environment.

INTRODUCTION

In recent years there has been much interest in research on submarine hot springs and the organisms associated with them (1, 2). Most of the work has been done on the deep-sea hydrothermal vents in the Pacific Ocean but recently some studies have been performed on submarine alkaline hot springs in Iceland (3, 4). In these studies salt tolerant stains of *Thermus* were isolated (3, 5) and a new species (and genus) of bacteria *Rhodothermus marinus*, was described (4). Bacteria of the genus *Thermus* are the most characteristic heterotrophs in terrestrial alkaline hot springs (6) and this also seems to be the case for submarine alkaline hot springs (4, 5). The salt tolerant *Thermus* strains were very similar to their terrestrial counterparts, except in being adapted to the submarine thermal environment (5).

The study presented here was done on submarine freshwater hot springs in the skerries of the Breidafjördur bay in West-Iceland. These hot springs are under sea-level except for a few hours when the tide is very low. Any thermophilic organisms growing in such submarine hot springs would be subjected to great fluctuations in the environment and stresses from the cold and salty seawater.

MATERIALS AND METHODS

Study Site.
Around the island Flatey in Breidafjördur bay in West-Iceland there are many small islands and skerries where hot springs can be found. Some of these hot springs are under sea-level, except for a few hours at extreme low tide. The skerries which we collected samples from were Drápssker, Nordursker, Reykey and Oddbjarnarsker. The water in the hot springs is alkaline freshwater (Conductivity 500-700 uS/cm). The temperature ranges from 70 to 100°C in the openings and the pH is 8 to 9.

Sampling.
Samples were collected at low tide directly from many different openings and the run-off water. Samples were also collected by scraping the rocks 1-3 cm above the hot water. The samples were taken in sterile bottles and kept at about 10°C until processed further in the laboratory. No seawater was mixed with the samples. Samples were also collected from the sea around the skerries.

Isolation of Bacteria.
Samples were filtered through 0,45 um membrane filter and these were then placed on nutrient agar 162 as previously described (5). The isolation medium had NaCl concentrations of 0, 1 and 3% (w/v). The plates were incubated at 20, 37, 50, 65 and 72°C for 4 days.

RESULTS AND DISCUSSION

The Macroflora.
This study showed a drastic but localized effect on all the organisms close to hot the springs. The visible effect on the macroflora was seen only close to the springs, usually only within 1/2 meter from the source. Typically there was a few centimeter zone of green cyanobacteria next to the hot water. After that came usually a clear zone with no visible growth. Then came a zone of barnacles, followed by an *Enteromorpha* zone and furthest away was an extensive zone of large brown algae.

The Hot Spring Bacteria.
Thermophilic bacteria were isolated at 65 and 72°C from all the hot springs. They could be isolated at NaCl concentrations up to 3% with the highest number of bacteria at 1% NaCl. When isolations were done at different temperatures (20-72°C) the highest viable counts were obtained at 37-50°C (Fig. 1 A and B).

In some cases the number of bacteria at 37-50°C were up to 50 times higher at 3% NaCl than at 0% NaCl (Fig. 1B). This shows that the population at the lower temperatures (37-50°C) is adapted to higher salt concentrations than the thermophiles growing at 72°C. This can be expected since increased mixing with seawater both increases the salt concentration and lowers the temperature. The fact that viable count at 20°C was highest at 0% NaCl (Fig. 1A) is probably due to post-sampling growth in the water samples, which were kept at ambient temperature (about 10°C) during the expedition (3-4 days). The growth of marine bacteria, (growing optimally at about 3% NaCl) would, however be suppressed in these samples.

When thermophilic marine strains which were isolated on 0% NaCl medium were tested for salt tolerance we found that more than 90% of them could grow in 3% NaCl, whereas only 20% of the thermophilic terrestrial strains could grow in 1% NaCl and not at higher concentrations (5). Most of the thermophiles isolated in this work were of the genus *Thermus* (5, 6). The recently described species *Rhodothermus marinus* (4) was not found in the submarine hot springs in Breidafjördur.

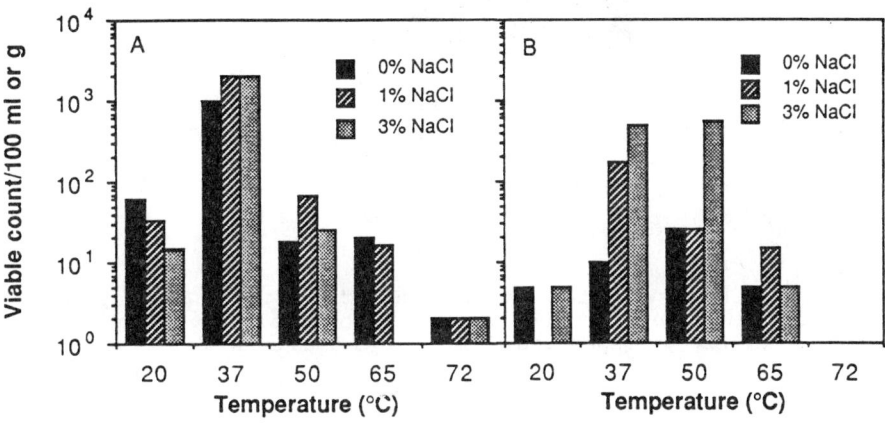

Figure 1. Viable counts at different temperatures and NaCl concentrations. A. In 100 ml water sample taken at 68°C in the runoff from hot spring NS-1 on Nordursker (Source Temp. 100°C). B. In 100 g wet wt. of algalmat scrapings taken at 55°C from rocks close to hot spring DR-5 on Drápssker (Source Temp. 95°C).

The Marine Bacteria.
Samples were also collected from the sea around the skerries and the viable counts compared to those in the hot springs. In these samples (which were processed within 24 hrs.) the highest viable counts were obtained at 20°C. Very few bacteria could be isolated from these seawater samples at 50°C and none at higher temperatures.

In all cases the highest number of bacteria were found at 3% NaCl but lowest at 0% NaCl (Fig. 2). The difference was about 10 fold at 20 and 37°C but 3 fold at 50°C. The 10 fold higher counts at 20°C as compared to 37°C is typical for seawater far from the coast. The opposite is usually only seen close to polluted areas (Unpublished results). Our study sites were several kilometers away from any such pollution and therefore the higher counts at 37°C can only be attributed the localized temperature increase due to the presence of the hot springs.

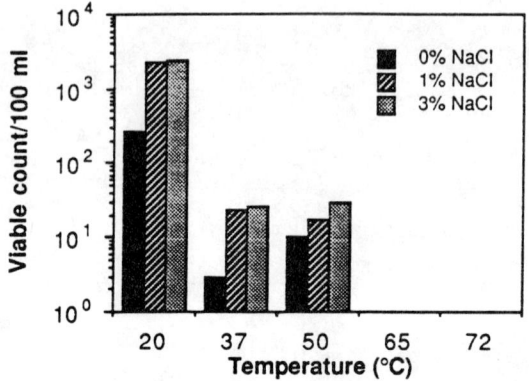

Figure 2. Viable counts at different temperatures and NaCl concentrations in 100 ml of seawater taken close to skerries (BS-4) in Breidafjördur bay.

ACKNOWLEDGEMENTS

This work was supported by the Icelandic Science Foundation and the Research Fund of the University of Iceland. We thank Gudmundur Ó. Hreggvidsson and Gudmundur Gudmundsson for assistance.

REFERENCES

1. Jannasch, H.W. and Nelson, D.C. Recent progress in the microbiology of hydrothermal vents. In Current Perspectives in Microbial Ecology, ed., M.J. Klug and C.A. Reddy, American Society for Microbiology, Washington, D.C., 1984, pp. 170-176.

2. Setter, K.O. Thermophilic archaebacteria occurring in submarine hydrothermal areas. In Planetary Ecology, ed., D.E. Caldwell, J.A. Brierley and C.L. Brierley, Van Nostrand Reinhold, New York, 1985, pp. 320-332.

3. Kristjansson, J.K. and Alfredsson, G.A. Life in Icelandic hot springs. Natturufraedingurinn, 1986, **56**, 49-68 (in Icelandic with English summary).

4. Alfredsson, G.A., Kristjansson, J.K., Hjörleifsdottir, S. and Stetter, K.O. Rhodothermus marinus, gen. nov., sp. nov., a thermophilic, halophilic bacterium from submarine hot springs in Iceland. J. Gen. Microbiol., 1988, **134**, 299-306.

5. Kristjansson. J.K., Hreggvidsson, G.O. and Alfredsson, G.A. Isolation of halotolerant Thermus spp. from submarine hot springs in Iceland. Appl. Environ. Microbiol., 1986, **52**, 1313 - 1316.

6. Kristjansson, J.K. and Alfredsson, G.A. Distribution of Thermus spp. in Icelandic hot springs and a thermal gradient. Appl. Environ. Microbiol.,1983, **45**, 1785-1789.

CATECHOL PRODUCTION BY BACILLUS STEAROTHERMOPHILUS

P. ORIEL, G. GURUJEYALAKSHMI
Department of Microbiology and Public Health,
Michigan State University, East Lansing,
Michigan 48824-1101, U.S.A.

SUMMARY

A phenol-resistant Bacillus stearothermophilus isolate accumulated catechol in the phenol growth medium following inhibition of catechol oxidation with tetracycline.

INTRODUCTION

It is recognized that enzymes possessing thermostability are more resistant to chemical denaturation as well (for a recent review, see Berquist et al., 1987). This interesting correlation between thermal and chemical stability suggests that if appropriate metabolic pathways are available, thermophiles might prove valuable in bioconversions of chemicals at levels which are toxic to mesophiles. As a test of this concept, we have chosen the oxidation of phenol, an environmental pollutant. Although oxidation of phenol has been documented for many mesophilic microorganisms, growth is usually inhibited at low phenol concentrations, requiring low level feeding or slow release "phenol sinks" for successful utilization (Ehrhardt and Rehm, 1985). This growth inhibition appears at least partly due to phenol inhibition of phenol hydroxylase, the first step in phenol utilization. This enzyme has been shown to be substrate inhibited above 0.3 mM phenol concentration both in Pseudomonas putida (Janke et al., 1981) and Trichosporon yeast (Neujahr and Gaal, 1983).

Following the early report of Buswell (1975) describing an isolated Bacillus stearothermophilus strain with the capability to degrade phenol via a meta pathway, we have isolated an even more phenol-tolerant strain BR219. This organism grows well on phenol at concentrations to 15 mM, and has an NADH-dependent phenol hydroxylase which is resistant to phenol at concentrations to at least 15 mM (Gurujayalakshmi and Oriel, submitted). In this report we describe that the phenol catabolism of BR219 can be inhibited by tetracycline, resulting in catechol accumulation in the medium. Because of its wide utilization in the plastic and photographic industries, efficient bioproduction of catechol could provide an interesting alternative to current chemical production processes.

METHODS AND MATERIALS

Measurement of Catechol

Catechol was measured colorimetrically using the 4-aminoantipyrene method of Irie et al. (1987). Independent identification of catechol was made using thin layer chromatography.

Organism and Media

BR219 was isolated from the Tittabawassee River near Midland, Michigan using phenol enrichment culture (Gurujayalakshmi and Oriel, submitted). It was grown on DP medium of Buswell (1975) containing phenol at the concentration indicated. Organism growth was measured using apparent optical density at 515 nm on a Cary 15 spectrophotometer, or by viable count of suitable dilutions on L agar plates. BR219 transconjugants containing the conjugating transposon Tn916 (Gawron-Burke and Clewell, 1982) were isolated at 55°C on L agar plates with 5 mg/ml tetracycline following 48°C liquid culture mating with Bacillus subtilis BS250 containing the transposon integrated in the chromosome (Sen, Gurujeyalakshmi and Oriel, submitted).

Intracellular 2-Hydroxymuconic Semialdehyde

The relative activity of catechol 2,3 dioxygenase was measured using the appearance of 2-hydroxymuconic semialdehyde following addition of known amounts of catechol to the culture. For this, catechol was added to 100 mgm/ml to log phase cells which had been grown in DP medium containing 10 mM phenol and resuspended in DP medium with no phenol to OD 0.5 at 515 nm in a Cary 15 spectrophotometer. The reference beam contained the same cell concentration with no catechol addition. Formation of 2-hydroxymuconic semialdehyde was observed by the appearance of an absorption band at 375 nm, which was recorded 15 minutes after catechol addition.

Assay of 2,3 Catechol Dioxygenase

Cellular extracts of BR219 were made of log phase cells using the method of Buswell, 1975. Assay of catechol 2,3 dioxygenase was made using the method of Kojima et al. (1961).

RESULTS AND DISCUSSION

Initial experiments indicated that although small amounts of catechol were formed in overnight flask cultures growing on phenol, significant enhancement of catechol accumulation with the addition of tetracycline to the culture at the end of the exponential growth phase (Table I). While this effect was significant with wild-type BR219, even higher accumulation occurred with BR219 carrying conjugating transposon Tn916. The location of the Tn916 insertion is not known, but does not appear to be in the phenol degradation pathway since phenol utilization was not altered in the absence of tetracycline (data not shown). BR219::Tn916 was therefore used in all further studies.

Table I

Catechol production (ugm/ml) by B. stearothermophilus BR219

	No tetracycline	5 mM tetracycline
BR219	2.2	7.2
BR219::Tn916	1.2	12.8

Tetracycline was added in late exponential phase, and catechol measured 16 hours thereafter.
Accumulation of catechol with addition of tetracycline suggested that tetracycline might act through direct inhibition of catechol 2,3 dioxygenase. This inactivation was measured by addition of catechol to resuspended cultures and measurement of intracellular 2-hydroxymuconic semialdehyde. As seen in Figure 1, intracellular accumulation of 2-hydroxymuconic semialdehyde following catechol addition was lost

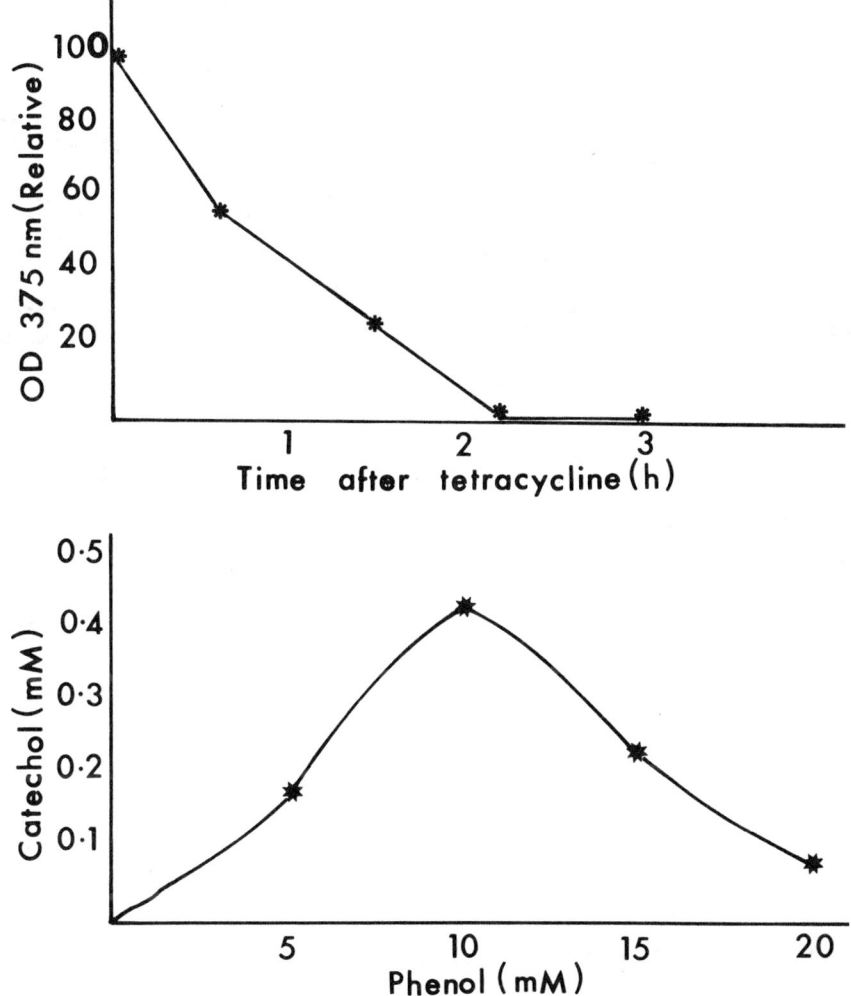

Figure 1. (Upper)   Inhibition of catechol 2.3 dioxygenase following addition of 5 mM tetracycline.

Figure 2. (Lower)   Extracellular catechol production with varied phenol concentrations.

rapidly after tetracycline treatment, with no activity evident after 2 hours. During this period, there was no loss in cell viability. Verification of direct catechol 2,3 dioxygenase inhibition by tetracycline was carried out using crude cell extracts, where complete inhibition of catechol 2,3 dioxygenase activity was observed at tetracycline concentrations above 1 mgm/ml.

It was of interest to determine the optimal concentration of phenol that could be used for catechol production. As seen in Figure 2, 10 mM phenol proved to be the optimum concentration for catechol formation in this system.

CONCLUSIONS

Availability of a phenol-tolerant B. stearothermophilus strain BR219 has allowed testing of characteristics of this thermophile in phenol bioconversion. Bioaccumulation of the metabolic intermediate catechol can be achieved through addition of tetracycline in the cells in culture, which inhibits catechol 2,3 dioxygenase in cell extracts and blocks intracellular oxidation of catechol to 2-hydroxymuconic semialdehyde. Introduction of transposon Tn916 further aids catechol production, possibly by countering antibiotic inhibition of protein synthesis. While conversion levels are in need of further optimization before consideration as a bioproduction method, we are encouraged by these initial results regarding the potential use of thermophiles in bioconversions of chemicals at levels toxic to mesophilic microorganisms.

REFERENCES

Berquist, P. L., D. R. Love, J. E. Croft, M. B. Streiff, R. M. Daniel, and W. H. Morgan. 1987. In: Biotechnology and Genetic Engineering Review, Volume 5, pp. 199-245. Intercept, Dorset.

Buswell, J. A. 1975. J. Bacteriol. **124**:1077-1083.

Kojima, Y., N. Itada, and O. Hyashi. 1961. J. Biol. Chem. **236**:2223-2228.

Ehrhardt, H. M., and H. J. Rehm. 1985. Appl. Microbiol. Biotechnol. **21**:32-36.

Gawron-Burke, C., and D. Clewell. 1982. Nature **300**:281-284.

Irie, S., K. Shirai, S. Doi, T. Yorifuji. 1987. Agr. Biol. Chem. **51**:1489-1493.

Janke, D., R. Pohl, and W. Fritsche. 1981. Z. Allgem. Mikrobiol. **21**:295-303.

Neujahr, H. Y., and A. Gaal. 1973. Eur. J. Biochem. **35**:386.

CELL ENVELOPES OF ARCHAEBACTERIA

HELMUT KÖNIG
Universität Ulm, Angewandte Mikrobiologie
Oberer Eselsberg M 23, 7900 Ulm, F.R.G.

ABSTRACT

Archaebacteria have quite diverse muramic acid-less cell envelopes. Cell walls of the gram positive archaebacteria are composed of pseudomurein, methanochondroitin or heteropolysaccharide. Some species possess an additional subunit layer (S-layer). The gram negative archaebacteria possess one or two surface layers (S-layers) composed of protein or glycoprotein subunits in a two-dimensional crystalline arrangement. Cell wall-less organisms are also present.

INTRODUCTION

The archaebacterial concept was initially based on 16S rRNA oligonucleotide comparisons [1]. Cell envelope structures were one of the first biochemical characteristics studied in detail [2, 3, 4].

During cell envelope investigations of *Halobacterium* Houwink [5] already established significant differences from typical bacterial cell walls in 1956 [6], which were confirmed by later work with other cell envelopes [6, 7, 8]. The heterogeneous archaebacterial groups possess a great variety of cell envelopes [2], but the lack of muramic acid is a common feature of all archaebacteria. In the following chapters a brief overview on archaebacterial cell envelopes (Fig. 1) is given.

Gram Reaction	Profile		Genus
negative	●●●●●●●●●●● ═══════════	SL CM	Acidianus Archaeoglobus Desulfurococcus Haloarcula Halobacterium Haloferax Methanococcoides Methanococcus Methanocorpusculum Methanogenium Methanolobus Methanomicrobium Methanoplanus Natronobacterium Pyrococcus Pyrodictium Staphylothermus Sulfolobus Thermococcus Thermofilum Thermoproteus
negative	●● ●● ●● ●● ●●●●●●●●●●● ═══════════	PS SL CM	Methanospirillum Methanothrix
negative	●●●●●●●●●●●	CM	Thermoplasma
positive or negative	▬▬▬▬▬▬▬▬▬ ●●●●●●●●●●● ═══════════	MC SL CM	Methanosarcina
positive	▬▬▬▬▬▬▬▬▬ ═══════════	PM,HP CM	Halococcus Methanobacterium Methanobrevibacter Methanosphaera
positive	●●●●●●●●●●● ▬▬▬▬▬▬▬▬▬ ═══════════	SL PM CM	Methanothermus

Figure 1. Schematic profiles of archaebacterial cell envelopes. CM = cytoplasmic membrane, HP = heteropolysaccharide, PM = pseudomurein, PS = protein sheath, SL = surface layer (S-layer), MC = methanochondroitin. *Methanosarcina* can lack one of the two layers. It has to be proofed, whether in the case of *Methanospirillum* and *Methanothrix* the layer surrounding the individual cells is composed of proteinaceous subunits.

METHANOGENIC BACTERIA

Pseudomurein

All species of the order Methanobacteriales possess electron-dense cell wall sacculi with a thickness of about 15 nm. These cell walls are composed of pseudomurein, a new type of peptidoglycan [9, 10, 11]. Significant differences were found between the eubacterial murein and the methanobacterial pseudomurein. The glycan strands consist of alternating $\beta(1\rightarrow3)$ linked N-acetyl-D-glucosamine and N-acetyl-L-talosaminuronic acid residues. The peptide moiety is composed of a set of the three L-amino acids: glutamic acid, alanine and lysine. Both the glycan and the peptide moiety can be modified [9, 12]. Some amino acids e.g. glycine are incorporated in the pseudomurein [13], when they occur in elevated concentrations in the culture medium. Under these conditions the rods become irregular in shape and the pseudomurein sacculi could be partially removed in some species.

The pseudomurein sacculi have a density of 1.39-1.46 g/cm^3 and the unit cell has a dimension of 4.5 Å x 10 Å x 21-22.5 Å [14, 15]. Conformational energy calculations suggested a similar secondary structure for murein and pseudomurein [16].

The proposed biosynthetic pathways of murein and pseudomurein are different [17]. In Methanobacteriales a UDP-activated disaccharide composed of N-acetyltalosaminuronic acid and N-acetylglucosamine is the main compound of the nucleotide pool. In addition, UDP-activated dipeptides, tripeptides and pentapeptides have been purified from cell extracts.

The degradation pathway of the pseudomurein in nature is unknown. Recently, an autolytic enzyme was described for *Methanobacterium wolfei* [18, 19]. It is a peptidase that splits the peptide bound between alanine and lysine. This enzyme is useful for the production of protoplasts for membrane studies [20] and for the isolation of nucleic acids [19].

The pseudoumurein shows antigenicity [21]. The glycan components N-acetylglucosamine, N-acetylgalactosamine and N-acetyltalosaminuronic acid and the C-terminal dipeptide γ-glutamyl-alanine were found to be the antigenic determinants.

Protein and glycoprotein S-layers

The cells of Methanococcales have only a single S-layer outside of the cytoplasmic membrane [22]. It is composed of nonglycosylated protein subunits in hexagonal arrangement. The S-layers have similar chemical and physical stabilities and amino acid compositions. The acidic proteins are predominantly composed of hydrophobic amino acids. The average hydrophobicity, however, is lower than in many cellular proteins.

Crystalline proteinaceous surface layers associated with pseudomurein sacculi are only found in the extremely thermophilic members of the genus *Methanothermus* [23]. *Methanothermus fervidus* possesses an S-layer glycoprotein with an apparent molecular weight of 92 500. Asx and Glx account for 32 mol%). The carbohydrate moiety is composed of mannose, galactose, 3-O-methylglucose, N-acetylglucosamine and N-acetylgalactosamine. In contrast to the eukaryotes the biosynthesis of the glycan strands proceeds via UDP-activated oligosaccharides, which are transferred to dolichyl C_{55} diphosphate [24]. Undecaprenyl derivatives occur also in *Methanothermus* [24].

Methanochondroitin

The cells of *Methanosarcina* are often closely packed into large "tissue"-like aggregates. Most species possess an outer cell wall layer with a thickness up to 200 nm. It consists of N-acetylgalactosamine, glucuronic or galacturonic acid, glucose and small amounts of mannose [25]. The shape-maintaining component is a fibrillar nonsulfated polymer composed of a uronic acid and two N-acetylgalactosamine residues. This polymer forms a compact or sometimes loose matrix, which is reminiscent of components of the eucaryotic connective tissue. It was therefore named "methanochondroitin" [25]. Some species of *Methanosarcina* lack the methanochondroitin matrix and possess single S-layers and other species have a subunit layer in addition to the methanochondroitin matrix.

Protein sheaths

The outer envelope of *Methanospirillum* and *Methanothrix* is formed by a tubular sheath composed of hoop like rings [26]. The sheath exhibits unusual resistance against chemical agents and lytic enzymes [9, 27]. The sheaths consist mainly of protein and carbohydrate [2, 9, 28]. A covalent linkage between the two components has not been demonstrated. Within the sheaths, the individual rod-shaped cells are surrounded by an additional layer. The rods are separated by pairs of spacer plugs. The hoop like rings of the sheaths consist of proteinaceous subunits in a two-dimensional crystalline arrangement [29, 30, 31, 32, 33]. Two structural models with a p1 [29] or a p2 [30, 33] symmetry have been proposed.

EXTREMELY HALOPHILIC BACTERIA

Glycoprotein S-layers

Gram negative halobacteria are surrounded by S-layers. The surface glycoprotein of *Halobacterium salinarium* was the first glycosylated protein detected in prokaryotes [34]. The analysis of the amino acid sequence and the carbohydrate moiety revealed a molecular weight of 130 000 [35]. The carbohydrate moiety is composed of three different glycopeptides with novel linkage types [36, 37, 38, 39, 40, 41, 42]. (1) A sulfated glycosaminoglycan of high molecular weight consisting of N-acetylglucosamine, N-acetylgalactosamine, galacturonic acid, galactose and 3-O-methylgalacturonic acid is connected via N-acetylgalactosamine to an asparagine residues. (2) Sulfated glycans of low molecular weight consisting of glucose, glucuronic acid and iduronic acid are bound via glucosyl-asparaginyl residues to the peptide moiety. (3) Disaccharides composed of glucose and galactose are connected to threonine residues. Dolichyl pyrophosphate serves as lipid carrier for the glycosaminoglycan. Dolichyl monophosphate is the lipid carrier for the second sulfated oligosaccharide. The acceptor peptide for the glycosaminoglycan is Asn-X-Ser(Thr). A transient methylation of a glucose residue is required for the transfer of the dolychyl oligosaccharides onto the protein [43, 44, 45].

The three-dimensional structure of the surface glycoprotein of *Halobacterium volcanii* has been studied recently [46]. The subunits are arranged on a p6 lattice. They are supposed to form dome-shaped structures, which are separated from the cytoplasmic membrane by spacers.

Heteropolysaccharide

Halococcus morrhuae is gram positive. The cells possess electron-dense cell walls with a thickness of about 50 nm [7], which prevent the lysis of the cocci in media of low ionic strength. The cell wall polymer is composed of a complex heteropolysaccharide consisting of N-acetylglucosamine, N-acetylgalactosamine, N-acetylgulosaminuronic acid, glucose, galactose, mannose, glucuronic acid, galacturonic acid, acetate, sulfate and glycine [47, 48, 49, 50]. The carbohydrate monomers are supposed to be arranged in three domains, which may be partly linked by N-glycyl-glucosaminyl bridges [51, 52].

THERMOPHILIC SULFUR AND SULFATE METABOLIZERS

Protein and glycoprotein S-layers

All extreme thermophilic sulfur and sulfate metabolizing archaebacteria (Sulfolobales, Thermoproteales, Thermococcales, Archaeoglobales) posses cell envelopes composed of proteinaceous subunits (S-layers). While the fine structure of several S-layers has been studied, little is known about the chemistry of these cell envelopes [53, 54, 55, 56, 57, 58, 59].

The S-layer of *Sulfolobus acidocaldarius* is composed of a glycoprotein [53, 54]. In contrast to *Halobacterium* and *Methanothermus* the polar amino acids serine and threonine predominate over Glx and Asx. The analysis of the fine structure revealed that the outer surface appears smooth, while the inner side is sculptured by the occurence of depressions at the sixfold and pedestals at the threefold axes [4, 55, 56]. The glycoprotein is arranged hexagonally (p6 symmetry) with a unit cell dimension of 22 nm. The glycoprotein subunits form three different domains. Large holes are created at the sixfold and the threefold axes. Due to the occurence of depressions and holes the S-layer possess a rather spongy structure. Only 30% is occupied by the glycoprotein.

S-layers with unusual high stability against chemical agents were isolated from *Thermoproteus tenax* [2, 54]. Ultrastructural analyses of the S-layers of *Thermoproteus tenax* and *T. neutrophilus* revealed [57, 58] that the outer surface of these S-layers is smooth, while long spikes protrude from the inner surface at the six-fould symmetry axes. The subunits are hexagonally (p6 symmetry) arranged with a center to center distance of 30 nm. Lattice faults were found at the hemispherical poles. The subunits form unusual rigid envelopes and thus determine the rod-shape of the cells.

Cell wall-less organisms
Despite the extreme environmental conditions, *Thermoplasma* lacks a cell wall [60]. The stabilization of the cells of this organism is most likely maintained by a combination of a mannose-rich glycoprotein [61] and a lipoglycan [62] located in the cytoplasmic membrane, which form most likely a glycocalyx.

CONCLUSIONS

Presently, the following conclusions can be drawn from the studies of archaebacterial cell walls:
(1) Archaebacteria possess diverse cell envelopes and they have no universal cell wall polymer like most eubacteria. However, all archaebacteria lack muramic acid. The lack of muramic acid can not be used as a general feature for distinguishing eu- and archaebacteria, since some genera of eubacteria with cell envelopes are also devoid of murein [63].
(2) The most common archaebacterial cell envelope is represented by single layers composed of protein or glycoprotein subunits. Cell envelopes composed of proteinaceous subunits

represent most probably an early invention of the archaebacteria. Only more progressive groups of the methanogenic branch have developed other polymers like pseudomurein, heteroplysaccharide or methanochondroitin. S-layers of nonglycosylated proteins may be the most primitive cell envelopes. Outside the cells no other "activated" compounds are required for the biosynthesis of the cell envelopes as with glycoproteins or peptidoglycans, since the information for the formation of stable crystalline layers is exclusively preserved in the amino acid sequence.

(3) Despite their structural similarities the methanobacterial peptidoglycan (pseudomurein) and the eubacterial peptidoglycan (murein) represent analogue and not homologue polymers as indicated by the replacement of MurNAc by NAcTalNUA, the sequence of the peptide subunit and the composition of the precursors.

(4) The structure of some polymers like methanochondroitin and the halobacterial S-layer glycoprotein are reminiscent of components of eucaryotic connective tissue.

(5) The biosynthesis of the archaebacterial cell envelope polymers exhibits eukaryotic and eubacterial characteristics, but also features not found in the other kingdoms so far. The usage of dolichol as lipid carrier for glycoprotein synthesis is a feature found in eukaryotes. *Methanothermus* possesses both, dolichol and undecaprenol. Undecaprenol, however, is a common lipid carrier of eubacteria. It seems most likely, that in general dolichol is involved in glycoprotein synthesis in pro- and eukaryotes and undecaprenol plays a role in other prokaryotic cell wall polymers e.g. pseudomurein or murein.

(6) The cell envelopes exhibit a high degree of resistance against cell wall specific antibiotics and lytic agents [64].

(7) In the case of *Thermoproteus, Thermofilum, Methanospirillum* and *Methanothrix* single S-layers have an unusual rigidity and a shape-determining function. However, the rigidity of the cell envelopes often does not parallel the extreme living conditions and cell envelopes with high rigidity are not required for life at high temperature and low

pH values. Organisms exposed to high temperatures and to low pH values lack rigid covalently linked cell wall sacculi composed of peptidoglycan or heteropolysaccharide. Layers consisting of protein or glycoprotein subunits surround such cells. In most cases these layers are easily disintegrated. *Thermoplasma* even demonstrates that cell envelopes are not required for life in extreme environments.

(8) Organisms with different cell envelope profiles display different modes of cell division. Cells with pseudomurein, heteropolysaccharide and methanochondroitin divide by septum formation. Bacteria with subunit envelopes divide by constriction. Cell wall less organisms multiply by budding. Division of the sheathed cells of *Methanospirillum* and *Methanothrix* proceeds by complicated fragmentation mechanisms.

, (9) Due to their high stability and isoporosity some S-layers can be used for the production of ultrafiltration and hyperfiltration membranes [65].

(10) The cell envelopes of archaebacteria are often directly exposed to extreme environments. They may serve as a model to elucidate survival strategies of these unusual bacteria and the molecular strategy of stability, shape-maintenance, resistance to heat and low pH and other functional or evolutionary aspects.

REFERENCES

1. Woese, C.R., Bacterial evolution. **Micorbiol. Rev.**, 1987, 51, 221-271.

2. Kandler, O. and König, H., Cell envelopes of archaebacteria. In **The Bacteria**, Vol. 8, eds., C.R. Woese and R.S. Wolfe, Academic Press, New York and London, 1985, pp. 413-457.

3. Sleytr, U.B., Messner, P., Sára, M. and Pum, D., Crystalline envelope layers in archaebacteria. **System. Appl. Microbiol.**, 1986, 7, 310-313.

4. Baumeister, W. and Engelhardt, H., Three-dimensional structure of bacterial surface layers. In **Electron Microscopy of Proteins**, Vol. 6, ed., J.R. Harris and R.W. Horne, Academic Press, London, 1987, pp. 109-154.

5. Houwink, A.L., Flagella, gas vacuoles and cell-wall struc-

ture in *Halobacterium halobium*; an electron microscope study. J. Gen. Microbiol., 1956, 15, 146-150.

6. Weiss, L.R., Subunit cell wall of *Sulfolobus acidocaldarius*. J. Bacteriol., 1974, 118, 275-284.

7. Brown, A.D. and Cho, K.J., The walls of the extremely halophilic cocci. Gram-positive bacteria lacking muramic acid. J. Gen. Microbiol., 1970, 62, 267-270.

8. Kandler, O. and Hippe, H., Lack of peptidoglycan in the cell walls of *Methanosarcina barkeri*. Arch. Microbiol., 1977, 113, 57-60.

9. Kandler, O. and König, H., Chemical composition of the peptidoglycan-free cell walls of methanogenic bacteria. Arch. Microbiol., 1978, 118, 141-152.

10. König, H., Kralik, R. and Kandler, O., Structure and modifications of pseudomurein in Methanobacteriales. Zbl. Bakt. Hyg., I. Abt. Orig., 1982, C 3, 179-191.

11. König, H., Kandler, O., Jensen, M. and Rietschel, T., The primary structure of the glycan moiety of pseudomurein from *Methanobacterium thermoautotrophicum*. Hoppe-Seyler's Z. Physiol. Chem., 1983, 364, 627-636.

12. König, H., Chemical composition of cell envelopes of methanogenic bacteria isolated from human and animal feces. System. Appl. Microbiol., 1986, 8, 159-162.

13. König, H., Influence of amino acids on growth and cell wall composition of Methanobacteriales. J. Gen. Microbiol., 1985, 131, 3271-3275.

14. Labischinski, H., Barnickel, G., Leps, B., Bradaczek, H., and Giesbrecht, P., Initial data for the comparison of murein and pseudomurein conformations. Arch. Microbiol., 1980, 127, 195-201.

15. Formaneck, H., Three-dimensional models of the carbohydrate moieties of murein and pseudomurein. Z. Naturforsch., 1985, 40c, 555-561.

16. Leps, B., Labischinski, H., Barnickel, G., Bradaczek, H. and Giesbrecht, P., A new proposal for the primary and secondary structure of the glycan moiety of pseudomurein. Eur. J. Biochem., 1984, 144, 279-286.

17. König, H., Kandler, O. and Hammes, W., Biosynthesis of pseudomurein: Isolation of supposed precursors from *Methanobacterium thermoautotrophicum*. Can. J. Microbiol., 1988, in press.

18. König, H., Semmler, R., Lerp, C. and Winter, J., Evidence for the occurence of autolytic enzymes in *Methanobacterium*

wolfei. Arch. Microbiol., 1985, 141, 177-180.

19. Kiener, A., König, H., Winter, J. and Leisinger, Th., Purification and use of Methanobacterium wolfei pseudomurein endopeptidase for lysis of Methanobacterium thermoautotrophicum. J. Bacteriol., 1987, 169, 1010-1016.

20. Mountfort, D.O., Mörschel, E., Beimborn, D.B. and Schönheit, P., Methanogenesis and ATP synthesis in a protoplast system of Methanobacterium thermoautotrophicum. J. Bacteriol., 1986, 168, 892-900.

21. Conway de Macario, E., Macario, A.J.L., Magarinos, M.C., König, H. and Kandler, O., Dissecting the antigenic mosaic of the archaebacterium Methanobacterium thermoautotrophicum by monoclonal antibodies of defined molecular specificity. Proc. Natl. Acad. Sci., 1983, 80, 6346-6350.

22. Nuβer, E. and König, H., Cell envelope studies on 3 species of Methanococcus living at different temperatures. Can. J. Microbiol., 1987, 33, 256-261.

23. Nuβer, E., Hartmann, E., Allmeier, H., König, H., Paul, G. and Stetter, K.O., A glycoprotein surface layer covers the pseudomurein sacculus of the extreme thermophile Methanothermus fervidus. 1987, In Proceedings of the Second International Workshop on S-layers in Procaryotes, Vienna, eds. U. B. Sleytr, P. Messner, D. Pum and M. Sara. Springer Verlag, Berlin, 1988, 21-25.

24. Hartmann, E. and König, H., Uridine and dolichyl diphosphate activated oligosaccharide intermediates in the biosynthesis of the S-layer glycoprotein of Methanothermus fervidus . Arch. Microbiol., 1988, submitted.

25. Kreisl, P. and Kandler, O., Chemical structure of the cell wall polymer of Methanosarcina. System. Appl. Microbiol., 1986, 7, 293-299.

26. Zeikus, J.G. and Bowen, V.G., Fine structure of Methanospirillum hungatei. J. Bacteriol., 1975, 121, 373-330.

27. Beveridge, T.J., Stewart, M., Doyle, R.J. and Sprott, G.D., Unusual stability of the Methanospirillum hungatei sheath. J. Bacteriol., 1985, 162, 738-737.

28. Sprott, G.D. and McKellar, R.C., Composition and properties of the cell wall of Methanospirillum hungatei. Can. J. Microbiol., 1980, 26, 115-120.

29. Shaw, P.J., Hills, G.J., Henwood, J.A., Harris, J.E. and Archer, D.B., Three-dimensional architecture of the cell sheath and septa of Methanospirillum hungatei. J. Bacteriol., 1985, 161, 750-757.

30. Stewart, M., Beveridge, T.J. and Sprott, G.D., Crystalline

order to high resolution in the sheath of *Methanospirillum hungatei*: A cross beta structure. J. Mol. Biol., 1985, 183, 509-515.

31. Beveridge, T.J., Patel, G.B., Harris, B.J. and Sprott, G.D., The ultrastructure of *Methanothrix concilii*, a mesophilic aceticlastic methanogen. Can. J. Microbiol., 1986, 32, 703-710.

32. Beveridge, T.J., Harris, B.J., Patel, G.B. and Sprott, G.D., Cell division and filament splitting in *Methanothrix concilii*. Can. J. Microbiol., 1986, 32, 779-786.

33. Patel, G.B., Sprott, G.D., Humphrey, R.W. and Beveridge, T.J., Comparative analyses of the sheath structures of *Methanothrix concilii* GP6 and *Methanospirillum hungatei* GP1 and JF1. Can. J. Microbiol., 1986, 32, 623-631.

34. Mescher, M.F. and Strominger, J.L., Purification and characterization of a procaryotic glycoprotein from cell envelope of *Halobacterium salinarium*. J. Biol. Chem., 1976, 251, 2005-2014.

35. Lechner, J. and Sumper, M., The primary structure of a procaryotic glycoprotein. Cloning and sequencing of the cell surface glycoprotein gene of halobacteria. J. Biol. Chem., 1987, 262, 9724-9729.

36. Wieland, R., Dompert, W., Bernhardt, G. and Sumper, M., Halobacterial glycoprotein saccharides contain covalently linked sulfate. FEBS Lett., 1980, 120, 110-114.

37. Wieland, F., Lechner, J., Bernhardt, G. and Sumper, M., Sulphation of a repetitive saccharide in halobacterial cell wall glycoprotein: occurence of a sulphated lipid-linked precursor. FEBS Lett., 1981, 132, 319-323.

38. Wieland, F., Lechner, J. and Sumper, M., The cell wall glycoprotein of halobacteria: Structural, functional and biosynthetic aspects. Zbl. Bakt. Hyg., I. Abt. Orig. C 3, 1982, 161-170.

39. Wieland, F., Heitzer, R. and Schaefer, W., Asparaginylglucose: novel type of carbohydrate linkage. Proc. Natl. Acad. Sci. (Wash.), 1983, 80, 5470-5474.

40. Wieland, F., Lechner, J. and Sumper, M., Iduronic acid: constituent of sulphated dolichylphosphate oligosaccharides in halobacteria. FEBS Lett., 1986, 195, 77-81.

41. Paul, G. and Wieland, F., Sequence of the halobacterial glycosaminoglycan. J. Biol. Chem., 1987, 262, 9587-9593.

42. Paul, G., Lottspeich, F. and Wieland, F., Asparaginyl-N-acetylgalactosamine: linkage unit of halobacterial glycosaminoglycan. J. Biol. Chem., 1986, 261, 1020-1024.

43. Lechner, J., Wieland, F. and Sumper, M., Biosynthesis of sulfated saccharides N-glycosidically linked to the protein via glucose. J. Biol. Chem., 1985, 260, 860-866.

44. Lechner, J., Wieland, F. and Sumper, M., Transient methylation of dolichyl oligosaccharides is an obligatory step in halobacterial sulfated glycoprotein biosynthesis. J. Biol. Chem., 1985, 260, 8984-8989.

45. Lechner, J., Wieland, F. and Sumper, M., Sulfated dolicholphosphate oligosaccharides are transient methylated during biosynthesis of halobacterial glycoproteins. System. Appl. Microbiol., 1986, 7, 286-292.

46. Kessel, M., Wildhaber, I., Cohen, S, and Baumeister, W., Three-dimensional structure of the regular surface glycoprotein layer of *Halobacterium volcanii* from the Dead Sea. EMBO J., 1988, 7, 1549-1554.

47. Reistadt, R., Cell wall of an extremely halophilic coccus. Investigation of ninhydrine-positive compounds. Arch. Microbiol., 1972, 82, 24-30.

48. Reistadt, R., 2-Amino-2-deoxyguluronic acid: a constituent of the cell wall of *Halococcus* sp., strain 24. Carbohydr. Res., 1974, 36, 420-423.

49. Reistadt, R., Amino sugar and amino acid constituents of the cell wall of the extremely halophilic cocci. Arch. Microbiol., 1975, 102, 71-73.

50. Steber, J. and Schleifer, K.H., *Halococcus morrhuae*: A sulfated heteropolysaccharide as the structural component of the bacterial wall. Arch. Microbiol., 1975, 105, 173-177.

51. Schleifer, K.H., Steber, J. and Mayer, H., Chemical composition and structure of the cell wall of *Halococcus morrhuae*. Z. Bakt. Hyg., I. Abt. Orig., 1982, C 3, 171-178.

52. Steber, J. and Schleifer, K.H., N-glycyl-glucosamine, a novel constituent in the cell wall of *Halococcus morrhuae*. Arch. Microbiol., 1979, 123, 209-212.

53. Michel, H., Neugebauer, D.-C. and Oesterhelt, D., The 2-d crystalline cell wall of *Sulfolobus acidocaldarius*: Structure, solubilization, and reassembly. In Electron Microscopy at Molecular Dimensions, eds., B. Baumeister and W. Vogell, Springer Verlag, Berlin, 1980, pp. 27-35.

54. König, H. and Stetter, K.O., Studies on archaebacterial S-layers. System. Appl. Microbiol., 1986, 7, 300-309.

55. Taylor, K.A., Deatherage, J.F. and Amos, L.A., Structure of the S-layer of *Sulfolobus acidocaldarius*. Nature, 1982, 299, 840-842.

56. Deatherage, J.F., Taylor, K.A. and Amor, L.A., Three-dimensional arrangement of cell wall protein of *Sulfolobus acidocaldarius*. J. Mol. Biol., 1983, 167, 823-852.

57. Messner, P., Pum, D., Sára, M., Stetter, K.O. and Sleytr, U.B., Ultrastructure of the cell envelope of the archaebacteria *Thermoproteus tenax* and *Thermoproteus neutrophilus*. J. Bacteriol., 1986, 166, 1046-1054.

58. Wildhaber, I. and Baumeister, W., The cell envelope of *Thermoproteus tenax*: three-dimensional structure of the surface layer and its role in shape maintenance. EMBO J., 1987, 6, 1475-1480.

59. Wildhaber, I., Santarius, U. and Baumeister W., Three-dimensional structure of the surface protein of *Desulfurococcus mobilis*. J. Bacteriol., 1987, 169, 5563-5568.

60. Darland, G., Brock. T.D., Samsonoff, W. and Conti, S.F., A thermophilic, acidophilic mycoplasma isolated from a coal refuse pile. Science, 1970, 170, 1416-1418.

61. Yang, L.L. and Haug, A., Purification and partial characterization of a procaryotic glycoprotein from the plasma membrane of *Thermoplasma acidophilum*. Biochim. Biophys. Acta, 1979, 556, 265-277.

62. Langworthy, T.A., Lipids of archaebacteria. In The Bacteria, Vol. 8, eds., R.S. Wolfe and C.R. Woese, Academic Press, Inc. New York and London, 1985, pp. 459-497.

63. Liesack, W., König, H., Hirsch, P., Chemical composition of the peptidoglycan-free cell envelopes of budding bacteria of the *Pirella/Planctomyces* group. Arch. Microbiol., 1986, 145, 361-366.

64. Böck, A. and Kandler, O., Antibiotic sensitivity of archaebacteria. In The Bacteria, Vol. 8, eds., C.R. Woese and R.S. Wolfe, Academic Press Inc., New York and London, 1985, pp. 525-544.

65. Sleytr, U.B. and Sara, M., Ultrafiltration membranes with uniform pores from crystalline bacterial cell envelope layers. Appl. Microbiol. Biotechnol., 1986, 25, 83-90.

LIPIDS OF ARCHAEBACTERIA: STRUCTURAL AND BIOSYNTHETIC ASPECTS

Mario De Rosa[a,b], Virginia Lanzotti[a], Barbara Nicolaus[a], Antonio Trincone[a] and Agata Gambacorta[a]

[a]Istituto per la Chimica di Molecole di Interesse Biologico CNR, Via Toiano 6, Arco Felice, Napoli, Italy

[b]Istituto di Biochimica delle Macromolecole Universita' di Napoli, I Facolta' di Medicina, Via Costantinopoli 16, Napoli, Italy

ABSTRACT

Archaebacteria, which at present comprise three phenotypes: halophiles, methanogens and thermophiles, have lipids characterized by unusual structural features, that can be considered as specific taxonomic markers and important factor in placing these organisms on a separate evolutionary path. Whereas all other hitherto known living organisms have membrane lipids based on ester linkages, archaebacterial lipids are based on ether linkages, formed by the condensation of glycerol or more complex polyols with isopranoid alcohols of 20, 25 or 40 carbon atoms. Furthermore the chirality of the β carbon of the glycerol in archaebacterial lipids is the opposite to that found in the glycerolipids of the eubacteria and eukaryotes. This presentation is meant to be a survey of the lipids occurring in archaebacteria and of the most important aspects of their biosynthesis.

INTRODUCTION

The archaebacteria have been recognized as a phylogenetically coherent separate group of microorganisms and differ from eubacteria and eukaryotes in a series of genetic and molecular aspects (1,2).

The archaebacteria are a collection of disparate phenotypes and thrive in environments that would normally kill many other known organisms. In fact, they are segregated into a few peculiar ecological niches, such as saturated brine for halophiles, strict anaerobiotic environment for methanogens, and thermal habitats for extreme thermophiles (1,2).

In view of the key role of the membrane in the ability of archaebacteria to survive in the face of drastic environmental stress, extensive studies have been undertaken to characterize the structural identity and the biogenetic origin of their membrane lipids (3-7).

Upon comparing molecular components of archaebacteria with prokaryotes and eukaryotes, one is struck by the deep chemical differences observed in the lipids, especially in comparison with the other components, whose essential chemical features appear to be preserved. In fact, the structure of membrane lipids remains one of the major distinctions between all archaebacteria and other organisms.

All the membrane lipids of the archaebacteria so far identified are characterized by unusual structural features, which can be considered to be specific taxonomic markers of this group of microorganisms. Unlike the eubacterial and eukaryotic lipids, based on ester linkages formed by condensation of alcohols and fatty acids, archaebacterial lipids are mainly isopranyl glycerol ethers. These ether linked lipids, common to all archaebacteria, are formed by the condensation of glycerol or more complex polyols with isopranoid alcohols containing 20, 25 or 40 carbon atoms.

Moreover, it is worth noting that all glycerol ethers in archaebacteria contain a 2,3-di-O-sn-glycerol, which is unusual, since the glycerol, in the naturally occurring glycerophosphatides or diacylglycerols, is known to have the 1,2-sn stereochemistry.

Although isopranoids with specific functions occur in the lipid membrane of most cells, the ether lipids dealt with here are the only such compounds that provide the major structural components of the membrane in which they occur (1-7).

We may wonder how archaebacterial lipids with a chemical composition so different from conventional ester lipids can manage to perform similar functions, and why that chemical variability is required for life.

This presentation is meant to be a survey of the lipid structure occurring in archaebacteria examined so far and of the most important aspects of archaebacterial lipids biosynthesis.

LIPID STRUCTURE OF ARCHAEBACTERIA

The complex lipids of archaebacteria are mainly based on two classes of isopranoid ether core lipids that are classified as diethers and tetraethers. These compounds are easily obtained by acid or alkaline hydrolysis of complex lipids.

Diether archaebacterial lipids

Different types of isopranoid diethers, basic structural elements of the membrane lipids occurring in different groups of archaebacteria, are reported in Figure 1. These molecules could be formally considered the archaebacterial counterparts of the conventional diglycerides of eukaryotic or eubacterial origin.

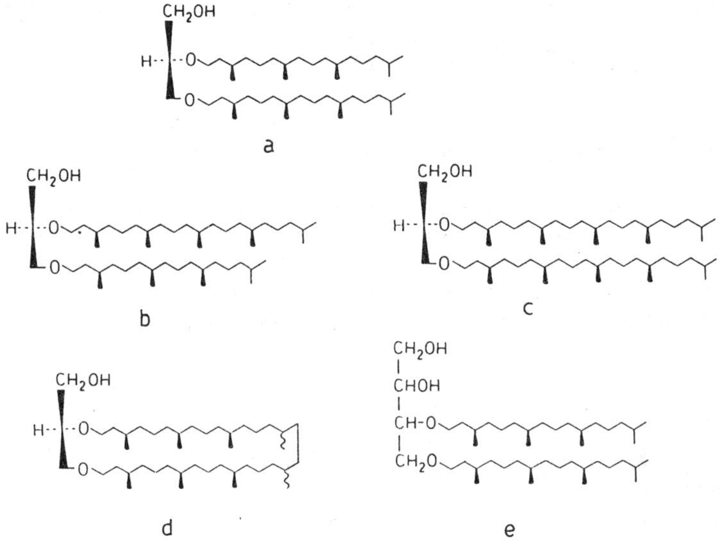

Figure 1. Diether archaebacterial lipids.

These compounds originate in the formation of two ether links between two vicinal hydroxyl groups of usually glycerol, occasionally tetritol, and C_{20}, C_{25} or C_{40} alcohols.

A general feature of all archaebacterial glycerol lipids is the unusual 2,3-sn glycerol stereochemistry, opposite to the one found in common diglycerides, that is probably related to the biochemical mechanism of ether linkage assembly.

The 2,3-di-O-phytanyl-sn-glycerol, a (Fig.1), could be considered a universal core lipid in such types of microorganisms. It may be present as the major component depending on the type of archaebacteria. This glycerol diether represents 100% of ether core lipids in the majority of neutral halophiles and in some coccoid forms belonging to the methanogens and thermophiles (7-9).

Structural types b-c (Fig.1) occur in the extremely alkaliphilic red halophiles of the genera Natronococcus and Natronobacterium, living at pH 10, and in a few strains of neutral halophiles of the genera Halobacterium and Halococcus. The 2-O-sesterterpanyl-3-O-phytanyl-sn-glycerol, b (Fig.1), represents about 80 % of total ether core lipids in the haloalkaliphiles (5-9).

Macrocyclic glycerol diether, d (Fig. 1), is characterised by the presence of a 36 membered ring, which formally originates from the condensation of a glycerol moiety with a C_{40} α-ωisopranyl diol. From a biosynthetic point of view, the macrocyclic structure probably derives from the unprecedented head-to-head condensation of two isopranoid residues of a 2,3-di-O-geranyl-geranyl-sn-glycerol precursor (6-10).

Diphytanyl tetritol diether, e (Fig.1), isolated from Methanosarcina barkeri and Methanosarcina mazei, represents up until now the only example, among archaebacteria, of core lipid in which glycerol is absent (7,8).

Although polar complex lipids based on 2,3-di-O-phytanyl-sn-glycerol are found in all archaebacterial phenotypes, they differ in the nature of the attached polar heads.

In the halophilic archaebacteria are found four basic complex lipid structures: the isoprenoid ether analogues of phosphatidylglycerol (PG), phosphatidylglycerophosphate (PGP), phosphatidylglycerosulfate (PGS), (Fig. 2) and a family of glycolipids and their sulfate derivatives, differing among them in the number and type of sugars (Fig.3), (4,11).

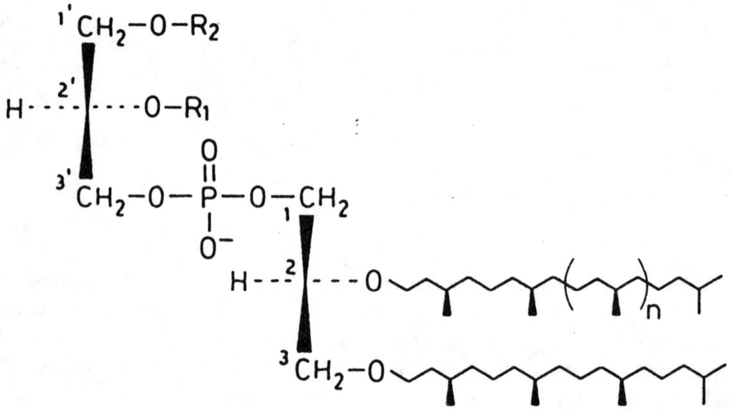

	PG	PGP	PGS	PL2
n	1,2	1,2	1	1,2
R_1	H	H	H	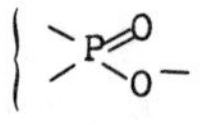 $>P{\scriptstyle\nwarrow}^{O}_{O^-}$
R_2	H	PO_3H_2	SO_3H	

Figure 2. Structures of phospholipids in halophilic archaebacteria.

The major phospholipid component present in the membrane lipids of halophiles is PGP, while PG and PGS usually occur in smaller amounts. However, the PGS is absent only in Haloferax genus (W.D. Grant, personal communication) of neutrophilic halophiles and in all alkaliphilic halophiles. It is worth noting that in these phospholipids both glycerols have a 2,3-sn stereochemistry (Fig. 2), (11).

The glycolipids appear to be derived from the basic structure diglycosyl diether, mannosyl-glucosyl-diphytanylglycerol (DGD), by substitution of sugar or sulfate group at 6 position of the mannose residue, giving rise to triglycosyl diethers and sulfate diglycosyl (S-DGD) diethers respectively (Fig. 3),(11).

Triglycosyl glycolipids differ between themselves in the more external sugar residue that can be galactose (TGD-1) or glucose (TGD-2). TGD-1 in turn originates a sulfate triglycosyl derivative when the sulfate group is linked to C-3 of the third sugar residues (S-TGD-1, Fig.3).

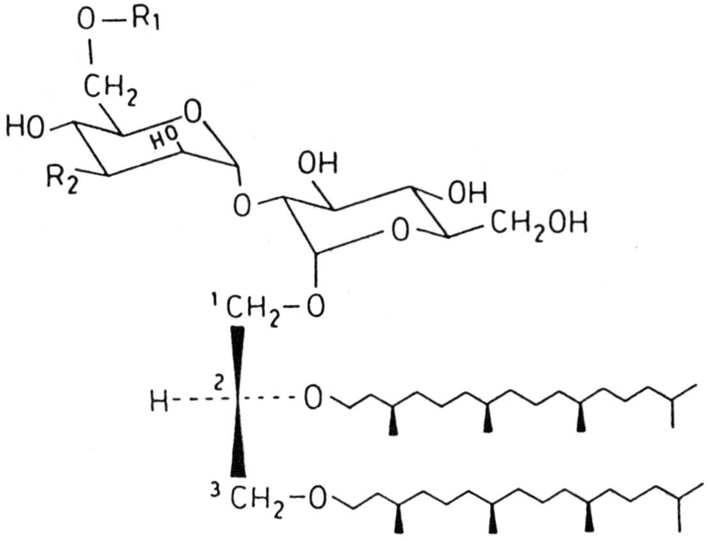

	R_1	R_2
DGD	H	OH
S-DGD	-SO$_3$H	OH
TGD-1	β-gal*p*	OH
TGD-2	β-glc*p*	OH
S-TGD-1	3-SO$_3$H- β-gal*p*	OH
TeGD	β-gal*p*	O- α-gal*f*
S-TeGD	3-SO$_3$H- β-gal*p*	O- α-gal*f*

Figure 3. Structures of glycolipids and their sulfate derivatives in halophilic archaebacteria.

Minor branched tetraglycosyl glycolipid (TeGD) and its sulfated derivative (S-TeGD) are also found. These glycolipids, having four sugars, show essentially the structure of the sulfate triglycosyl glycolipid with galactose residue (TGD-1), with the addition of α-galactose to the 3 position of the mannose residue (Fig. 3), (11,12).

The haloalkaliphilic archaebacteria of the genera Natronobacterium and Natronococcus have a relative simple polar lipid composition in comparison with neutrophilic halophiles, in that the glycolipids are completely absent and that the major species of polar lipids are PGP and PG. However, several minor unidentified phospholipids are also present in examples of these genera. In particular Natronococcus occultus has a significant amount of PGP derivative with a 1',2'-cyclic phosphate (PL2, Fig. 2), (13,14).

At the moment complex lipids based on $C_{20}C_{25}$ diether are found only as PG, PGP, and PL2 and are present essentially in haloalkaliphilic species (7,8,13).

A survey of the lipids of a wide range of extreme halophiles shows a correlation between the lipid composition and their taxonomy (11).

The knowledge of polar lipids in methanogens is quite incomplete both because many of these molecules are not fully characterised and few species of methanogens are analysed in detail for their lipid composition.

The complex lipids in methanogens are classified by the use of specific staining tests into five groups: mean phospholipids, aminophospholipids, aminophosphoglycolipids, phosphoglycolipids and glycolipids (15).

Aminophospholipids were found to be widely distributed in methanogens in which they prevail in contrast with the other two phenotypes of archaebacteria. For example, diphytanyl ether analog of phosphatidylserine (PNL2b, 1 Fig.4) occurs as a major constituent in the lipids of Methanobacteriales (16,17). In a genus of such an order is also found a phosphoethanolamine derivative of diphytanyl glycerol diether (2,Fig. 4), (18).

In species belonging to the Methanococcus genus has been identified a novel 2,3-di-O-phytanyl-1-(phosphoryl-2-acetoamido-2-deoxy-β-D-glucopyranosyl)-sn-glycerol (3, Fig. 4), (19).

Kushwaha et al. have been found in Methanospirillum hungatei, 2,3- di-O- phytanyl-1- (phosphoryl-1'-sn-glycerol)-sn- glycerol the diastereoisomer of PG found in extreme halophiles (20). In contrast to this result, Ferrante et al. did not find in M. hungatei this phospholipid but two new aminophospholipids identified as 2,3-di-O-phytanyl-1-[phosphoryl-2'-(1'-N,N-dimethylamino)-2',3',4',5'- pentane-tetrol]-sn-glycerol (PPAD,4 Fig. 4) and 2,3-di-O-phytanyl-1-[phosphoryl-2'-(1'-N,N,N,trimethylamino)-2',3',4',5'-pentanetetrol]-sn-glycerol (PPTAD, 5 Fig. 4), (21).

Glycolipids identified until now in methanogens are monoglycosyl or diglycosyl derivatives of 2,3-di-O-phytanyl-sn-glycerol. The glycosidic residues are glucose for monoglycoside, MGD, (6,Fig.4) and glucose and/or galactose for diglycosides DGD, also named GL1b, DGD-I,DGD-II,(7-9,Fig.4),(16,19,20).

Also identified in methanogens is a phosphatidylinositol derivative of 2,3-di-O-phytanyl-sn-glycerol (PL2b, 10,Fig.4),(16).

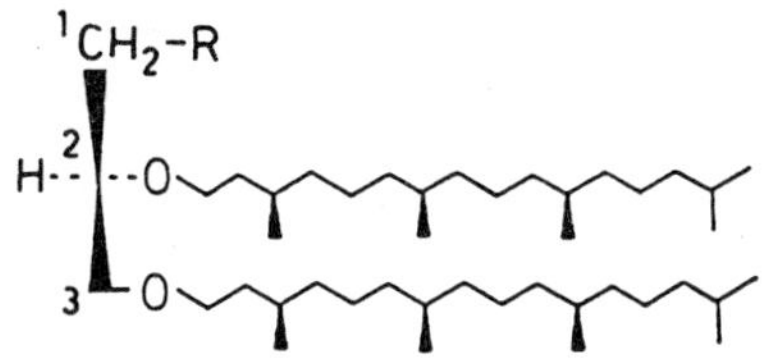

	R		R
1	$P-CH_2-CH(NH_3^+)-COOH$	6	β-D-glc$\underline{p}$
2	$P-(CH_2)_2-NH_3^+$	7	β-D-glc$\underline{p}$-(1→6)- β-D-glc$\underline{p}$
3	$P-1-[2-(NHAc)-2-deoxy]- \beta$-D-glc$\underline{p}$	8	α-glc$\underline{p}$-(1→2)- β-gal$\underline{f}$
4	$P-CH[CH_2-NH^+-(CH_3)_2]-(CHOH)_2-CH_2OH$	9	β-gal$\underline{f}$-(1→6)- β-gal$\underline{f}$
5	$P-CH[CH_2-N^+-(CH_3)_3]-(CHOH)_2-CH_2OH$	10	P-inositol
		11	α-D-glc$\underline{p}$-3-P

P= phosphate group

Figure 4. Structures of complex lipids based on 2,3-di-O-phytanyl-sn-glycerol in methanogenic and thermophilic archaebacteria.

Although the presence or absence of some of the complex lipids identified could be considered to reflect the taxonomy of the methanogens, comparative studies carried out on the basis of TLC profiles, as was done with the halophilic archaebacteria, are at present not possible because of the higher complexity of lipid pattern in methanogens.

Few species of thermophilic archaebacteria, all belonging to the Thermococcales order, have complex lipids based essentially on 2,3-di-O-phytanyl-sn-glycerol. In Thermococcus and in Pyrococcus is found a single phospholipid (about 85% of total complex lipids), in which the P-myoinositol is the polar head attached to the free hydroxyl of the glycerol moiety (10,Fig. 4), (22 and personal communication of the authors).

In addition to this type of phospholipid, that accounts for 40% of total complex lipids, AN1 isolate (7) has a novel phosphoglycolipid, 2,3- di-O-phytanyl-1- (3-phosphoryl-α-D-glucopyranosyl) -sn-glycerol, which represents the first example in archaebacteria of glycolipid with a phosphorylated sugar, (11,Fig.4), (personal communication of the authors).

In the other thermophilic species, complex lipids based on 2,3-di-O-phytanyl-sn-glycerol, are minor compounds which remain, until now, unidentified.

Furthermore complex lipids based on the diethers d and e (Fig. 1) are not yet characterized.

Tetraether archaebacterial lipids

Figure 5 shows a family of macrocyclic tetraethers which are the basic components of the membrane lipids of many methanogens and, with few exceptions, of thermophilic archaebacteria (4-9).

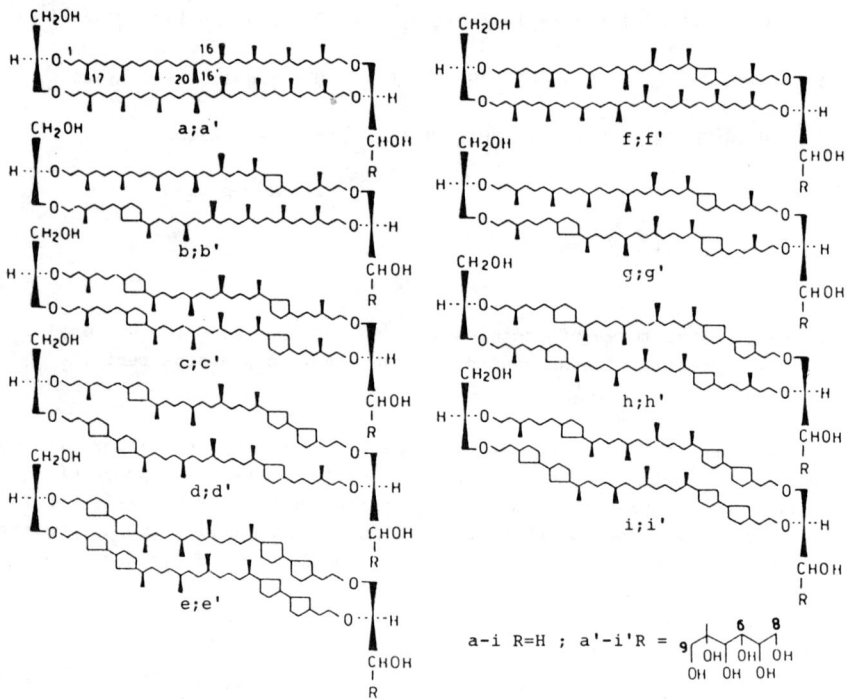

Figure 5. Tetraether archaebacterial lipids: structures a-i are named glycerol dialkyl glycerol tetraethers (GDGTs) and structures a'-i', glycerol dialkyl nonitol tetraethers (GDNTs).

The unusual molecular architecture of these bipolar amphipatic molecules has no counterpart in eubacterial and eukaryotic lipids with the exception of bipolar lipids of Butyrrovibrio based on the diabolic acids (23).

The tetraether a (Fig. 5) has a 72-member ring with 18 stereocenters, formed by two 2,3-sn-glycerol moieties, bridged through ether linkages by two isopranoid C_{40} diols. The series of tetraethers b-i (Fig.5) can be considered to derive from the first one by a process of cyclization resulting in the formation of up to eight cyclopentane in the isopranoid components. All these structures are simple named glycerol dialkyl glycerol tetraethers (GDGTs). The second series a'-i' (Fig.5) has a structural organization similar to that of GDGTs, but a more complex polyol with nine carbon atoms replaces one of the glycerol. By analogy this compound are named glycerol dialkyl nonitol tetraethers (GDNTs),(5-9). A high degree of structural specificity characterizes these tetraethers originated from two polyols and five different C_{40} isopranoids with up to four cyclopentane rings per chain. In all the structures identified so far, the two polyols are always antiparallel and the two C_{40} chains are either identical or differ by one cycle. Moreover in tetraethers with asymmetric end-to-end C_{40} chain (mono and tricyclic C_{40} isopranoids) the more cyclized end is always linked to primary carbinol of the polyols (24).

GDNT lipids are found only in microorganisms belonging to the Sulfolobales and seem to be a specific taxonomic marker for this order (7-9).

The tetraether a (Fig.5) together with the 2,3-di-O-phytanyl-sn-glycerol, a (Fig.1), in different proportions, are the basic components of complex lipids of methanogenic archaebacteria belonging to the genera Methanobacterium, Methanobrevibacterium, Methanospirillum, Methanoplanus, Methanogenium, Methanolobus and Methanothermus. Only in Methanothermus sociabilis and in Methanosarcina barkeri tetraethers with cyclopentane rings in the isopranoid chains have been detected (7,9).

Tetraethers are the main components of lipids of thermophilic archaebacteria with the exception of Pyrococcus woesei, Thermococcus celer and AN1 strain previously reported (22 and personal communication of the authors).

Complex lipids based on tetraethers, absent in the microorganisms belonging to halophilic phenotype, are found in methanogenic and thermophilic species.

Although relatively few polar lipid structures of archaebacteria based on tetraethers have been fully established, a series of structural constraints seems to be recurring: (i) tetraether polar lipids may contain polar groups attached on one or both glycerol moieties; (ii) there are no examples of symmetrical tetraether polar lipids; (iii) with few exceptions the tetraether polar lipid structures contain carbohydrate residues.

Kushwaha et al.have been elucidated the structure of complex lipids in Methanospirillum hungatei,in which the polar complex lipids based on tetraethers are limited to two glycolipids and two phospholipids. In the glycolipids, DGT-I and DGT-II (1,2, Fig.6), one of the free hydroxyl groups of GDGT a (Fig. 5) is linked glycosidically to a disaccharide residue. From the two glycolipids derive the phosphoglycolipids PGLI and PGLII (3,4, Fig.6) in which a 1-sn-glycerophosphate residue is attached in the opposite side of the tetraether molecule (20).

In Methanobacterium thermoautotrophicum were elucidated the structures of three of the major polar lipids present. The first one, an aminophospholipid, PNL1a (5, Fig.6) is the GDGT analog of phosphatidylethanolamine . The second, GL1a (6, Fig.6) is a glycolipid in which one of the free hydroxyl groups of the GDGT a (Fig.5) is linked glycosidically to a glucose disaccharide residue. The third, PNGL1, (7, Fig.6), an aminophosphoglycolipid, has the polar head groups of both PNL1 and GL1a (25).

Koga et al. (16) suggest, on the basis of TLC analysis that the occurrence of some polar lipids in the methanogens could be used for grouping these microorganisms at the family level.

Identities of the polar lipid structures among the thermophile phenotype are still incomplete, since the structures of complex lipids have been defined in only a few species.

The complex lipids of the thermophiles belonging to the group of sulfur-dependent archaebacteria have now been defined in Desulfurococcus, Thermoproteus, Sulfolobus and Desulfurolobus species.

In Thermoproteus have been found three type of glycolipids, based on GDGT, which have a mono- (8, Fig. 6),di- (9,Fig. 6) and tri- (10,Fig. 6) glucosyl residue as polar head group, named by the authors GLA,GLC and GLD. Phospholipids of this microrganism, named PLC and PLE (11,12, Fig.6), are phosphatidylinositol derivatives of GLA and GLD (8,10, Fig.6), respectively (26).

In the Desulfurococcus species have been identified three complex lipids based on GDGT. The first one is a glycolipid, in which one of the free hydroxyl groups of GDGT is linked glycosidically to a glucose-galactose disaccharide (13,Fig. 6). The second is a phosphoglycolipid in which one of the free hydroxyls of GDGT is linked glycosidically to galactose and the other one is esterified with P-myoinositol residue (14, Fig.6).The last compound is a phosphoglycolipid derived from the glycolipid 13 (Fig. 6) in which the free hydroxyl of GDGT moiety is esterified with P-myoinositol residue (15,Fig. 6),(27).

In the Sulfolobus species the complex lipids are based both on GDGTs and GDNTs differently cyclized (Fig. 5). The glycolipids are both disaccharide and monosaccharide derivatives of tetraethers. The glycolipid, based on GDGTs, has glucose-galactose disaccharide glycosidically linked to one of the free hydroxyls of the glycerol moiety (16,Fig 6). The GDNT glycolipid has glucose residue attached to the C-6 of the nonitol moiety (17, Fig.6). From this glycolipid derives

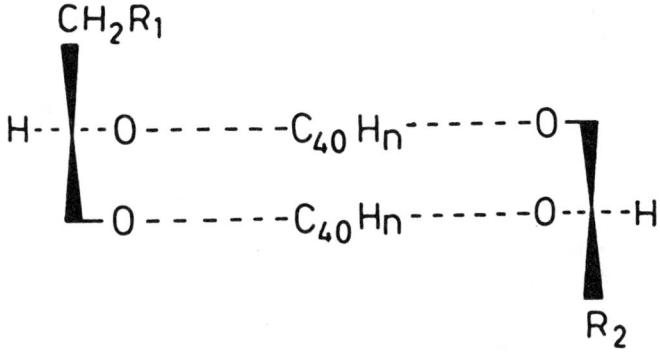

	n	R₁	R₂
1	80	OH	CH_2O-β-gal<u>f</u>-(1→6)-β-gal<u>f</u>
2	80	OH	CH_2O-α-glc<u>p</u>-(1→2)-β-gal<u>f</u>
3	80	P-1-<u>sn</u>-glycerol	CH_2O-β-gal<u>f</u>-(1→6)-β-gal<u>f</u>
4	80	P-1-<u>sn</u>-glycerol	CH_2O-α-glc<u>p</u>-(1→2)-β-gal<u>f</u>
5	80	P-$(CH_2)_2$-NH_3^+	CH_2OH
6	80	OH	CH_2O-β-glc<u>p</u>-(1→6)-β-glc<u>p</u>
7	80	P-$(CH_2)_2$-NH_3^+	CH_2O-β-glc<u>p</u>-(1→6)-β-glc<u>p</u>
8	76-80	OH	CH_2O-β-glc<u>p</u>
9	76-80	OH	CH_2O-glc<u>p</u>-(1→6)-glc<u>p</u>
10	76-80	OH	CH_2O-glc<u>p</u>-(1→6)-glc<u>p</u>-(1→6)-glc<u>p</u>
11	76-80	P-inositol	CH_2O- β-glc<u>p</u>
12	76-80	P-inositol	CH_2O-glc<u>p</u>-(1→6)-glc<u>p</u>-(1→6)-glc<u>p</u>
13	80	OH	CH_2O- α-glc<u>p</u>-(1→4)- β-gal<u>p</u>
14	80	P-myoinositol	CH_2O- β-gal<u>p</u>
15	80	P-myoinositol	CH_2O- α-glc<u>p</u>-(1→4)- β-gal<u>p</u>
16	72-80	OH	CH_2O- β-gal<u>p</u>- β-glc<u>p</u>
17	72-80	OH	*$C_7H_{14}O_7$- β-glc<u>p</u>
18	72-80	OH	*$C_7H_{14}O_7$- β-glc<u>p</u>-sulfate
19	72-80	P-myoinositol	CH_2OH
20	72-80	P-myoinositol	CH_2O- β-gal<u>p</u>- β-glc<u>p</u>
21	72-80	P-myoinositol	*$C_7H_{14}O_7$- β-glc<u>p</u>
22	76-80	P-3-<u>sn</u>-glycerol	CH_2O-monoglycosyl

P = phosphate group   * for structure and numbering see Figure 5

Figure 6. Structures of complex lipids based on tetraethers in methanogens and thermophilic archaebacteria.

the sulfoglycolipid 18 (Fig. 6) in which a sulfate group esterifies the glucose moiety. This tetraether sulfolipid is the only example of sulfate lipid in archaebacteria in addition to those described in halophilic phenotype based on diether. The three phospholipids identified in the Sulfolobus species all contain a myo-inositol phosphate residue. The first is derived by GDGT, in which one of the free hydroxyls of glycerol moiety is esterified by P-myoinositol (19,Fig.6). The last two (20,21, Fig.6) are derivatives of the glycolipids (16,17, Fig. 6) in which the P-myoinositol residue esterifies the free hydroxyl group of the glycerol moiety (5,28).

In Sulfolobus grown autotrophically two polar unidentified lipids also occur (5).

In Desulfurolobus, grown both aerobically and anaerobically, are identified the same complex lipids found in the Sulfolobus species (16-21, Fig. 6), (7 and personal communication of the authors).

Summing up, in microorganisms belonging to a sulfur-dependent archaebacteria group, such as Sulfolobus, Desulfurolobus, Desulfurococcus and Thermoproteus, the largest percentage of complex lipids occurs as phosphoglycolipids in which a sugar residue and a phosphate group are linked to opposite sides of the tetraether molecules. Polar head groups are restricted to galactose or glucose or both and P-myoinositol. Different structures originate from the stereochemistry of the glycosidic bond and location of the interglycosidic linkage.

The complex lipids of Thermoplasma acidophilum are based only on GDGTs (Fig.5) differently cyclized. At least six different glycolipids and seven phosphorus-containing lipids have been found, but none of these have been fully characterized. The major component is a phosphoglycolipid derivative, in which a 3-sn-glycerol-phosphate and an unidentified monosaccharide are attached to the free hydroxyl groups of the GDGT (22, Fig.6),(5). Other minor phosphoglycolipids appear to possess up to three carbohydrate residues along with free amino groups, possibly as amino sugars (29). Moreover on the cell surface of $\underline{T}$. acidophilum has been found an unusual lipoglycan that resembles the lipopolysaccharide of Gram-negative bacteria. This polymer, (Man$\underline{p}$-(α 1 → 2)-Man$\underline{p}$-(α 1→4)-Man$\underline{p}$-(α 1→3)$_8$-Glc$\underline{p}$-(α 1→1)-O-(diglyceryl tetraether)-OH, contains 24 mannose residues and one glucose, and the polysaccharide chain is attached to only one side of the tetraether molecule (5).

Minor archaebacterial lipids

In Figure 7 are reported two minor core lipids occurring in archaebacteria. Compound a (Fig.7) is a glycerol trialkyl glycerol tetraether based upon two glycerol linked by four ether bonds with two C_{20} and one C_{40} isopranoid chains. The absence in Sulfolobus solfataricus of tetraethers with this type of molecular architecture, but with cyclopentane rings in the isopranoid chains, as in GDGT and in GDNT (Fig.5), could indicate that the enzyme system devoted to cyclopentane ring formation operates only if the 72 membered macrocycle is closed (7).

The structure b (Fig.7) has been isolated from lipids of Methanosarcina barkeri and, in trace amounts, also in S. solfataricus (7).

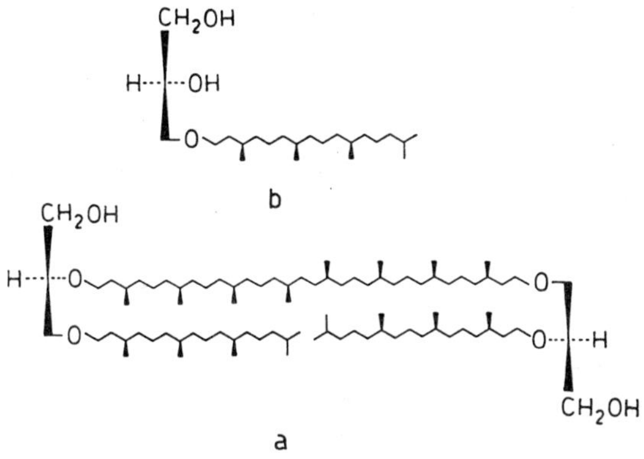

Figure 7. Minor archaebacterial lipids.

Up until now complex lipids based on these two last structures have not been isolated.

Nomenclature

Nishihara et al. (25) propose a new systematic nomenclature of archaebacterial polar lipids, because the alternative names are too lengthy and laboratory designations of these lipids are not at all systematic. Such laboratory designations are in fact merely symbols or numbers and do not have general significance. The new nomenclature starts with giving the names archaeol and caldarchaeol to dialkyl diethers and dibisphytanyl tetraethers, respectively.

Phospholipids with a phosphodiester bond were named as derivatives of archaetidic acid or caldarchaetidic acid by analogy with phosphatidic acid. For example, archaetidylserine (PNL2b 1, Fig. 4), found in Methanobrevibacter arboriphilus is a diether analogue of phosphatidylserine and caldarchaetidylethanolamine (PNL1a 5, Fig. 6) is a GDGT analogue of phosphatidylethanolamine.

By analogy the phosphoglycolipid PNGL1 (7, Fig. 6) is named diglucosyl caldarchaetidylethanolamine.

Although the suggestions of Nishihara et al. are useful to describe the general features of the archaelipids characterized up until now, we suggest that a nomenclature proposal could be deferred to a more complete analysis of this new class of lipid molecules.

BIOSYNTHESIS OF ARCHAEBACTERIAL ETHER LIPID

Experiments with labeled acetate or mevalonate have indicated that diether and tetraether lipids of some archaebacteria are derived from acetate via mevalonate by the same general biosynthetic pathway operating in eubacteria and eukaryotes (6, 30). However in a recent report Ekiel and collaborators (31) show that, in <u>Halobacterium cutirubrum</u> and <u>Hb. halobium</u> methyl and methyne carbons in phytanyl chains derive from lysine.

The characteristic steps of archaebacterial ether lipid biosynthesis are as follows: (i) ether linkage formation; (ii) head-to-head coupling of two geranylgeranyl residues, with reduction to form biphytanyls; (iii) cyclization within coupled geranylgeranyl residues, with reduction to form five membered cyclic biphytanyls; and (iiii) biogenesis of calditol, the branched chain nonitol occurring in GDNTs (a'-i', Fig.5) of <u>Sulfolobaceae</u>.

Ether linkage formation

The ether linkage formation is the common step in the biosynthesis of all archaebacterial lipids. It was first investigated by Kates in <u>Halobacterium cutirubrum</u>, using differently labelled glycerols as precursors. In particular the loss of tritium from the β-carbon of [U-^{14}C,2-^{3}H]glycerol, during the glycerol ether formation, was regarded as a significant feature of the etherification step. On the basis of this and other evidence the authors suggested that a more likely precursor would be dihydroxyacetone, which could be formed by the action of glycerol dehydrogenase, an enzyme known to be active in <u>Hb. cutirubrum</u> (30).

Different results were obtained with similar experiments in <u>Sulfolobus solfataricus</u>. Particularly in the experiment with [U-^{14}C,2-^{3}H]glycerol and [U-^{14}C,1(3)-^{3}H]glycerol, the distribution of radioactivity in tetraethers shows that the glycerol moieties of GDGT lipids (a-i, Fig.5) incorporate the radioactivity without any change in the ^{3}H/^{14}C ratio and with a high efficiency for both precursors. On the basis of these results glycerol ether assembly in GDGTs of <u>S. solfataricus</u> occurs without loss of hydrogen from any of the glycerol carbons and therefore without the intervening formation of oxidized intermediate derivative of the glycerol (6, 32).

More recently biogenetic experiments with 5-^{13}C,5,5-^{2}H mevalonolactone, performed on this archaebacterium indicate that during the alkylation step of glycerol and nonitol in GDGTs and GDNTs (Fig.5), both deuterium atoms of the precursor are retained on the first carbons of the isoprenic chains. This result supports a prenyl transfer mechanism for attachment of the isoprene residues to the polyol moiety without double bond isomerization in the first isoprene unit (33, personal comm. of authors). More recently Poulter and co-workers report data on the incorporation of C_{20} isoprenoidic alcohols with different degree of unsaturation in Methanospirillum hungatei, a strict anaerobic methanogenic archaebacterium. Geranylgeraniol was readily incorporated into diethers and tetraethers of this microorganism, while phytol (with one double bond in the first isoprene unit) and phytanol (without double bonds) were poorly and not incorporated, respectively (34).

Given the well demonstrated ability of prenyl pyrophosphates to act as alkylating agents in other biosynthetic mechanisms, the body of these results supports a direct ether formation from glycerol (or, facilitated by neighbouring group deprotonation, from glycerolphosphate). According to this hypothesis, the unusual configuration of the chiral centre in the glycerol moiety of archaebacterial lipids, would depend on the stereospecific nature of the alkylation step. In this respect studies on chirality of C-2 of the glycerol in ether lipids of archaebacteria, have been recently carried out by Kakinuma and co-workers with feeding experiments of (RS)-,(R)- and (S)-(1-^{2}H$_2$) glycerol to the culture of Hb. halobium. In these experiments the sn-C-1 of the glycerol moiety of the 2,3-di-O-phytanyl-sn-glycerol appears to be derived from the sn-C-3 carbon of glycerol (35).

Head-to-head coupling

The process of head-to-head coupling is particularly striking and has no parallel in other fields of terpene biochemistry. Up until now there has been no direct information about this mechanism. Incorporation experiments of ^{13}C$_2$ acetate (6) and 2-^{13}C,2,2-^{2}H mevalonolactone in Sulfolobus solfataricus establish that the coupling occurs between the two carbons derived from C-2 of mevalonate and that the metabolic fate of the C-2 hydrogen of the double labelled mevalonolactone is almost identical both in head-to-tail elongation process and in head-to-head C_{20} coupling (36, personal comm. of the authors). There is no direct information as to whether the coupling reaction is between C_{20} chains ether linked to glycerol or between C_{20} precursors themselves. Indirect evidence favors the former: in fact, the structural regularities of GDGTs and GDNTs (Fig.5) are in accord with the supposition that cyclization, to form cyclopentane rings in C_{40} chains of these tetraethers, occurs in the axially symmetric tetraethers rather than in the free C_{20} or C_{40} components (37). Recently some experimental results on the incorporation of ^{14}C diether in Methanospirillum hungatei, seem to give an indication that the direct precursor for head-to-head coupling could be an isoprenic C_{20} chain ether linked to glycerol (34).

Cyclopentane ring formation

In the lipids of Sulfolobus solfataricus, the proportion of differently cyclized isoprenoid tetraethers GDGTs and GDNTs (Fig.5) shows a relationship to the growth temperature, increasing the percentage of the more cyclized products at higher temperatures (7, 8, 33, 38).This evidence has been regarded as an indication that cyclization of the isoprenoid components acts as a buffer against the effects of the external temperature increase. Probably additional cyclopentane rings in the C_{40} isoprenoids could decrease the available modes of flexing and rotating of the chains and increase the inertial moment of the molecules.

The high structural specificity of GDGTs (Fig. 5) originating from two glycerines and five differently cyclized isoprenoids, with regard to the fact that only 9 combinations occur out of the large number of possibilities, suggests that some biosynthetic hypothesis would be more plausible than others (37). For example, if the tetraethers were formed from a pool of mixed C_{40} diols or from variously-cyclized diphytanyl glycerol, we would expect a much wider range of products. Those, actually found, all show the same symmetry with each of the antiparallel C_{40} chains having the same steric relationships to each of the antiparallel sn-glycerol units, and cyclizations seem to have been introduced in conformity with that symmetry, probably as a consequence of enzyme-substrate interaction. The regular disposition of cyclopentane rings in these tetraethers could indicate that cyclopentanes were closed in an ordered way by a mechanism that operates in a concerted manner on both alkyl chains, starting from the middle of the isoprenoid system towards ether bonds. Similar considerations can be performed for GDNTs (Fig.5).

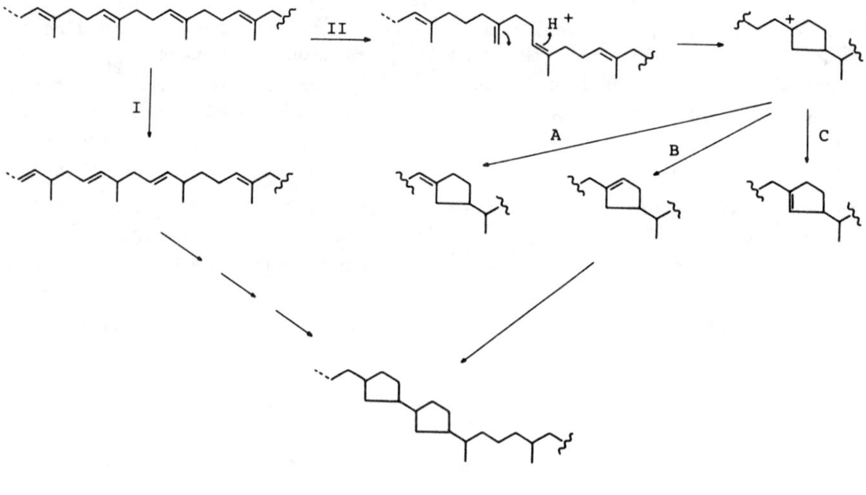

Scheme 1. Possible mechanisms of cyclopentane ring formation in tetraether lipids in S. solfataricus.

The mechanism of cyclopentane ring formation has been investigated using as precursors of tetraether lipids in S. solfataricus $^{13}C, ^{2}H$ strategically labelled mevalonolactones. Results of these experiments allowed the discrimination of a series of mechanisms based on the initial presence of an unsaturation and possibly concerted with a hydride reduction step (Scheme 1). From the labelling pattern in these experiments the more plausible mechanism of cyclopentane ring formation is depicted in A (personal comm. of the authors), but other mechanisms based, for example, on hydroxylation of isoprenic chains could not be excluded.

Biogenesis of nonitol
GDNT lipids (Fig.5) have been found only in the Sulfolobales and seem to be a specific taxonomic marker for this order. In particular the presence of these lipids in the recently isolated microorganisms belonging to the Desulfurolobus and Acidianus genera, has been a help in ascribing these microorganisms to the Sulfolobales order (7, 39). The studies of the biogenetic origin and assembly of the carbon skeleton of the nonitol are based on the incorporation in Sulfolobus solfataricus of labelled precursors such as $[U-^{14}C, 2-^{3}H]$ and $[U-^{14}C, 1(3)-^{3}H]$ glycerols and $D-[1-^{14}C, 6-^{3}H]$ glucose and fructose. The body of the results, provided by the labelling pattern of the nonitol in GDNTs, leads to conclude that, without regarding implications as to stereochemistry or phosphorylation, the biosynthesis of calditol occurs via an aldolic condensation between dihydroxyacetone and fructose giving rise to a 2-keto-intermediate, that is in turn reduced and alkylated to yield GDNTs (6).

Pathway for ether lipid assembly in archaebacteria.
Biogenetic studies and structural constraints of isoprenoid ether lipids permit to propose for the assembly of these molecules a metabolic pattern (Fig.8) that appears more plausible and economical than do alternative pathways, without considering implications as to activation of the intermediate.

Ether bond formation between low molecular weight alcohols (glycerol, tetritol or calditol) and geranyl geranyl pyrophosphate or geranyl farnesyl pyrophosphate is the common biosynthetic step in archaebacteria lipid biogenesis. In halophiles hydrogenation of the unsaturated intermediate leads to the formation of glycerol diether lipids, while in methanogens and thermophiles, in competition with the reduction step, it effects head-to-head coupling between C_{20} unsaturated isoprenoid chains and cyclopentane ring formation, to give rise to a variety of structural types of lipids occurring in these phenotypes.

ACKNOWLEDGEMENTS

The authors thank Mr. R. Turco for drawing the figures.

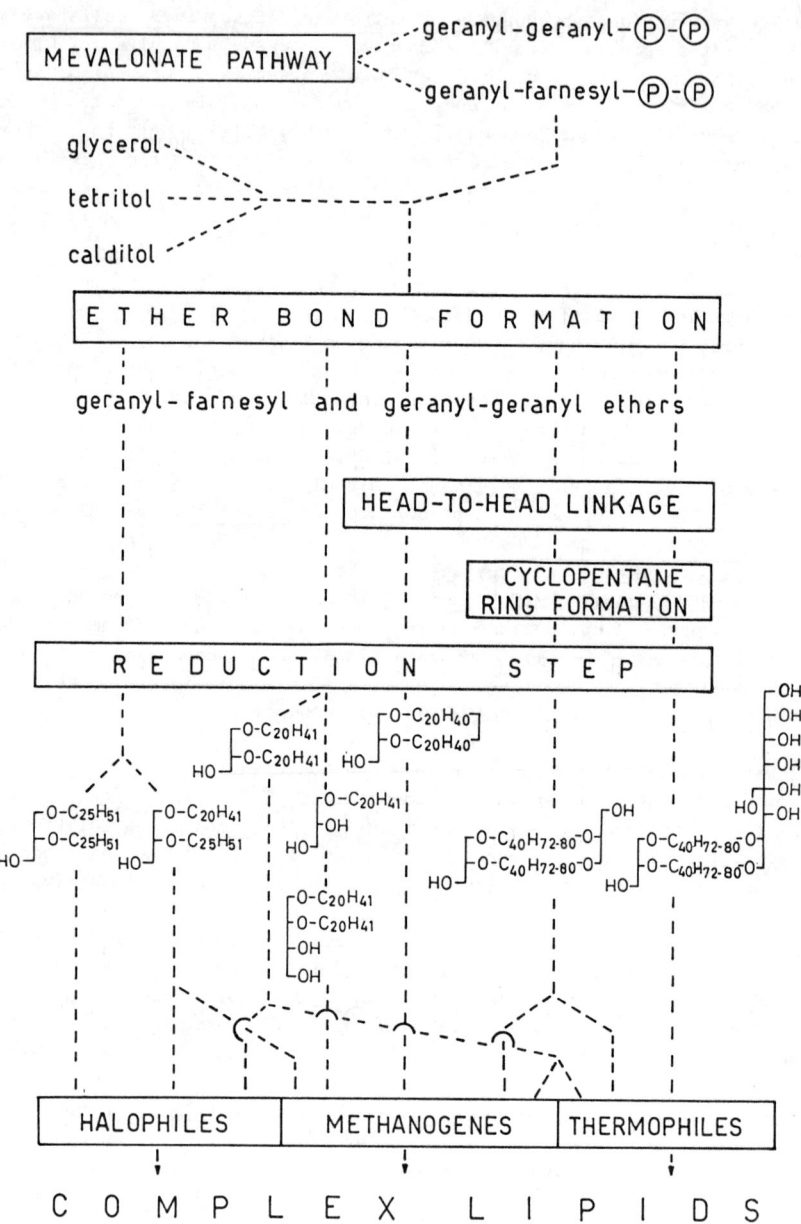

Fig. 8. Pathway for ether lipid assembly.

REFERENCES

1. Woese, C.R. and Wolfe, R.S., The Bacteria, Archaebacteria, Academic Press, Inc, New York, 1985, **8**.
2. Woese, C.R., Bacterial evolution. Microbiol. Rev., 1987, **51**, 221-71.
3. Kates, M., Ether-linked lipids in extremely halophilic bacteria. In Ether Lipids: Chemistry and Biology, ed., F. Snyder, Academic Press, New York, 1972, pp. 351-98.
4. Kates, M. Kushwaha, S.C. and Sprott, D.G., Lipids of purple membrane from extreme halophiles and of methanogenic bacteria. In Methods in Enzymology, ed., L. Parker, Academic Press, New York, 1982, **88**, pp. 98-111.
5. Langworthy, T.A., Lipids of archaebacteria. In The Bacteria, eds., C.R. Woese and R.S. Wolfe, Academic Press, New York, 1985, **8**, pp. 459-91.
6. De Rosa, M. and Gambacorta, A., Lipid biogenesis in archaebacteria. In Archaebacteria '85, eds., O. Kandler and W. Zillig, Gustav Fischer Verlag, Stuttgart, 1986, pp. 278-85.
7. De Rosa, M. and Gambacorta, A., The lipids of archaebacteria. In Progress in Lipid Research, Pergamon Press, Oxford, 1988, in press
8. De Rosa, M. Gambacorta, A. and Gliozzi, A., Sructure, biosynthesis and physicochemical properties of archaebacterial lipids. Microbiol. Rev., 1987, **50**, 70-80.
9. Langworthy, T.A. and Pond, J.L., Archaebacterial ether lipids and chemotaxonomy. In Archaebacteria '85, eds., O. Kandler and W. Zillig, Gustav Fischer Verlag, Stuttgart, 1986, pp. 253-7.
10. Comita, P.B., Gagosian, R.B., Pang, H. and Costello, C.E., Structural elucidation of a unique macrocyclic membrane lipid from a new, extremely thermophilic, deep-sea hydrothermal vent archaebacterium, Methanococcus jannaschii. J. Biol. Chem., 1984, **259**, 15234-41.
11. Kates, M., Influence of salt concentration on membrane lipids of halophilic bacteria. FEMS Microbiol. Rev., 1986, **39**, 95-101.
12. Smallbone, B.W. and Kates, M., Structural identification of minor glycolipids in Halobacterium cutirubrum. Biochim. Biophys. Acta, 1981, **665**, 551-8.
13. De Rosa, M., Gambacorta, A., Grant, W.D., Lanzotti, V. and Nicolaus, B., Polar lipids and glycine betaine from haloalkaliphilic archaebacteria. J. Gen. Microbiol., 1988, **134**, 205-211.
14. Lanzotti, V., Nicolaus, B., Trincone, A., De Rosa, M., Grant, W.D. and Gambacorta, A., A complex lipid with a cyclic phosphate from the archaebacterium Natronococcus occultus. Biochim. Biophys. Acta, in press.
15. Nishihara, M. and Koga, Y., Extraction and composition of polar lipids from the archaebacterium, Methanobacterium thermoautotrophicum: effective extraction of tetraether lipids by an acidified solvent. J. Biochem., 1987, 101, **997**-1005.

16. Koga, J., Ohga, M., Nishihara, M. and Morii, H., Distribution of a diphytanyl ether analog of phosphatidylserine and an ethanolamine-containing tetraether lipid in methanogenic bacteria. System. Appl. Microbiol.,1987, **9**, 176-82.
17. Morii, H., Nishihara, M., Ohga, M. and Koga, Y., A diphytanyl ether analog of phosphatidylserine from a methanogenic bacterium, Methanobrevibacter arboriphilus. J. Lipid. Res., 1986, **27**, 724-30.
18. Kramer, J.K.G., Sauer, F.D. and Blackwell, B.A., Structure of two new aminophospholipids from Methanobacterium thermoautotrophicum. Biochem. J., 1987, **245**, 139-43.
19. Ferrante, G., Ekiel, I. and Sprott, G.D., Structural characterization of the lipids of Methanococcus voltae, including a novel N-acetylglucosamine 1-phosphate diether. J. Biol. Chem., 1986, **261**, 17062-6.
20. Kushwaha, S.C., Kates, M., Sprott, G.D. and Smith, I.C.P., Novel polar lipids from the methanogen Methanospirillum hungatei GP1. Biochim. Biophys. Acta, 1981, **664**, 156-73.
21. Ferrante, G., Ekiel, I. and Sprott, G.D., Structures of diether lipids of Methanospirillum hungatei containing novel head groups N,N-dimethylamino and N,N,N-trimethylaminopentanetetrol. Biochim. Biophys. Acta, 1987, **921**, 281-91.
22. De Rosa, M., Gambacorta, A., Trincone, A., Basso, A., Zillig, W. and Holz, I., Lipids of Thermococcus celer, a sulfur-reducing archaebacterium: structure and biosynthesis. System. Appl. Microbiol., 1987, **9**, 1-5.
23. Klein, A., Hazlewood, G.P., Kemp, P. and Dawson, M.C., A new series of long-chain dicarboxylic acid with vicinal dimethyl branching found as major components of the lipids of Butyrivibrio spp. Biochem. J., 1979, **183**, 691-700.
24. De Rosa, M., Gambacorta, A., Nicolaus, B., Chappe, B. and Albrecht, P., Isoprenoid ethers backbone of complex lipids of the archaebacterium Sulfolobus solfataricus. Biochim. Biophys. Acta, 1983, **753**, 249-56.
25. Nishihara, M., Morii, H. and Koga, Y., Structure determination of a quartet of novel tetraether lipids from Methanobacterium thermoautotrophicum. J. Biochem. 1987, **101**, 1007-15.
26. Thurl, S., Uber lipide aus archaebakterien. Thesis, Ludwig-Maximilians-Universitat, Munchen, 1986.
27. Lanzotti, V., De Rosa, M., Trincone, A., Basso, A., Gambacorta, A. and Zillig, W., Complex lipids from Desulfurococcus mobilis, a sulfur-reducing archaebacterium. Biochim. Biophys. Acta., 1987, **922**, 95-102.
28. De Rosa, M., Gambacorta, A. and Nicolaus, B., A new type of cell membrane in thermophilic archaebacteria, based on bipolar ether lipids. J. Membrane Sci., 1983, **16**, 287-94.
29. Langworthy, T.A.,Special features of Thermoplasmas. In The Micoplásma, eds., M.F. Barile and R. Razin, Academic Press, Inc., New York, 1979, pp. 495-513.

30. Kates, M. and Kushwaha, S.C., Biochemistry of the lipids of extremely halophilic bacteria. In Energetics and Structure of Halophilic Microorganisms, eds. S.R. Caplan and M. Ginzburg, Elsevier, North-Holland, 1978, pp. 461-80.
31. Ekiel, I., Sprott, G.D. and Smith, I.C.P., Mevalonic acid is partially synthesized from amino acids in Halobacterium cutirubrum: a ^{13}C nuclear magnetic resonance study. J. Bacteriol., 1986, 166, 559-64.
32. De Rosa, M., Gambacorta, A., Nicolaus, B. and Sodano, S., Incorporation of labelled glycerols into ether lipids in Caldariella acidophila. Phytochemistry, 1982, 21, 595-9.
33. Scolastico, C., Sidymov, A., Potenza, D., De Rosa, M., Gambacorta, A. and Trincone, A., Cyclopentane ring formation in isoprenoid ether lipids of archaebacteria. In Archaebacteria '85, eds. O. Kandler and W. Zillig, Gustav Fischer Verlag, Stuttgart, 1986, pp.417-8.
34. Poulter, C.D., Aoki, T. and Daniels, L., Biosynthesis of isoprenoid membranes in the methanogenic archaebacterium Methanospirillum hungatei. J. Am. Chem. Soc., 1988, 110, 2620-4.
35. Kakinuma, K., Yamagishi, M., Fujimoto, Y. and Oshima, T., Stereospecific incorporation of glycerol into sn-2,3-di-O-phytanyl glycerol, a membrane lipid of archaebacteria Halobacterium halobium. 16th International Symposium on the Chemistry of Natural Products, Kyoto, Japan, May-June 1988, PD19.
36. De Rosa, M., Gambacorta, A. and Nicolaus, B., Regularity of isoprenoid biosynthesis in the ether lipids of archaebacteria. Phytochemistry, 1980, 19, 791-3.
37. De Rosa, M., Gambacorta, A., Nicolaus, B., Sodano, S. and Bu'Lock, J.D., Structural regularities in tetraether lipids of Caldariella and their biosynthetic and phyletic implications. Phytochemistry, 1980, 19, 833-6.
38. De Rosa, M., Esposito, E., Gambacorta, A., Nicolaus, B. and Bu'Lock, J.D., Effects of temperature on ether lipid composition of Caldariella acidophila. Phytochemistry, 1980, 19, 827-31.
39. Segerer, A., Neumer, A., Kristjansson, J.K. and Stetter, K.O., Acidianus infernus gen. nov., sp. nov., and Acidianus brierleyi comb. nov.: facultatively aerobic, extremely acidophilic thermophilic sulfur-metabolizing archaebacteria. Int. J. Syst. Bacteriol., 1986, 36, 559-64.

DNA POLYMERASES AND TOPOISOMERASES IN ARCHAEBACTERIA: POTENTIAL APPLICATIONS.

PATRICK FORTERRE
Institut de Microbiologie, Université Paris-Sud,
Centre d'ORSAY, Bat.409, 91405 ORSAY Cedex, FRANCE

ABSTRACT

DNA polymerases and topoisomerases have been isolated from thermoacidophilic archaebacteria. *Sulfolobus acidocaldarius* contains a thermostable DNA polymerase which can be used for DNA amplification by the PCR method. The major DNA topoisomerase activity of this archaebacterium is a novel type I DNA topoisomerase, reverse gyrase, which introduces positive superturns into the DNA. In contrast, the major DNA topoisomerase activity of *Thermoplasma acidophilum* is an enzyme which relaxes negatively supercoiled DNA. Both positively and negatively supercoiled DNA have been found in archaebacteria. Antitumoral drugs and antibiotics which inhibit type II DNA topoisomerases in eubacteria and eukaryotes change the topology of plasmids from halobacteria indicating that both type I and type II topoisomerases exist in archaebacteria. The plasmid pGRB-1 from *Halobacterium* GRB could be used to prescreen new antitumoral drugs and new antibiotics.

INTRODUCTION

The studies of enzymes from extremophilic archaebacteria have several interests : beside the obtention of fundamental informations on the evolution of molecular mechanisms they permit to isolate new enzymes with potential applications in biotechnology and to discover new enzymatic activities unique to the archaebacterial urkingdom. These two points have been illustrated in ours studies of thermophilic archaebacterial DNA polymerases and DNA topoisomerases. In addition, the phylogenetic position of archaebacteria at an intermediate stage between eukaryotes and eubacteria offers an opportunity to study in the same organism different molecular mechanisms otherwise considered to be specific for eukaryotes or prokaryotes. An example is provided by the studies on the mode of action of type II DNA topoisomerase inhibitors using plasmids from halophilic archaebacteria.

RESULTS AND DISCUSSION

DNA polymerases
We have purified to near homogeneity thermostable and thermophilic DNA polymerases from two thermoacidophilic archaebacteria:*Sulfolobus acidocaldarius* (1) and *Thermoplasma acidophilum* (2). These enzymes are monomers of 100 kDa and 85 kDa, respectively. Antibodies raised against *Sulfolobus* DNA polymerase did not crossreact with *Thermoplasma* DNA polymerase, consistent with the large phylogenetic distance between these two archaebacteria. The two enzymes are optimally active at 70°C when they are tested with activated calf thymus DNA as a template. The properties of *Sulfolobus* DNA polymerase have been studied in more details (3 and Salhi *et al*, manuscript in preparation). This enzyme can use a 20 mer-primed single-stranded DNA as template at temperatures well above its Tm (60°C). *Sulfolobus* DNA polymerase is thermostable up to 80°C and can replicate completely a 20 mer-primed M13 DNA at this temperature. At higher temperatures, the reaction stops prematurely but the purified DNA polymerase can still polymerize 100 to 200 nucleotides at 100°C.

Recently, we have found that the *S. acidocaldarius* DNA polymerase can perform DNA amplification by the polymerase chain reaction (PCR) method (3 and S. Salhi *et al*, manuscript in preparation). This methodology has been developped by the scientists from the Cetus Corporation first using Klenow DNA polymerase and later on DNA polymerase from the eubacterium *Thermus aquaticus* (4). The PCR method (see Fig. 1) allows to amplify specific DNA regions 1 000 000 folds or even more. This is extremely useful for medical diagnosis (detection of pathological mutations or DNA sequences from pathogens), for the identification of individuals or species by DNA typing and in the day to day practice of molecular biology by allowing direct DNA cloning and sequencing. A thermostable and thermophilic DNA polymerase improves the specificity and the speed of the method, reduces its cost and permits to use automatic instruments. *S. acidocaldarius* DNA polymerase can also be used for DNA sequencing by the method of Sanger on either single-stranded or double-stranded DNA.

DNA topoisomerases.
Reverse gyrase : The isolation of new enzymatic activities in archaebacteria was illustrated four years ago by the discovery of reverse gyrase, a novel type of DNA topoisomerase, in the sulfur-dependent extremely thermophilic archaebacterium *Sulfolobus acidocaldarius* (5-8). This enzyme introduces positive superturns at high temperature into covalently closed circular DNA at the expense of ATP (instead of negative ones in the case of eubacterial DNA gyrase). In eubacteria, DNA gyrase is responsible for the negative supercoiling of plasmids and chromosomal DNA; in agreement with the discovery of a reverse gyrase in *Sulfolobus*, we

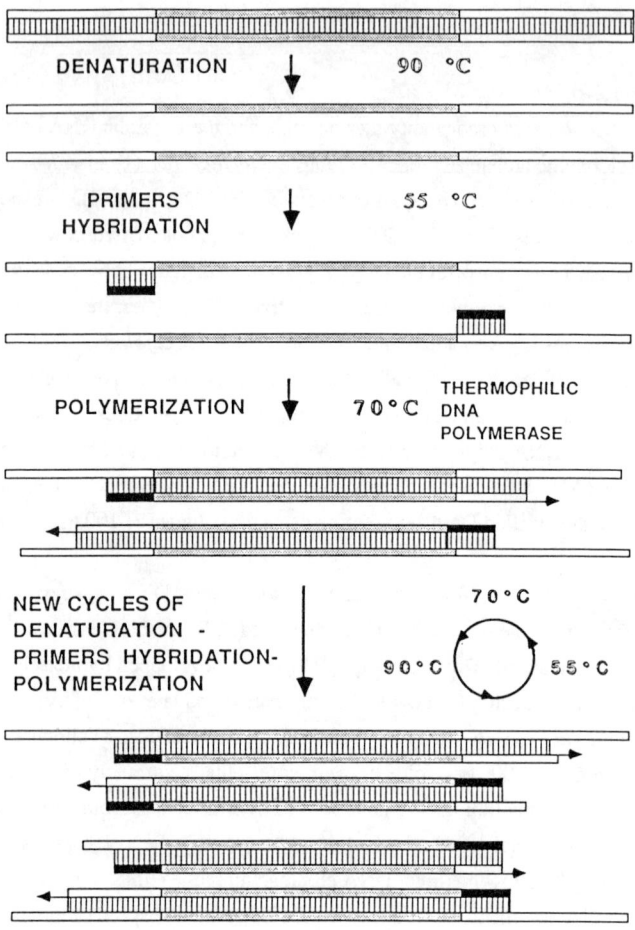

Figure 1. DNA amplification by the polymerase chain reaction (PCR) method.

have found that the genome of the *Sulfolobus* virus-like particle SSV1 is positively supercoiled (9). Remarkably, SSV1 is the only organism discovered so far with a positively supercoiled genome.

Positive supercoiling could help to stabilize the DNA double helix at extremely high temperatures (Fig. 2). Nevertheless, it does not seem to be a general principle for extremely thermophilic organisms. A reverse gyrase has been purified recently from the anaerobic thermophilic archaebacterium *Desulfuroccocus amylolyticus* (10) but we have not found reverse gyrase activity in *Thermoplasma acidophilum*. This archaebacterium contains only a DNA topoisoimerase which relaxes negative superturns (2). Recently, Collin *et al* have

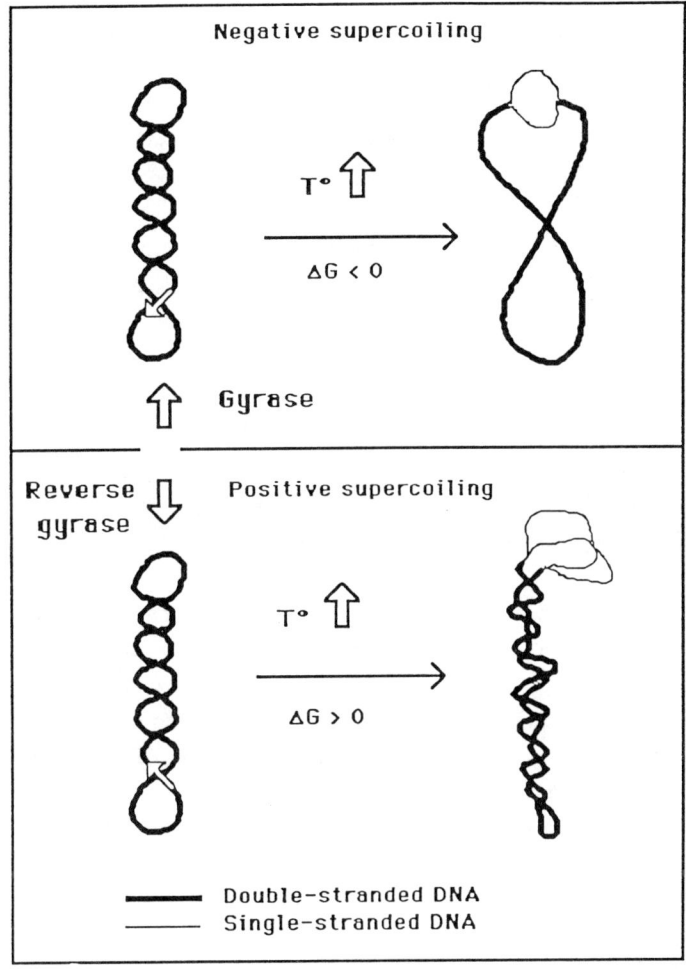

Figure 2. Thermal denaturation of negatively or positively supercoiled DNA. Opening of the DNA double helix via thermal denaturation reduces the number of turns (T) of the helix. In a covalently closed circular DNA, this reduction should be compensated by an increase in the absolute number of superturns (W) in order to keep constant the linking number of the DNA molecule according to the formula $L = T + W$. If the DNA is negatively supercoiled ($W < 0$), this increase leads to relaxation of the molecule, an energetically favorable process ($\Delta G < 0$). If the DNA is positively supercoiled, this increase leads to additionnal supercoiling, an energetically unfavourable process ($\Delta G > 0$). Accordingly, a positively supercoiled DNA resists to thermal denaturation.

detected reverse gyrase activity in crude extracts of *Thermoproteus* and a*Thermococcus* species but not in thermophilic methanogens, *Thermoplasma acidophilum* and an extremely thermophilic eubacterium of the genus *Thermotoga* (11). All these results suggest that reverse gyrase could be restricted to one of the two main archaebacterial branches.

Unlike eubacterial gyrase, which is a type II DNA topoisomerase, reverse gyrase is a type I DNA topoisomerase (6,7) i.e. its changes the linking number of the DNA by step of 1 by introducing transient single-stranded breaks in the DNA double helix. Reverse gyrase is the first type I DNA topoisomerase which is ATP dependent. Previously, ATP dependence was considered to be a hallmark of type II DNA topoisomerases. In the absence of ATP, stoechiometric amonts of reverse gyrase produce single-stranded DNA breaks with the enzyme covalently linked to the 5' end of the breaks (C. Jaxel *et al*, manuscript in preparation). This suggests that reverse gyrase could be evolutionarily related to the eubacterial type I DNA topoisomerase (protein ω) and not to the eukaryotic type I DNA topoisomerase which binds at tha 3' ends of DNA breaks. Reverse gyrase may be used to produce positively supercoiled DNA for studies of the properties of this type of molecules. In addition, this enzyme could be useful in future manipulations of DNA at high temperature.

<u>Topoisomerases II</u>: In contrast to the genome of the *Sulfolobus* virus-like particle SSV1 DNA, the small circular plasmids of halophilic archaebacteria are negatively supercoiled (12). These plasmids are probably supercoiled *in vivo* by a DNA gyrase-like enzyme related to the eubacterial and eukaryotic type II DNA topoisomerase since inhibitors of these enzymes change their topology (13-14). In contrast to DNA topoisomerases I, DNA topoisomerases II introduce transient *double-stranded* breaks into the DNA in the course of the topoisomerization reaction. The archaebacterial plasmids became positively supercoiled by treatment of halobacteria with the anti-gyrase inhibitor novobiocin (14). This can be explained either by the existence of both a gyrase (sensitive to novobiocin) and a reverse gyrase in halobacteria or by asymetric transcriptional supercoiling in the absence of DNA gyrase activity as reported in *Escherichia coli* (2). Positive supercoiling of halobacterial plasmids after novobiocin treatment provides an alternative strategy to reverse gyrase for the large scale preparation of positively supercoiled DNA.

Antitumor drugs active on eukaryotic type II DNA topoisomerases (epipodophyllotoxines) and quinolone antibiotics active on DNA gyrase (ciprofloxacin) produce *in vivo* both DNA relaxation and DNA cleavage of small plasmids from halophilic archaebacteria. This indicates that these drugs have the same mechanism in archaebacteria as in the other two urkingdoms: they stabilize "cleavable complexes" in which type II DNA topoisomerases are covalently linked to DNA breaks. In the case of ciprofloxacin, these effects are only detected when the culture medium is depleted of magnesium at the time of drug addition. Single and double-stranded breaks produced by antitumoral drugs and antibiotics active against type II DNA topoisomerases can be monitored easily *in vitro* on agarose gel in the multicopy plasmid pGRB-1 of the strain *Halobacterium* GRB (Fig. 3-a). In addition antitumoral drugs which are also DNA intercalators, such as the anthracyclins daunorubicin, can be detected in this system because they change the electrophoretic mobility of the plasmid (Fig 3-b). Therefore, the archaebacterial plasmid pGRB-1 could be used to prescreen *in vivo* both antitumoral drugs which are DNA intercalators and/or which are active on eukaryotic DNA topoisomerase II or antibiotics active on eubacterial DNA gyrase. In addition, the mixture of eubacterial and eukaryotic features of the archaebacterial type II DNA topoisomerase at the level of antibiotic sensitivity gives a unique opportunity to investigate in a single system the mode of action of clinically useful drugs which were previously studied independently.

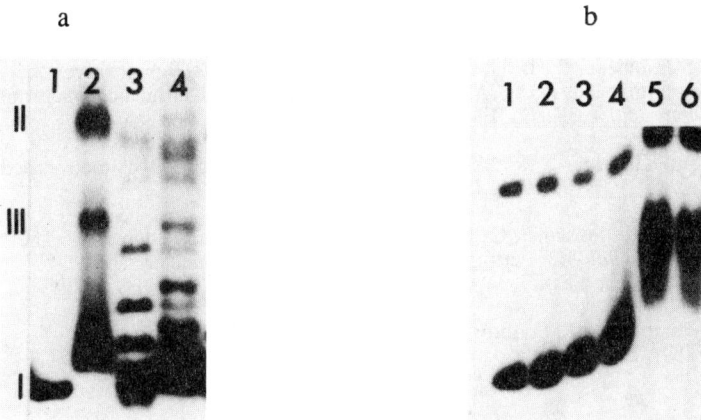

Figure 3. Effect of antibiotics and antitumoral drugs inhibitors of type II DNA topoisomerases on the topology of the archaebacterial plasmid pGRB-1. Panel a, lane 1: control corresponding to the negatively supercoiled form of the plasmid; lane 2 : the DNA gyrase inhibitor ciprofloxacin produces nicked (form II) and linear plasmid DNA (form III) through single and double-stranded DNA breaks; lane 3 : the DNA gyrase inhibitor novobiocin produces a ladder of positively supercoiled topoisomers; lane 4 : the antitumoral drug etoposide (inhibitor of eukaryotic type II DNA topoisomerase) produces a double ladder of positively and negatively supercoiled DNA as well as forms II and III. Panel b : the antitumoral drug daunorubicin (a DNA intercalator) reduces the electrophoretic mobility of the plasmids forms I and II. Lane 1 : control, lanes 2 to 6 : plasmids isolated from cells treated with increasing concentrations of daunorubicin (0, 2.5, 5, 10, 20 µg/ml). For Material and Methods see references 11-13.

REFERENCES

1. Elie, C., De Recondo, A.M. and Forterre, P. Thermostable DNA polymerase from the archaebacterium *Sulfolobus acidocaldarius*; Purification, characterization and obtention of polyclonal antibodies. Eur. J. Biochem. 1989, in press.

2. Forterre, P. Elie, C., Sioud, M.and Hamal, A. Studies on DNA polymerases and topoisomerases on archaebacteria. Can. J. Biochem., 1989, in press.

3. Elie, C., Salhi, S., Rossignol, J.M., Forterre, P. and De Recondo, A.M. DNA polymerase from a thermoacidophilic archaebacterium: evolutionary and technological interest. Biophysica Biochemica Acta, 1988, in press.

4. Saiki, R.K., Gelfand, D.H., Stoffel, S., Scharf, S.J., Higuchi, R., Horn, G.T., Mullis, K.B. and Erlich, H.A. Primer-directed enzymatic amplification of DNA with a thermostable DNA polymerase. Sciences, 1988, **239**, 487-491.

5. Kikuchi, A. and Asai, K., Reverse gyrase-a topoisomerase which introduces positive superhelical turns into DNA. Nature, 1984, **309**, 677-681

6. Forterre, P., Mirambeau, G., Jaxel, C., Nadal, M. and Duguet, M. High positive supercoiling in vitro catalysed by an ATP and polyethylene glycol-stimulated topoisomerase from *Sulfolobus acidocaldarius*. EMBO J., 1985,**4**, 2123-2128.

7. Nakasu, S., and Kikuchi, A.. Reverse-gyrase: ATP dependent type I DNA topoisomerase from *Sulfolobus*. EMBO J., 1985, **4**, 2705-2710.

8. Nadal, M., Jaxel, C., Portemer, C., Forterre, P., Mirambeau, G. and Duguet, M. The reverse gyrase of *Sulfolobus*: Purification to homogeneity and characterization. Biochemistry 1988, in press.

9. Nadal, M., Mirambeau, G., Forterre, P., Reiter, W.D. and Duguet, M. Positively supercoiled DNA in a virus-like particle of an archaebacterium. Nature, 1986, **321**, 256-258.

10. Slezarev, A.I.. Positive supercoiling catalysed in vitro by ATP dependent topoisomerase from *Desulfurococcus amylolyticus*. Eur. J. Biochem., 1988, **173**, 395-399.

11. Collin, R.G., Morgan, H.W., Musgrave, D.R., and Daniel, R.M. Distribution of reverse gyrase in representative species of eubacteria and archaebacteria. FEMS letters. 1988, **55**, 235-239

12. Sioud, M., Possot, O., Elie, C., Sibold, L. and Forterre, P. Coumarin and quinolone action in archaebacteria: evidence for the presence of a DNA gyrase-like enzyme. J. Bacteriol., 1988, **170**, 946-953.

13. Sioud, M., Baldacci, G., Forterre, P. and De Recondo, A.M. Effects of the antitumor drug VP16 (etoposide) on the archaebacterial *Halobacterium* GRB 1.7 kb plasmid *in vivo*. Nucl. Acids Res., 1987, **15**, 8217-8234.4

14. Sioud, M., Baldacci, G., De Recondo, A.M. and Forterre, P. Novobiocin induces positive supercoiling of small plasmids from halophilic archaebacteria *in vivo*. Nucl. Acids Res., 1988, **16**, 1379-1391.

DEEP-SEA HYDROTHERMAL VENTS OF THE 13°N (East Pacific Rise): PRELIMINARY BACTERIAL SURVEY AND BIOTECHNOLOGICAL POTENTIAL.

DANIEL PRIEUR
CNRS,LP 4601, Station Biologique,
Place G. Teissier, 29211 Roscoff, FRANCE.

ABSTRACT

Several deep-sea hydrothermal vents have been discovered in the Eastern Pacific, and the 13°N site was explored by the French in 1982,1984 and 1987. These marine ecosystems are almost independent of solar energy, and the invertebrate communities associated with the vents are supported by an abundant local primary production of chemolithotrophic bacteria.

During the last "Hydronaut" cruise, a microbial survey was conducted to detect bacteria involved in sulfur and nitrogen cycles and showing resistance against heavy metals. Heterotrophic bacteria using different carbon sources were also cultivated. Preliminary results showed that vent ecosystems were inhabited by complex and diversified bacterial communities which had adapted to the extreme conditions of this unique environment.

For these reasons, some of these bacteria could be suitable for biotechnological applications in the following fields: marine invertebrate nutrition, thermostable enzymes, polysaccharide production, heavy metal depuration or recovery.

INTRODUCTION

Since the first discovery of deep-sea hydrothermal vents near the Galapagos Islands in 1977 (1), several sites of the Eastern Pacific have been explored by American, French and Canadian submersibles and teams (2). These vent sites were inhabited by rich animal communities which had several features in common (3): high biomass, dependence on active vents, dominance of large new species, low specific diversity and relatively constant composition. These animal communities were supported by an abundant, local primary production of chemolithotrophic bacteria which could be found in seawater, on rock surfaces and in close associations (epibiotic and endobiotic) with primary consumers (4,5,6,7).

Although sulfur-oxidizing bacteria were found to play a primordial role in the deep-sea hydrothermal food chain (8), other bacterial metabolisms were noted or expected, taking into account available electron donnors and acceptors (9). A major feature of Hydrothermal vent environments was featured by the presence of large amounts of heavy metals due to the discharge of vent fluids (1). As a result, the role of invertebrate associated bacteria in the detoxication of metals was suspected, particularly in the case of Alvinellid epibionts (10).

During the "Hydronaut" cruise (october to december 1987), we aimed at cultivating bacterial communities of different origins, belonging to various metabolic types and showing resistance agains heavy metals.First the collection of strains constituted after the cruise will be studied from a physiological point of view to improve the knowledge on vent bacterial communities. Secondly this collection will be screened for potential biotechnological utilizations, some strains being interesting candidates because of the extreme conditions (hydrostatic pressure, temperature gradients, toxic compounds) of the environment they are stemming from.

MATERIAL AND METHODS

Different samples (hydrothermal waters, rocks and surfaces, invertebrates) were collected during the "Hydronaut" cruise by the man operated "Nautile" submersible. Rock and surface samples were treated with a vortex, and invertebrate tissues were ground with an ultra-turax homogenizer to obtain bacterial suspensions for inoculation in liquid media or plating.

Sulfur-oxidizing, sulfate reducing, nitrifying, denitrifying, nitrogen fixing, manganese oxidizing bacteria were cultivated on specific media (11). Heterotrophic bacteria were cultivated on 2216E medium under aerobic and anaerobic conditions, on TCBS medium for Vibrionaceae enrichment, and with amino acids or carbohydrates as carbon sources. Heavy metal resistant bacteria (Silver, Copper, Cadmium, Zinc, Arsenic, Selenium) were enriched on 2216E medium supplemented with metal salts (12).

The cultures were incubated under different temperature conditions (according to the origin of the samples) and under both atmospheric and "in situ" hydrostatic pressure.

RESULTS

Sulfur bacteria

Sulfur-oxidizing bacteria were cultivated from many samples, including invertebrate tissues, under both atmospheric and hydrostatic pressure, mostly at 20°C, on thiosulfate medium. Although sulfur-oxidizing bacteria were cultivated from the tube worm *Riftia pachyptila* and the bivalve *Bathymodiolus thermophilus*, further studies are needed to conclude if symbionts have been obtained or not.

Sulfate-reducing bacteria were also cultivated from many samples, showing that this metabolism is rather common in the vent environment. The product of this metabolism being hydrogen sulfide, some interactions between these two kinds of sulfur bacteria could be suspected, particularly in microenvironments like the Polychaete *Alvinella* and its tube. The occurence of sulfate reducers in hydrothermal waters or in sulfide chimneys, after incubation at 80°C and 250 atmospheres, implies a possible biological origin of part of the hydrogen sulfide.

Nitrogen bacteria

The results concerning nitrogen bacteria proved to be rather unbalanced. Nitrifying and nitrogen fixing bacteria were detected in few samples of the Polychaete *Alvinella* and its tube. Moreover, nitrate respiration was found in almost all the samples. Denitrification was detected from *Alvinella*, hydrothermal waters and non living surfaces, mostly after incubation at 40°C. Some strains showed a high denitrification activity, compared to coastal strains previously studied in our laboratory. Nitrate respiration was also detected after incubation at 80°C under hydrostatic pressure.

Heterotrophic bacteria

Heterotrophic bacteria have been studied a little in hydrothermal environments, autotrophic bacteria being considered as primordial. However, in regard to the high biomass of invertebrates, the occurence of heterotrophic bacteria using the dissolved and organic matter produced by the invertebrates is rather normal. Almost all the samples allowed us to cultivate aerobic and facultative anaerobic heterotrophic bacteria. Many isolated strains will be studied in the next future. The distribution of Vibrionaceae, as indicated by TCBS enrichment cultures is worth being commented. These bacteria were found in water, chimney and surface samples, sometimes in Polychaete samples, and non-existent in bivalve tissues. A study of pure strains is required for a better understanding of these results.

Manganese oxidizing and heavy metal resistant bacteria

Manganese oxidizing bacteria were cultivated from many samples of various origins. The oxydation of manganese has been checked on pure strains isolated from the Polychaete *Alvinella*, but the process of manganese oxydation have not yet been considered.

Heavy metal resistant heterotrophic bacteria were also very common, particularly in *Alvinella* samples, where Arsenic resistant strains are particularly abundant. The ability of these strains to concentrate arsenic could be very important for the detoxication of the Polychaete environment.

DISCUSSION AND CONCLUSIONS

During the "Hydronaut" cruise, 1000 enrichment cultures were made, and more than 2000 strains (mostly metal resistant) were isolated from 200 positive cultures.

Sulfur-oxidizing bacteria, which are known to play a primordial role in the vent ecosystem food chain, were found in many samples, which constitutes an expected result. These bacteria could be checked as a cultivated food source for marine invertebrate aquaculture as previously suggested (13).

Sulfate-reducing, nitrate respiring and denitrifying bacteria were found in many samples, while nitrifying and nitrogen fixing bacteria were detected in very few samples. This apparently unbalanced nitrogen cycle must be confirmed by further studies.

Heterotrophic bacteria are common in many samples, but need to be studied in details tso that we understand their role within the ecosystem. Most of them are resistant to heavy metals and are presently studied for their possible role in detoxication.

Thus, the vent ecosystem is inhabited by abundant and complex bacterial communities, which are adapted to extreme conditions: hydrostatic pressure, sharp temperature gradients, toxic compounds. For these reasons, some strains could represent interesting candidates in biotechnology.

Most of these bacteria are attached to non living or living surfaces, and exopolymers are probably involved in the adhesion process (14). Because of the extreme conditions of the vent environment, those exopolymers could have interesting new properties for industrial application.

Only moderate thermophilic cultures have been obtained (80°C was the highest temperature tested), and we failed in the isolation of sulfate reducers. Anyway, the results obtained by several authors on other vent sites, clearly indicate that thermpophilic (including new genus) bacteria and archaebacteria do exist in the deep-sea vent environment and require more attention for the next vent expeditions. These bacteria could produce thermostable enzymes for industrial applications.

In conclusion, deep-sea hydrothermal vents constitute unique ecosystems where marine bacteria of usual metabolic types, but adapted to extreme conditions, and new bacterial types exist. They are worth more studying in the future from an applied point of view, all the more because usual aquatic and marine bacteria have yet showed interesting properties (15).

ACKNOWLEDGEMENTS

This work was supported by CNRS (GRECO "ECOPROPHYCE" and AIP 034808). The cruise "Hydronaut" was organised by IFREMER (D. Desbruyères and A-M. Alayse-Danet). D. Prieur, E. Jacq and C. Jeanthon participated to the cruise; enrichment cultures and strain isolation were done by also S. Chamroux, P. Durand, G. Erauso;, Ph. Fera, L.Le Borgne, A. Marhic and G. Mével.

REFERENCES

1. Corliss,J.B., Dymond,J., Gordon, L.I., Edmond, J.M., Von Herzen, R.P., Ballard, R.D., Green, K., Williams, D., Bainbridge, H., Crane, K. and Van Handel, T.H., Submarine thermal springs on the Galapagos Rift. Science, 1979, 203, 1073-83.

2. Prieur, D., Jeanthon, C. and Jacq, E., Les communautés bactériennes des sources hydrothermales profondes du Pacifique oriental. Vie et Milieu, 1987,37,149-64.

3. Laubier, L. and Desbruyères, D., Les oasis du fond des océans. La Recherche, 1984, 15, 1506-17.

4. Jannasch, H.W. and Wirsen, C.O., Chemosynthetic primary production at East Pacific sea floor spreading centers. Bioscience, 1979, 29, 592-98.

5. Jannasch, H.W. and Wirsen, C.O., Morphological survey of microbial mats near deep-sea thermal vents. Appl. Environ. Microbiol., 1981, 41,428-38.

6. Cavanaugh,C.M., Symbiotic chemoautotrophic bacteria in marine invertebrates from sulphide-rich habitats. Nature, 1983, 302, 58-61.

7. Gaill, F., Desbruyères, D. and Prieur, D., Bacterial communities associated with "Pompei worms" from the East Pacific Rise hydrothermal vents: SEM, TEM observations.Microb.Ecol., 1987, 13, 129-39.

8. Felbeck, H. and Somero, G.N., Primary production in deep-sea hydrothermal vent organisms: role of sulfide-oxidizing bacteria. Trends in Biochemical Sciences, 1982, 7, 201-04.

9. Jannasch, H.W. and Mottl, M.J., Geomicrobiology of deep-sea hydrothermal vents. Science, 1985, 229,717-25.

10. Prieur, D. and Jeanthon, C., Preliminary studies of heterotrophic bacteria isolated from deep-sea hydrothermal vent invertebrates. Symbiosis, 1987, 4, 87-98.

11. Prieur, D. Etude des peuplements bactériens associés aux communautés animales des sites hydrothermaux actifs (13°N). Programme National d'Etude de l' Hydrothermalisme Océanique, Rapport AIP CNRS 034808, Brest, May 1988.

12. Jeanthon, C. and Prieur, D., Resistance to heavy metals of heterotrophic bacteria isolated from the deep-sea hydrothermal vent Polychaete *Alvinella pompejana*. Proc. 5th Deep-sea Biology Symposium, June 1988 (Brest, France), submitted.

13. Berg, C.J. and Alatalo, P., Potential of chemosynthesis in molluscan mariculture. Aquaculture, 1984, 39, 165-79.

14. Geesey, G., Microbial exopolymers: ecological and economic considerations. A.S.M. news, 1982, 48, 9-13.

15. Staley, J.T. and Stanley, P.M., Potential commercial applications in aquatic microbiology. Microb. Ecol., 1986, 12, 79-100.

DIVERSITY OF ARCHAEBACTERIA IN BIOTECHNOLOGICALLY IMPORTANT ENVIRONMENTS REVEALED BY ANTIBODY PROBES

EVERLY CONWAY de MACARIO AND ALBERTO J.L. MACARIO
Wadsworth Center for Laboratories and Research
New York State Department of Health, School of Public Health
SUNY-NYSDOH, Albany, New York 12201-0509 USA

ABSTRACT

Immunologic strategies for elucidating the microbial flora of complex ecosystems by direct examination of minute samples are reviewed. The reactivities of antibodies with thermo- and halophilic methanogens, and halophiles are analyzed to reveal distinctive and shared antigens and species diversity. Also discussed are the applications of antibodies in the exploration of extreme environments and in biotechnology involving microbes.

INTRODUCTION

Recently poly- and monoclonal antibodies have been generated against archaebacteria and immunologic strategies have been developed for identifying and quantifying these organisms [1]. Application of these tools has revealed the antigenic organization of archaebacteria and has begun to unravel the structure of their antigens [2]. Immunotypes and antigenic clusters have been delineated which are useful as reference for identifying unknowns by simple and rapid antibody-based tests. This facilitates exploration of a variety of ecosystems, including extreme environments that would otherwise remain largely unpenetrated. A considerable diversity of organisms is emerging. Illustrative data are reviewed in this report.

MATERIALS AND METHODS

Antibody probes, 31 reference methanogens, halophiles, techniques, and antigenic fingerprinting have been described [1-4].

RESULTS AND DISCUSSION

Comparative Analysis of Antigenic Fingerprints.

Meso- and thermophilic methanogens. Reference and non-reference organisms were studied, Table 1. Four reference thermophilic methanogens (11,12,20 and 24) were found to share antigens with at least one of these reference mesophilic strains: 1,3,4,6,10,16,18 and 19. Likewise two of the three non-reference thermophilic methanogens examined shared antigens with mesophilic organisms, Table 1.

Halophilic and other methanogens. No antigen sharing was found between three halophilic methanogens examined and 31 reference strains, Table 2 (to save space only four reference methanogens representing different genera are shown).

Halophiles and methanogens. Eight halophiles were found to elicit antibodies in rabbits and mice and were studied comparatively with respect to 31 reference methanogens, Table 3. No antigen sharing was found between these two groups although antigen sharing was observed within each group [3].

TABLE 1
Antigenic fingerprints reveal that meso and **thermophilic** methanogens share structural markers

No.	Strain	ref.	1	2	3	4	6	10	11	12	16	18	19	20	23	24	31
11	Mb. thermoautotrophicum GC1	[1]			1				4	2							
12	Mb. thermoautotrophicum ΔH	[1]	1			1		1	2	4							
20	Ms. thermophila TM1	[1]		1		3					3	2	4	4			
24	Mt. fervidus V24S	[1]	2													4	
--	Methanobacterium FTF	[5]							3	3							
--	Mg. thermophilicum Ratisbona	[6]													1		
--	Methanosarcina CHTI55	[7]	2		2						2	4	2	4			

[a] See complete list in ref. 4. Blanks: 0 (zero)

TABLE 2

Antigenic fingerprints show that halophilic and other methanogens do not share structural markers

No. Strain	ref.	1	3	4	5	6	7	10	13	14	16	17	18	19	20	23	31
4 Mb. bryantii MoH	[1]	1		4	3		1										
13 Mc. vannielii SB	[1]								4	1							
17 Mg. cariaci JR1	[1]				2						4					1	
18 Ms. mazei S6	[1]	1		3							3		4	3	1		
- Halophilic SF1	[8]	0	0	0	0	0	0	0	0	0	0	0	0	0	0	0	0
- M. coccus DOII	[9]	0	0	0	0	0	0	0	0	0	0	0	0	0	0	0	0
- M. halophilus mahii SLP	[10]	0	0	0	0	0	0	0	0	0	0	0	0	0	0	0	0

[a]See complete list in ref. 4. Blanks in the fingerprints of methanogens No. 4, 13, 17, and 18 mean 0 (zero).

TABLE 3

Antigenic cohesiveness and distinctiveness of halophiles as compared with methanogens demonstrated by antigenic fingerprinting analysis

No. Halophile	Halophile No.								Methanogen No.[a]
	1	2	3	4	5	6	7	8	1-31
1 Hb. mediterranei	4			1	2	1			0
2 Hb. halobium		4	2	2		3			0
3 Hb. salinarium		3	4	2	2	3			0
4 Nb. pharaonis				4					0
5 Hb. volcanii		1	1	2	4	1			0
6 Hb. cutirubrum	2	2	2			4			0
7 Hb. saccharovorum							4		0
8 Hc. morrhuae								4	0
1-31 Methanogens	0	0	0	0	0	0	0	0	

[a]See list in ref. 4. Blanks in fingerprints of halophiles No. 1-8 mean 0 (zero) (ref. 3).

Detection, Identification, and Quantification of Microorganisms in Extreme Environments using Antibody Probes.

Rapid identification of archaebacteria by direct examination of minute samples -avoiding culture-isolation procedures- from complex microbial communities is now possible [4]. Key tools are calibrated antibody probes of predefined specificity spectra. Using comprehensive panels of these probes, microbial identification is done by antigenic fingerprinting [1]. A blown up portrait of microbes is obtained with the probes using the slide immunoenzymatic assay (SIA) constellation [4]. This consists of a modular set of complementary micromethods, all carried out on adjacent small solid phases on a single support. The geometry of the assay components allows for visual and microscopic observations, and spectrophotometric measurements. Microbes are identified and counted, and are quantified by measuring a antibody-mediated enzymatic reaction. Applying this immunotechnology, the antigenic mosaics of many archaebacteria and associated microbes (syntrophs) have been resolved into components. Some of the latter structural markers have been elucidated further using monoclonal antibodies [2]. A considerable diversity of organisms has been revealed in ecosystems presumed monotonous according to data from conventional methods [4]. A broad spectrum of immunotypes and molecular markers is unfolding. For example, various immunotypes of M. thermophila have been found in digestors. Information thus far indicates that the immunologic approach will help investigating extreme environments by providing data hardly obtainable by other means as quickly.

ACKNOWLEDGEMENT

This work was supported in part by Grant #706IERBEA85 (GRI-NYSERDA-NY gas).

REFERENCES

1. Macario, A.J.L. and Conway de Macario, E., Syst. Appl. Microbiol. 1983, **4**, 451-458.
2. Conway de Macario, E. and Macario, A.J.L., Syst. Appl. Microbiol. 1986, **7**, 320-324.
3. Conway de Macario, E., Konig, H. and Macario, A.J.L., J. Bacteriol. 1986, **168**, 425-427.
4. Macario, A.J.L. and Conway de Macario, E., Appl. Environ. Microbiol. 1988, **54**, 79-86.
5. Touzel, J.P., Petroff, D., Maestrojuan, G.M., Prensier, G. and Albagnac, G., Arch. Microbiol. 1988, **139**, 291-296.
6. Zabel, H.-P., Konig, H. and Winter, J., Syst. Appl. Microbiol. 1985, **6**, 72-78.
7. Touzel, J.P., Petroff, D. and Albagnac, G., Syst. Appl. Microbiol. 1985, **6**, 66-71.
8. Mathrani, I.M. and Boone, D.R., Appl. Environ. Microbiol. 1985, **50**, 140-143.
9. Yu, I.K. and Kawamura, F., J. Gen. Appl. Microbiol. 1987, **33**, 303-310.
10. Paterek, R.J. and Smith, P.H., Int. J. Syst. Bacteriol. 1988, **38**, 122-123.

LIPID STRUCTURES IN THERMOTOGA MARITIMA

Mario De Rosa[a,b], Agata Gambacorta[a], Robert Huber[c], Virginia Lanzotti[a], Barbara Nicolaus[a], Karl O. Stetter[c] and Antonio Trincone[a]

[a]Istituto per la Chimica di Molecole di Interesse Biologico del C.N.R., Via Toiano 6, Arco Felice, Napoli, Italy

[b]Istituto di Biochimica delle Macromolecole, Universita' di Napoli I Facolta' di Medicina e Chirurgia, Via Costantinopoli 16 Napoli, Italy

[c]Lehrstuhl fur Mikrobiologie, Universitat Regensburg, Federal Republic of Germany

ABSTRACT

The lipid structures of Thermotoga maritima, a new anaerobic eubacterium, growing up to 90° C have been characterized. Acid methanolysis of complex lipids of this thermophilic eubacterium afforded a mixture of myristic, palmitic and stearic acid methyl esters; 15,16-dimethyltriacontanedioic acid dimethyl ester (the free acid of which has been named diabolic acid); and a new type of ether core lipid. This last compound, on the basis of chemical and spectroscopic evidence, was identified as 15,16-dimethyl-30-glyceryloxytriacontanoic acid methyl ester. Complex lipids of the microorganism consist mainly of two glycolipids and three phospholipids.

INTRODUCTION

The genus Thermotoga comprises the most extremely thermophilic eubacteria presently known. Up to now, the species Thermotoga maritima and Thermotoga neapolitana were described, growing at temperatures between 55 and 90°C with an optimum around 80°C (1-3). The type species T. maritima was isolated from a submarine hydrothermal system at Vulcano, Italy (1). Members of Thermotoga are strictly anaerobic fermentative marine bacteria. Cells are Gram-negative rods, exhibiting the "toga", a characteristic sheath-like structure overballooning at both ends. Phylogenetically, Thermotoga represents the deepest branch-off within the eubacteria (4). Besides their extremely high growth temperatures, members of this group are unique by their 16S rRNA, their rifampicin-resistant RNA polymerase and their sensitivity against aminoglycoside antibiotics (1,5,6). Taxonomically, Thermotoga represents a separate order, the "Thermotogales" (Stetter et al. manuscript in preparation).

It is well known that all the membrane lipids of the eubacteria are based on ester linkages formed by condensation of alcohols and fatty acids, whereas archaebacteria possess isopranyl glycerol ether lipids. However some species of eubacteria may have unique lipids which are neither of archebacterial type, nor are known in other eubacteria. Examples are the lipids of Thermodesulfotobacterium commune, based on alkyl (straight and iso or anteiso branched chain of variable carbon number) ether of glycerol (7), and the lipids of Thermomicrobium roseum, in which a series of straight chain and internally methyl branched 1,2 diols (C_{18} to C_{23}), are found to replace glycerolipids. Lipids of this last microorganism also contain fatty acids but they are ester-linked to the diols or amide-linked to polar head groups and not to the glycerol which is essentially absent in this thermophilic eubacterium (7). Non conventional lipid molecules also occur in some members of the Butyrivibrio genus, whose lipids are based on a series of long-chain dicarboxylic acids with a vicinal dimethyl branching; the trivial name for this series of compounds is diabolic acid (8).

This paper reports data on lipid structures of Thermotoga maritima, a new example of eubacterium with unusual lipid structures.

MATERIALS AND METHODS

Microorganism and culture conditions
Thermotoga maritima DSM 3109, was grown in MMS medium at 85°C (1). Large scale cultures were grown in HTE-Fermentor 300 l (Bioengineering, Wald, Switzerland) under N_2 (2 l/min) with gentle stirring (200 rev/min). Cells were harvested in the late exponential growth phase and lyophilized (the yield was about 60 g wet weight per 100 l).

Extraction and methanolysis of lipids
Lyophilized cells were extracted continuously (Soxhlet) for 12 h with $CHCl_3$/MeOH (1:1 v/v). The extract, taken to dryness under vacuum (7-8% of dried cells), was treated with methanol/HCl (9:1 v/v) for 6 h, at reflux.

Purification of methanolyzed lipids
The chloroform soluble fraction of the methanolysis mixture was chromatographed on a silica gel column. Light petroleum/Et_2O (99:1 v/v) eluted fatty acid methyl esters (ca 36% of the $CHCl_3$ soluble fraction); light petroleum/Et_2O (98:2 v/v) eluted 15,16-dimethyltriacontanedioic

acid dimethyl ester (33%); $CHCl_3$ eluted the new ether core lipid 1 (Fig. 1)(30%) and trace amounts of an unidentified compound; $CHCl_3$/MeOH (80:20 v/v) eluted a more polar yet unidentified compound (1%).

Chromatographic procedures

Thin layer chromatography (TLC) was performed on 0.25 mm layers of silica gel F254, Merck, activated by heating at 100° C for 2 h. Solvents included $CHCl_3$/MeOH/H_2O (65:25:4 v/v) for complex lipids (solvent A); $CHCl_3$/MeOH (95:5 v/v) for compound 1 (Fig. 1) (solvent B); $CHCl_3$ for acetylated derivative of 1 (compound 2, Fig. 1) (solvent C); light petroleum/Et_2O (98:2 v/v) for dimethyl ester of diabolic acid and for methyl ester of fatty acids (solvent D). All compounds were detected by exposure to I_2 vapour, or by spraying with $Ce(SO_4)_2$/sulfuric acid. Staining test analyses for complex lipids were performed using specific reagents for phospho, glyco and aminolipids. GLC of fatty acid methyl esters was performed on a SE-30 capillary column, N_2 (20 ml/min) was used as carrier gas, with temperature program 6°C/min from 165 to 220°C.

Chemical procedures

Acetylation of compound 1 (Fig. 1) was performed with Ac_2O/pyridine (9:1 v/v) at 60° C overnight. Compound 1 was reduced (8) with $NaAlH_2(OCH_2CH_2OCH_3)_2$, obtaining the alcohol 3 (Fig. 1), that in turn was converted into the parent C_{32} hydrocarbon by HI/$LiAlH_4$ degradation (9).

Instrumental

The MS measurements in electron impact (EI-MS) were performed at 70 eV with a Kratos MS-30 instrument. NMR spectra were run on a Bruker WH-500 spectrometer in $CDCl_3$ as solvent and TMS as internal standard.

RESULTS

This paper characterizes the core lipids of <u>Thermotoga maritima</u>, a new hyperthermophilic eubacterium, recently isolated in Italy and the Azores (1-3).

TLC analysis of complex lipids of the microorganism showed two glycolipids and three phospholipids, as major components, whereas aminolipids were absent.

Chloroform soluble fraction of the methanolysis mixture of complex lipids, chromatographed on silica gel column, was resolved in three main fractions: fatty acid methyl esters, diabolic acid dimethyl ester and a new type of ether core lipid (compound 1, Fig. 1).

Fatty acid methyl esters

This fraction (36% of chloroform soluble fraction of methanolyzed complex lipids), eluted with light petroleum/Et_2O (99:1 v/v), gave on

TLC a single spot, with Rf 0.8 in the solvent D.
^{1}H NMR spectrum was consistent with the structure of methyl esters of n-fatty acid.

By GLC the fraction was resolved in three peaks identified as methyl myristate (12% of the mixture), methyl palmitate (86%) and methyl stearate (1%).

Diabolic acid dimethyl ester

The fraction (33%), eluted with light petroleum/Et$_2$O (98:2 v/v), gave a single spot on TLC with Rf 0.4 in solvent D.

^{1}H NMR signals (δ 0.72, d, C$\underline{H}_3$-CH; 1.28, bs, -(C$\underline{H}_2$)$_n$-; 1.63, bm, -C$\underline{H}_2$CH$_2$-CO-; 2.3, t, -C$\underline{H}_2$-CO-; 3.65, s, -COOC$\underline{H}_3$) and ^{13}C NMR signals (δ 14.4, q, $\underline{C}$H$_3$-CH-; 25.0 to 35.0, t, -($\underline{C}$H$_2$)$_n$-; 36.7, d, $\underline{C}$H-CH$_3$; 51.4, q, $\underline{C}$H$_3$O-; 174.4, s, -$\underline{C}$OOCH$_3$) were consistent with the structure of a methyl branched fatty acid methyl ester.

EI-MS spectrum resulted identical to that of 15,16-dimethyl triacontanedioic acid (diabolic acid) dimethyl ester, previously described (8), with diagnostically important peaks at m/z 538 (M$^+$), 506 and 474 (loss of one and two methanol fragments, characteristic of this class of dimethyl branched dicarboxylic acid dimethyl esters).

New ether core lipid

Compound 1 (Fig.1) (30%), eluted with CHCl$_3$, gave a single spot in TLC with Rf 0.4 in the solvent B.

The acetylation of 1 afforded the diacetylated derivative 2 (Fig.1) (Rf 0.6 in solvent C).

^{1}H NMR signals of 1 and 2, reported in Table 1, indicated the presence of a glyceryloxy group ether linked to a dimethyl branched C$_{30}$ carboxylic acid methyl ester. In particular in the ^{1}H NMR spectrum of 2 the singlets of the two acetyl groups at δ 2.06 and 2.08, the doublet at 3.54, due to the 1'-protons, and the ABX system, centred at 4.25 and 5.18, due to the 3' and 2' protons, pointed out that the glycerol moiety was ether linked at 1' position.

The EI-MS spectrum of 2 besides M$^+$ at m/z 668 (0.1%), included diagnostically important peaks at m/z 637 (M$^+$ - CH$_3$O, 3.1%), 523 (M$^+$ - CH(OAc)-CH$_2$OAc, 7.1%), 493 (M$^+$ - C$_7$H$_{11}$O$_5$; C(30)-O-cleavage, 3.9%), 339 (M$^+$ - AcOH and C$_{17}$H$_{33}$O$_2$; C(15)-C(16) cleavage, 32.9%) and 159 (loss of C$_{33}$H$_{65}$O$_3$; C(1')-O cleavage, 100%), confirming the ether linkage at the 1' position of glycerol and localizing a vic-dimethyl branching at the 15-16 position of the C$_{30}$ aliphatic chain. The ^{13}C NMR data of 1 (Table 1) confirmed the dimethyl branching of alkyl chain ether linked to the glyceryloxy group. In particular the steric compression shift of methyl resonance to higher field (δ 14.5), also observed for diabolic acid (8), indicated a higher probability of gauche environment for these methyl groups. A final confirmation of the vic-dimethyl branching of aliphatic chain was obtained by degradation of 1 to the parent C$_{32}$ hydrocarbon, identified as 15,16-dimethyltriacontane by comparison with EI-MS data previously reported (8). Studies on the stereochemistry of the glycerol of compound 1 (Fig. 1) are in progress.

TABLE 1

^{1}H and ^{13}C NMR data for the compounds 1 and 2

Carbon atom	1		2
	δ_H	δ_C	δ_H
1'	3.53 d	72.5 t	3.54 d
2'	3.88 bm	70.4 t	5.18 X part of ABX
3'	3.72 bd	64.3 t	4.25 AB part of ABX
1		174.2 s	
2	2.30 t	34.1 t	2.29 t
3	1.62 m	25.0 t	1.60 m
4,27	1.26 bm	29.4 t	1.26 bm
5-11,20-26	1.26 bm	29.7 t	1.26 bm
12,19	1.26 bm	30.0 t	1.26 bm
13,18	1.26 bm	27.7 t	1.26 bm
14,17	1.26 bm	34.9 t	1.26 bm
15,16	1.40 m	36.6 d	1.39 m
28	1.26 bm	24.7 t	1.26 bm
29	1.56 m	33.5 t	1.54 m
30	3.46 t	71.9 t	3.43 t
31,32	0.73 d	14.5 q	0.73 d
CH$_3$-O-C(1)	3.67 s	51.4 q	3.66 s
CH$_3$-CO-O-C(2')			2.08 s
CH$_3$-CO-O-C(3')			2.06 s

Chemical shifts are in ppm downfield from TMS.
d=doublet; bm=broad multiplet; bd=broad doublet; t=triplet; s=singlet.

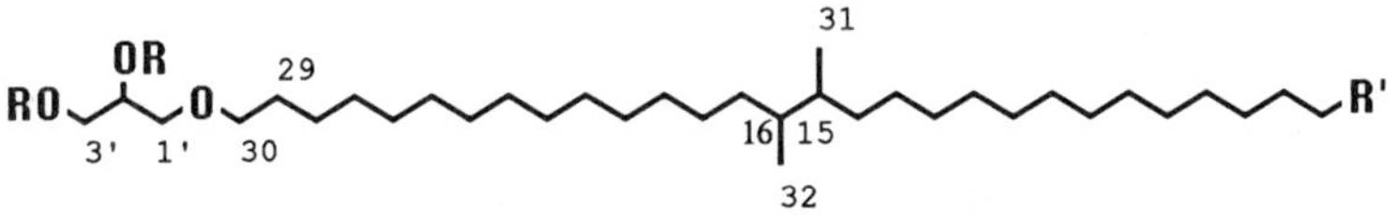

Figure 1. 15,16-dimethyl-30-glyceryloxytriacontanoic acid methyl ester (compound 1); R = H, R' = -COOCH$_3$. Diacetylated derivative of 1 (compound 2); R = CH$_3$-CO-, R' = -COOCH$_3$. Compound 3; R = H, R' = -CH$_2$OH. Numbering of carbon skeleton in Figure was according to Table 1.

CONCLUSIONS

Membrane lipids of Thermotoga maritima are based on n-fatty acids, diabolic acid (15,16-dimethyltriacontanedioic acid) and on a new glycerol ether lipid (15,16-dimethyl-30-glyceryloxytriacontanoic acid). The presence of ether lipid, with an unprecedented structure, accords with the proposed phylogenetic segregation of Thermotogales, representing the deepest branch-off within eubacteria (4), and provides further evidence that membrane lipid structure could be a useful taxonomic marker (10).

The complex lipids of Thermotoga maritima based on the 15,16-dimethyl-30-glyceryloxytriacontanoic acid could have two different types of molecular architecture: a) they are amphipatic monopolar molecules if the two free hydroxyl groups of the glycerol moiety are linked to a polar head and to the carboxylic group of triacontanoic chain respectively, giving rise to a hairpin structure similar to that reported in the thermophilic, methanogenic archaebacterium Methanococcus jannaschii (11); b) they are bipolar amphipatic molecules if two different polar heads are linked to the glycerol moiety and to the carboxylic group of the triacontanoic chain, giving rise to a molecular organization similar to that occurring in thermophilic and methanogenic archaebacteria and in eubacteria of Butyrivibrio spp. (8, 12).

Both the lipid organization and the role of such molecules in resistance to the high growth temperature of this microorganism can be clarified only when the structure of complex lipids is entirely known.

ACKNOWLEDGEMENTS

The authors thank E. Esposito and R. Turco for technical assistance.

REFERENCES

1. Huber R., Langworthy, T.A. Konig, H., Thomm, M. Woese, C.R., Sleytr, U.B. and Stetter, K.O., Thermotoga maritima sp. nov. represents a new genus of unique extremely thermophilic eubacteria growing up to 90°C. Archiv. Microb., 1986, **144**, 324-33.
2. Belkin, S., Wirsen, C.O., Jannasch, H.W., A new sulfur-reducing extremely thermophilic eubacterium from a submarine thermal vent. Appl. Environ. Microbiol., 1986, **51**, 1180-5.
3. Jannasch, H.V. Huber, R., Belkin, S. and Stetter, K.O., Thermotoga neapolitana sp. nov. of the extremely eubacterial genus Thermotoga. Arch. Microbiol., 1988, **150**, 103-4.
4. Woese, C.A., Bacterial evolution. Microbiol. Rev., 1987, **51**, 221-71.
5. Achenbach-Richter, L., Gupta, R., Stetter, K.O. and Woese, C.R., Were the original eubacteria thermophiles? System. Appl. Microbiol. 1987, **9**, 34-9.

6. Londei, P., Altamura, S., Huber, R., Stetter, K.O. and Cammarano, P., Ribosomes of the extremely thermophilic eubacterium Thermotoga maritima are uniquely insensitive to the miscoding-inducing action of aminoglycoside antibiotics. J. Bacteriol., in press.
7. Langworthy, T.A. and Pond, J.L., Archaebacterial ether lipids and chemotaxonomy. In Archaebacteria '85, eds., O. Kandler and W. Zillig, Gustav Fischer Verlag Stuttgart, 1986, pp. 253-7.
8. Klein, A., Hazlewood, G.P., Kemp, P. and Dawson, M.C., A new series of long-chain dicarboxylic acid with vicinal dimethyl branching found as major components of the lipids of Butyrivibrio spp. Biochem. J., 1979, 183, 691-700.
9. De Rosa, M., De Rosa, S., Gambacorta, A., Minale, L. and Bu' Lock, J.D., Chemical structure of the ether lipids of thermophilic acidophilic bacteria of the Caldariella group. Phytochemistry, 1977, 16, 1961-5.
10. Goodfellow, M. and Minnikin, D. E., Chemical Methods in Bacterial Systematic, Academic Press, New York, 1985.
11. Comita, P.B.,Gagosian, R.B., Pang, H. and Costello, C.E., Structural elucidation of a unique macrocyclic membrane lipid from a new, extremely thermophilic, deep-sea hydrothermal vent archaebacterium Methanococcus jannaschii. J. Biol. Chem., 1984, 259, 15234-41.
12. De Rosa, M., Gambacorta, A. and Gliozzi, A., Structure, biosynthesis and physicochemical properties of archaebacterial lipids. Microbiol. Rev., 1986, 50, 78-80.

BEHAVIOUR OF THE THERMOSTABLE SULFOLOBUS SOLFATARICUS MALIC ENZYME IN WATER-MISCIBLE ORGANIC SOLVENTS: BIOTECHNOLOGICAL PROSPECTS

ANNAMARIA GUAGLIARDI[1], MOSE' ROSSI[1,2], SIMONETTA BARTOLUCCI[1]
[1] Dipartimento di Chimica Organica e Biologica, Università di Napoli, Italia
[2] Istituto di Biochimica delle Proteine ed Enzimologia, CNR, Napoli, Italia

ABSTRACT

This paper reports the effects of certain water-miscible organic solvents on both the stability and activity of the malic enzyme (EC 1.1.1.40) from the extreme thermoacidophilic archaebacterium Sulfolobus solfataricus. The enzyme exhibited a remarkable resistance to solvents: after 12 hours' incubation at 25 °C, it was completely active in 50% dimethylformamide and had lost 15% of its initial activity in 50% methanol or 15% ethanol. The Sulfolobus solfataricus malic enzyme was able to function in the presence of an organic solvent; moreover, a number of solvents caused an increase of the initial reaction rate. The biotechnological potentials of this thermostable, NAD(P)-dependent oxidoreductase are discussed.

INTRODUCTION

Traditionally, the biological catalysts have been employed in water, but many enzymes have recently been discovered to function in non-strictly aqueous solutions. The enzymatic catalysis in such media can offer a number of advantages: decrease of the product's inhibition, increase of the enzyme's activity because of a better solubility of hydrophobic substrates or of an altered conformation of the protein molecule, and the ability to carry out useful reactions unfeasible in aqueous buffer [1]. However, the strong inhibitory and denaturing effects of relatively high concentrations of water-miscible organic solvents on most mesophilic enzymes account for the search for more resistant catalysts. The finding that enzymes from thermophilic sources exhibit a high tolerance towards organic solvents as compared to similar enzymes from mesophiles [2] represents an attractive prospect to perform enzyme-catalyzed reactions in several industrial processes and assess the structure-function-stability relationship of these proteins.

Here we report the effects of certain water-miscible organic solvents on both the stability and activity of the malic enzyme (L-malate: NADP oxidoreductase (oxalacetate - decarboxylating) EC 1.1.1.40) previously purified [3] from the extreme thermoacidophilic archaebacterium Sulfolobus solfataricus, strain MT-4, grown aerobically at 87 °C and pH 3.0 [4] and collected after 40 hours' growth. In addition to the reversible oxidative decarboxylation of L-malate (malic activity), the S. solfataricus enzyme catalyzes the decarboxylation of oxalacetate at a rate comparable to the L-malate's decarboxylation [5]. All the data reported are related to the malic activity. The enzyme, a dimer with a native molecular weight of 105,000 ± 2,000 and apparently identical subunits, requires divalent metal cations and displays maximal activity at 85 °C and pH 8.0. NAD can replace NADP although with a lower efficiency. Amino acid composition was determined and found to be significantly higher in the tryptophan content than the malic enzyme from E. coli; three cysteine residues per subunit were evaluated by amino acid analyses (as carboxymethylcysteine) and by titration with 5,5'-dithiobis (2-nitrobenzoic acid).

MATERIALS AND METHODS

The malic enzyme was purified to homogeneity by using ion exchange chromatography, ammonium sulfate fractionation, affinity chromatography and gel filtration [3]. The protein concentration was determined by the BIO-RAD method using bovine serum albumin as standard.

The enzyme activity was assayed at 60 °C in 20 mM Tris/HCl buffer, pH 8.0, 1 mM L-malate, 0.05 mM NADP, 0.1 mM $MnCl_2$, 50 mM ammonium sulfate and the enzyme (5-10 µg) (final volume of 1 ml). One unit of the enzyme activity was defined as the amount of enzyme reducing 1 µmol of NADP per min under the assay conditions. The effects of increasing concentrations of organic solvents on the malic activity were investigated at 25 °C under the above assay conditions, by adding the solvent to the reaction mixture in the place of an equivalent volume of buffer; the percentages are volumes/100 volumes (v/v).

The stability of the malic activity in the presence of an organic solvent was tested by incubating at 25°C in sealed glass tubes a mixture containing the enzyme (150 µg) and the solvent in 20 mM Tris/HCl buffer, pH 8.0 (final volume 0.5 ml). At convenient time intervals, 30-µl aliquots were removed from each incubation mixture and assayed for the malic activity as described above.

RESULTS

Enzyme Stability

Table 1 summarizes the data referring to the stability of the malic activity after 12 hours' incubation in the presence of certain water-miscible organic solvents. It is noteworthy that no further activity decrease was detected after 24 hours in dimethylformamide, ethanol or methanol (not reported in the Table). Hence, the S. solfataricus malic enzyme displayed an excellent resistance to certain miscible solvents at a concentration that most enzymes inactivate.

Enzyme Activity

No inhibition was detected when the reaction was carried out in the presence of certain miscible solvents even at high concentrations. Furthermore, a number of solvents were found to stimulate the malic activity independently of time (Fig.1).

TABLE 1
Effect of water-miscible organic solvents on the stability of the S. solfataricus malic enzyme

Solvent	Concentration %	Residual Activity * %
Dimethylformamide	50	100
Ethanol	15	85
Methanol	50	85
2-Propanol	30	48

* After 12 hours of incubation.

Maximal activation was observed with 60% methanol, which was able to double the velocity of the reaction. Studies with methanol revealed that the effect of the solvent was exclusively on the K_{cat} of catalysis, the K_m's for both substrates remaining unvaried.

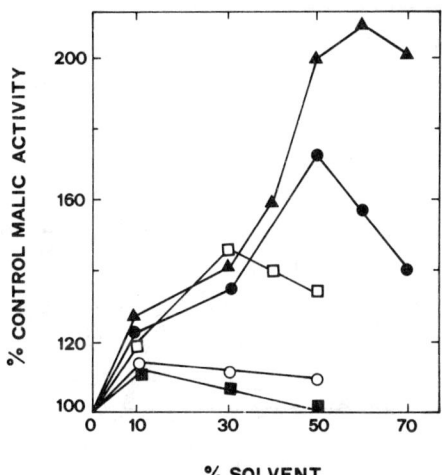

Figure 1. Effect of methanol (▲), ethanol (●), acetone (□) 2-propanol (○) and dimethylformamide (■) on the malic activity.

DISCUSSION

Some peculiar features of the S. solfataricus malic enzyme are summarized in Table 2.

TABLE 2
Kinetic properties of the S. solfataricus malic enzyme

			Reference
Temperature Optimum		85 °C	[3]
Stability at 85 °C		5 hours*	[3]
Resistance to 7.5 M urea		10 hours*	[6]
Resistance to 0.075% SDS		5 hours*	[6]
NADP**	K_m V_m	3 µM 1.3 U/mg	[3]
NADP-(2AE)-N^1-PEG**	K_m V_m	32 µM 6.65 U/mg	[7]

* 50% of the initial activity was assayed after this incubation time.
** Determined at 45 °C.

The thermal stability and resistance to protein denaturants displayed by the S. solfataricus malic enzyme are uncommon even when compared to those of other dehydrogenases from thermophilic source.

Interest is growing in NAD(P)-dependent oxidoreductases for application in the synthesis of fine chemicals because of their high stereo - specificity. In this respect, the macromolecular NAD(P) derivatives have potential for use in the systems of ultrafiltration membrane reactors with cofactor regeneration [8]. NADP macromolecularized by covalent coupling to polyethylenglycol MW 17,000 (a gift from Dr. A.F. Bückmann, Braunschweig, F.R.G.) was found to have a good cofactor activity for the malic enzyme. Moreover, the enzyme displayed a 5-fold higher reaction rate with NADP-PEG as coenzyme. The remarkable resistance of the malic enzyme in miscible solvents, in addition to its dual cofactor specificity, prompts us to determine the kinetic constants for PEG-derivatives of NAD(P) in organic media.
On the other hand, the S. solfataricus enzyme may constitute a model to evaluate the structural features that a protein requires for stability in non aqueous solutions.

ACKNOWLEDGEMENTS

This work was partially supported by the Commission of the European Communities, Contract BAP.0052.I.

REFERENCES

1. Zaks, A. and Klibanov, A.M., Enzymatic catalysis in nonaqueous solvents. J. Biol. Chem., 1988, 263, 3194-3201.

2. Fontana, A., Structure and stability of thermophilic enzymes. Studies on thermolysin. Biophys. Chem., 1988, 29, 181-193.

3. Bartolucci, S., Rella, R., Guagliardi, A., Raia, C.A., Gambacorta, A., De Rosa, M., and Rossi, M., Malic enzyme from archaebacterium Sulfolobus solfataricus. Purification, structure and kinetic properties. J. Biol. Chem., 1987, 262, 7725-7731.

4. Cacace, M.G., De Rosa, M., and Gambacorta, A., DNA-dependent RNA polymerase from the thermophilic bacterium Caldariella acidophila. Purification and basic properties of the enzyme. Biochemistry, 1976, 15, 1692-1696

5. Guagliardi, A, Moracci, M., Manco, G., Rossi, M. and Bartolucci, S., Oxalacetate decarboxylase and pyruvate carboxylase activities, and effect of sulfhydryl reagents in malic enzyme from Sulfolobus solfataricus. Biochim. Biophys. Acta, 1988, in press.

6. Bartolucci, S., Guagliardi, A., Moracci, M., Gambacorta, A., De Rosa, M. and Rossi, M., Sulfolobus solfataricus malic enzyme. Abstr.Book XIV Internatl. Congress of Microbiology, Manchester, September 1986, p.106.

7. Rossi, M., Rella, R., Bartolucci, S., De Rosa, M., Gambacorta, A., Iorio, G., Catapano, G., Drioli, E., Pensa, M., Guagliardi, A., and Lama, L., Potential applications of cells and enzymes of the thermophilic bacterium Sulfolobus solfataricus in biotechnology. Abstr. Book Research Programme in Biomolecular Engineering, Second Generation Bioreactor, Compiègne, April 1986, pp. 38-39.

8. Bückmann, A.F., Kula, M.-R., Wichmann, R., and Wandrey, C., An efficient synthesis of high molecular weight NAD(H)-derivatives suitable for continuous operation with coenzyme depending enzyme systems, J. Appl. Biochem., 1981, 3, 301-315.

MECHANISM OF CYCLOPENTANE RING FORMATION IN TETRAETHER LIPIDS OF SULFOLOBUS SOLFATARICUS

Antonio Trincone[a], Agata Gambacorta[a] and Mario De Rosa[a,b]
[a]Istituto per la Chimica di Molecole di Interesse Biologico Via Toiano, 6 Arco Felice, Napoli, Italy
[b]Istituto di Biochimica delle Macromolecole, Universita' di Napoli I Facolta' di Medicina e Chirurgia, Via Costantinopoli 16, Napoli, Italy

Carlo Scolastico[c], Atanas Sydimov[c] and Donatella Potenza[c]
[c]Dipartimento di Chimica Organica, Universita' di Milano, Via Venezian 21, Milano, Italy

ABSTRACT

Lipids of Sulfolobales are essentially based on two classes of tetraethers, differing both in the nature of polar heads, glycerol or nonitol, and in the number of cyclopentane rings in the isopranoid chains.
In order to contribute to the elucidation of the mechanism of the cyclopentane ring formation, $^{13}C, ^{2}H$ labelled mevalonic lactones were synthesized and fed to Sulfolobus solfataricus as precursors. Results obtained from the analysis of ^{13}C NMR spectra of labelled tetraether lipids permit the discrimination among a series of biogenetic hypotheses in which the mechanisms of cyclopentane ring formation require the initial presence of an unsaturation and are possibly concerted with a hydride reduction step.

INTRODUCTION

The eubacterial and eukaryotic lipids are based on fatty esters, while archaebacterial lipids are mainly isopranyl glycerol ethers. These molecules are obtained by condensation of glycerol (or more complex polyols) with isopranoid alcohols of 20, 25 or 40 carbon atoms. (1,2). The complex lipids of archaebacteria are mainly based on two classes of isopranoid ethers, that are classified as diethers and tetraethers. These latter, with unprecedented molecular architecture for a lipid, are bipolar amphipatic molecules characterized by the presence of two (equivalent or not) polar heads, linked by two C_{40} alkyl components, which are practically twice the average length of the aliphatic components of classic ester lipids. The tetraethers can be divided into two structural types: the glycerol dialkyl glycerol tetraethers (GDGTs) and the glycerol dialkyl nonitol tetraethers (GDNTs) (1,2). The C_{40} alkyl components in these lipids differ in the additional feature containing up to four cyclopentane rings. The cyclization degree of the ispranoid chain is sensitive to enviromental parameters such as temperature; it has been shown that the number of cyclopentane rings in the C_{40} components increases when some thermophilic archaebacteria are grown at increasing temperatures (3,4).

There is no direct information as to wheter the cyclopentane ring formation occurs at level of a free isoprenic precursor or on C_{40} chains already ether linked. However, indirect evidence favors the latter alternative. In fact the process of cyclopentane ring formation leads to remarkably specific products (5). The structural regularities observed in tetraether biosynthesis are in accord with the hypothesis that cyclizations occur in the axyally symmetric tetraethers rather than in the free C_{20} or C_{40} components. The regular disposition of cyclopentane rings in tetraethers could indicate that cyclopentanes were closed in an ordered way, by a mechanism that operates in a concerted manner on both alkyl chains, starting from the middle of the isoprenoid system towards the ether bonds. Studies on the mechanism of cyclopentane ring formation have been carried out with <u>Sulfolobus solfataricus</u>, a member of the thermophilic archaebacteria group, since the lipids of this microorganism are based essentially on tetraethers differently cyclized.

In order to contribute to the elucidation of the mechanism of the cyclopentane ring formation, $^{13}C, ^{2}H$ mevalonolactones strategically labelled were synthesized and fed to this archaebacterium.

MATERIALS AND METHODS

Synthesis of labelled precursors
$(5-^{13}C;5,5-^{2}H)$, $(3-^{13}C;3',3',3'-^{2}H)$, $(3'-^{13}C;3',3',3'-^{2}H)$ and $(2-^{13}C;2,2-^{2}H)$ mevalonolactones were synthesized according to the known procedures (6, 7, 8, 9).

Microorganism and culture conditions
<u>Sulfolobus solfataricus</u>, strain MT-4, isolated from an acid hot spring in Agnano, Naples, was grown in the standard medium (10) at 87°C in 90 l batch cultures (pH controlled 3.5, low mechanical agitation, aeration at 3 l/min) inoculated with 9 l of 12 h broth cultures. Labelled precursors were added slowly over a period of 10 h, in the logarithmic phase of growth. Cells were harvested in the stationary phase by continuous flow

centrifugation and lyophilized (yield approx 0.2 g lyophilized cells/l).

Extraction and methanolysis of lipids
Lyophilized cells were extracted continuously (Soxhlet) for 12 h with $CHCl_3/MeOH$ (1:1 v/v). The total lipid extract (approx 8% of dry cells) was treated with methanol/HCl (9:1 v/v) for 6 h under reflux and the methanolysis mixture dried in vacuo.

Purification of tetraether isopranoid lipids
The chloroform soluble fraction of the methanolysis mixture was chromatographed on a silica gel column. $CHCl_3-Et_2O$ (9:1,v/v) eluted the glycerol dialkyl glycerol tetraether mixture (GDGTs) (approx 33 % of chloroform soluble fraction) and $CHCl_3-MeOH$ (95:5, v/v) eluted the glycerol dialkyl nonitol tetraether mixture (GDNTs) (approx. 52 %). The GDGT fraction, was further resolved into single components by HPLC using as solvent n-hexane/ ethyl acetate (6:4 v/v) on a Microporasil column (3.9x30 cm, flow rate 0.5 ml/min for analytical work; 7.8 mm x 30cm, flow rate 5 ml/min for preparative work). Eluants from HPLC were detected with a differential refractometer.

NMR analysis
NMR spectra were recorded on a Bruker spectrometer WH-500. Samples for NMR analysis were prepared by dissolving purified GDGT in 0.4 ml of C^2HCl_3. DEPT (Distorsion Less Enhancement by Polarization Transfer) and Inverse gated experiments were performed using commercially available microprograms. Tetramethylsilane was used as internal standard.

RESULTS AND DISCUSSION

The membrane lipids of Sulfolobus solfataricus are mainly based on glycerol dialkyl glycerol tetraethers (GDGTs) and glycerol dialkyl nonitol tetraethers (GDNTs) that are organized as covalently bound bilayers in which each molecule, when fully stretched, anchors the two polar heads to the inner and outer faces of the membrane array (11).

In this paper the operativity of a series of possible mechanisms of cyclopentane ring formation, as illustrated in Scheme 1, has been evaluated. All the hypotheses taken into consideration require the initial presence of an unsaturation and are possibly concerted with a hydride reduction step.

A series of preliminary experiments with $2-^{14}C$ mevalonolactone, supplied to the cultures of S. solfataricus at concentrations ranging from 2.5 to 21.2 mg/l, allowed us to define optimal conditions for the $^{13}C,^2H$ mevalonolactone incorporation in the lipids. Maximum incorporation (6-8%) was obtained when the precursor was added to the beginning of logarithmic growth phase at a final concentration ranging from 8 to 10 mg/l.

The labelled lipids, extracted from the microorganism grown in the presence of differently $^{13}C,^{2}H$ mevalonolactones, were methanolyzed to obtain GDGTs and GDNTs. Glycerol dialkyl glycerol tetraethers were resolved by silica-gel column and differently cyclized compounds were obtained by preparative HPLC. Tetraether 1 (Fig.1) was choosen for the ^{13}C NMR studies because of structural features and aboundance.

5-^{13}C,5,5-^{2}H mevalonolactone incorporation

The incorporation of 5-^{13}C,5,5-^{2}H mevalonolactone permits the evaluation of the operativity of biogenetic mechanism I, hypothesized in Scheme 1. Carbons 1,1',5,5',9,9' and 13,13' of tetraether 1 are labelled. Multiplicity and one-bond isotopic shift (12) of signals 5', 9 and 9' (Table 1) indicate that cyclopentane ring formation on the unsaturated isoprenic intermediate (Scheme 1) occurs without isomerization of double bonds having taken place on the labelled site of the mevalonate precursor. On the basis of this evidence the operativity of path I (Scheme 1) can be excluded.

3-^{13}C,3',3',3'-^{2}H and 3'-^{13}C,3',3',3'-^{2}H mevalonolactone incorporations

The two biogenetic experiments with these precursors permit us to consider the operativity of II C mechanism (Scheme 1) in the cyclization process.

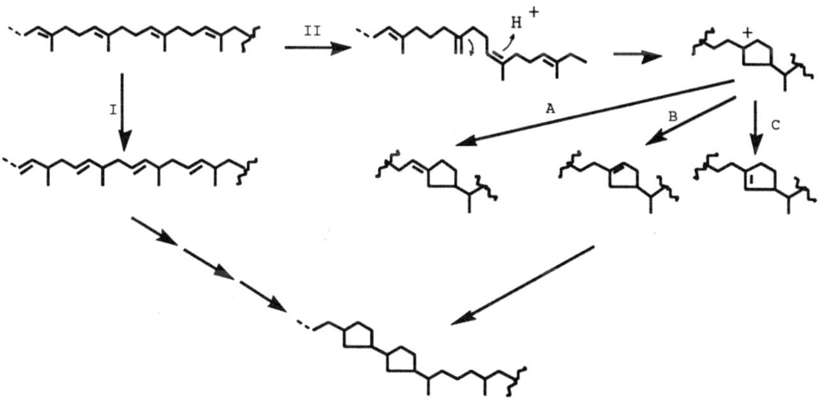

Scheme 1. Possible mechanism of cyclopentane ring formation in tetraether lipids of <u>Sulfolobus solfataricus</u>.

As reported in Table 1 in the experiment with $3-^{13}C, 3',3',3'-^2H$ mevalonolactone, carbons 3,3',7,7' 11,11' and 15,15' are labelled with ^{13}C while carbons 17,17',18,18',19,19',20,20' are deuterated; particularly the two-bond isotopic shift of the carbons 3', 7 and 7' shows a retention of two deuterium atoms on the geminal carbon atoms. However this result is ambiguous if 1,3 shift of deuterium takes place during double bond formation when the mechanism II C (Scheme 1) occurs. An experiment with $3'-^{13}C,3',3',3'-^2H$ mevalonolactone gives more direct information derived from the analysis of one-bond isotopic shift and multiplicity of labelled signals. Positions 17,17',18,18',19,19', 20,20' are labelled with ^{13}C and deuterium. Multiplicity and one-bond isotopic shift of carbons 17', 18 and 18' indicate the presence of two deuterium atoms on these carbons, thus excluding the operativity of mechanism II C.

$2-^{13}C,2,2-^2H$ incorporation

The incorporation experiment carried out with $2-^{13}C,2,2-^2H$ mevalonolactone permits the evaluation of the operativity of route II B. With this precursor, carbons 4,4',8,8',12,12',16,16' are labelled. One bond isotopic shift of these signals and their multiplicities accord with the loss from ^{13}C labelled positions of one or both deuterium atoms. The analysis of Inverse gated ^{13}C NMR spectra indicates, for all labelled positions, that the signals of monodeuterated and non-deuterated enriched carbons (13) are in 1:1 ratio.

The occurrence of an equivalent labelling pattern, both for carbons involved in the cyclopentane ring formation and for carbons in chain, does not support the formation of a double bond as hypothesized in route II B. The loss of one or both deuterium atoms from all labelled sites of the isoprenic chain is due to the biogenetic events from the mevalonate precursor to the isoprenic intermediate, rather than to the cyclization process (14).

Final remarks

Results of the incorporation of $^{13}C,^2H$ differently labelled mevalonolactones allowed the discrimination among the hypotheses taken into consideration in Scheme 1, based on the initial presence of an unsaturation. In fact, from the labelling pattern in these experiments the paths I, II B, and II C are excluded. The remaining route II A can be regarded as a more probable mechanism for cyclopentane ring formation; however, other mechanisms based, for example on hydroxylation of isoprenic chains, can not be "a priori" excluded.

TABLE 1

Labelling pattern and ^{13}C NMR data for tetraether 1, see Figure 1, using strategically labelled ^{13}C,^{2}H mevalonolactones as precursors of tetraether lipids in Sulfolobus solfataricus

Precursor	Labelled carbons	δ(ppm)	$^{n}\Delta$C(D) (ppm)	M*
1)	5	25.04	0.81	p
5-^{13}C,5,5-^{2}H	5'	29.66	0.77	p
mevalonolactone	9	30.43	0.83	p
	9'	31.02	0.78	p
	13, 13'	23.53	0.83	p
2)	3	29.73	0.19	s
3-^{13}C,3,3,3-^{2}H	7	39.07	0.17	s
mevalonolactone	11, 11'	38.02	0.23	s
	15, 15'	32.97	0.24	s
	3'	36.74	0.18	s
	7'	45.49	0.16	s
3)	17, 20, 20'	19.03	0.90	h
3'-^{13}C,3',3',3'-^{2}H	17'	38.73	0.78	p
mevalonolactone	18	35.17	0.72	p
	18'	33.98	0.74	p
	19, 19'	16.83	0.91	h
4)	4	36.37	0.40	t
2-^{13}C,2,2-^{2}H	4'	31.44	0.40	t
mevalonolactone	8	32.89	0.41	t
	8'	31.95	0.41	t
	12, 12'	35.30	0.43	t

n= number of bonds from ^{2}H to ^{13}C. n=1, for precursors 1, 3 and 4; n=2 for precursor 2.

* M= multiplicity (s= singlet, t=triplet, p=pentet, h=heptet)

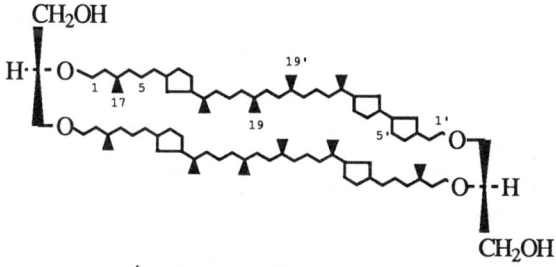

Fig. 1. Chemical structure of tetraether 1.

ACKNOWLEDGEMENTS

The authors thank I. Romano and R. Turco for technical assistance.

REFERENCES

1. De Rosa, M. and Gambacorta, A., Lipid Biogenesis in Archaebacteria. In Archaebacteria '85, eds., O. Kandler and W. Zillig, Gustav Fischer Verlag, Stuttgart, 1986 pp. 278-85.
2. Langworthy, T.A., Lipids of archaebacteria. In The Bacteria,eds., C.R. Woese and R.S. Wolfe, Academic Press, New York, 1985, **8**, pp. 459-91.
3. De Rosa, M., Esposito, E., Gambacorta, A., Nicolaus, B. and Bu' Lock, J.D., Effects of temperature on ether lipid composition of Caldariella acidophila. Phytochemistry, 1980, **19**, 827-31.
4. Furuya, T., Nagumo, T., Itoh, T. and Kaneko, H., The effect of growth temperature on the lipids in extremely thermoacidophilic bacterium, TA-1. Agric.Biol. Chem., 1980, **44**, 517-21.
5. De Rosa, M. Gambacorta, A., Nicolaus, B., Sodano, S. and Bu' Lock, J.D., Structural regularities in tetraether lipids of Caldariella and their biosynthetic and phyletic implications. Phytochemistry,1980, **19**, 833-6.
6. Lawson, J.A., Colwell, T.W., Degraw, J.I., Peters, R.H.,Dehn, R.L. and Tanabe, M., The synthesis of singly and doubly ^{13}C labelled mevalonolactone. Synthesis, 1975, 729-30.
7. Bardshiri, E. and Simpson, T.J.,Studies on a synthesis of (RS) mevalonic acid lactone J. Chem. Soc. Perkin I, 1984, 1765-7.
8. Rosseau, B., Beaucourt, J.P. and Pichat, L.,Syntheses de (R,S) mevalonolactones marquees au ^{14}C ou au ^{13}C: (R,S) mevalonolactone (^{14}C-4,5); (^{14}C-5);(^{14}C-3'). TELE, 1982, 2183-6
9. Ellison, R.A. and Bhatanagar, P.K.,A hight-yield synthesis of DL-mevalonolactone. Synthesis, 1974, 719.
10. De Rosa, M., Gambacorta, A., and Bu' Lock, J.D.,Extremely thermophilic acidophilic bacteria convergent with Sulfolobus acidocaldarius. J. Gen. Microbiol., 1975, **86**, 156-64.
11. Gliozzi, A., Paoli, G., De Rosa, M. and Gambacorta, A., Effect of isoprenoid cyclization on the transition temperature of lipids in thermophilic archaebacteria. Biochim. Biophys. Acta, 1983,**735**, 234-42.
12. Hansen, P.E.,Isotope effects on nuclear shielding. In Annual Report on NMR Spectroscopy Academic Press London 1983, **15**, pp. 108-225.
13. Stothers, J.B., ^{13}C NMR Studies Reaction mechanism and reactive intermediates. In Topics in Carbon-13 NMR Spectroscopy ed., G.C. Levy, John Wiley and Sons, New York, 1974, **1**, pp. 230-78.
14. Caspi, E.,The mode of incorporation of C-2 hydrogen atoms of mevalonic acid into protosterols and sterols. Tetrahedron, 1986, **42**, 3-50.

THE MICROBIOLOGY OF METHANE PRODUCTION IN LANDFILLS

DAVID B. ARCHER AND MICHAEL W. PECK[1]

A.F.R.C. Institute of Food Research, Colney Lane,
Norwich, NR4 7UA, U.K.

[1] Present address: Bioprocess Engineering Research Laboratory
University of Florida, Gainesville, FL 32611, U.S.A.

ABSTRACT

Refuse is commonly disposed of by landfilling and, in the U.K. alone approximately 20Mt is deposited every year. Landfills can cause local nuisance due to odours, hazardous seepage of explosive gas and pollution of groundwaters by acidic leachates. Alternatively, landfilling can be considered as a potentially cost-effective method of contained refuse disposal which could provide up to 5% of the U.K.'s energy needs through methane production. Improved management strategies, or reappraisal of refuse disposal methods, are required to realise this potential. Account must be taken of the microbiology of waste degradation in a heterogeneous environment of high solids content. From microbiological studies of samples from a variety of landfills taken at various depths many functional groups of bacteria have been identified including cellulolytic, sulfate-reducing, methane oxidising and methanogenic species. Our studies have concentrated upon the methanogenic bacteria in landfills. Species of <u>Methanosarcina, Methanothrix, Methanobacterium</u> and a coccoid methanogen have been detected. Techniques based either on serology or assay of coenzyme F_{420} are being developed for the indirect identification and quantification of methanogens in landfill samples.

INTRODUCTION

Landfilling is presently the most common method of domestic refuse disposal in the UK, with about 90% of refuse (~ 20Mt) deposited each year (1). It is also a common method of refuse disposal in other countries; for example, member countries of the European Community dispose an average of approximately 65% of their refuse in controlled landfills (2). Successful landfilling should provide a safe and economic means of refuse disposal but all too often the local environment can be polluted by offensive odours, litter, vermin and acidic leachates. Uncontrolled emissions of methane, often at considerable distances from the landfill itself, introduce a hazard which has had tragic consequences. Nonetheless, with such large quantities of waste being deposited, of which more than 50% is generally regarded to be biodegradable (3,4), there is considerable potential for exploiting such a resource. The potential methane yield is attractive (1) and methane is abstracted and utilised at several landfill sites (2). The fact remains, however, that despite the considerable potential in terms of methane production, yields of methane are often low and erratic. Acidic leachates constitute a loss of reducing equivalents as well as being pollutants of local groundwaters. There is a need to improve landfill management practices and an improved understanding of the microbiology of refuse degradation is first necessary.

Landfills are extreme environments, albeit man-made ones. They are environments of high solids content : emplaced refuse has a moisture content of approximately 20-25% but, within a landfill, there will be relatively dry and wet regions. Waste degradation has been shown in some landfills to be enhanced in regions below the water table (5). Aerobic and anaerobic zones exist within landfills. Upon emplacement, the initial breakdown of refuse is aerobic but oxygen is soon depleted and the environment becomes anaerobic. High temperature (70°C) have been recorded during the initial aerobic phase of degradation (5) but temperatures within the anaerobic regions of landfills are generally in the mesophilic range (20-40°C). From the ecological view, landfills house a variety of environments with gradients of pH, Eh, moisture levels, temperature, nutrient status etc.

Such heterogeneity and the scale of landfills are daunting prospects for microbiological investigation. The areas to be investigated therefore need to be carefully selected and dictated by the needs of the industry.

A major concern must be the survival of pathogens within landfills. High temperatures during the initial degradation of refuse under aerobic conditions are effective at removing pathogenic activity (6, 7). Other aspects have been less well investigated and most questions remain open. For example, there is very little information on the anaerobic flora of landfills, species present, their metabolic roles, the source and size of inoculum, or their yields. Such information could be exploited in devising strategies for enhanced gas production either within exisiting landfills or purpose-built reactors where more control over the process is feasible (8-12). Although the prospect of producing cultures for inoculating landfills is probably unrealistic there is every possibility of designing cultures for treating particular co-disposed wastes.

At present, methods for monitoring the survival or spread of anaerobes in landfills, and the response of anaerobes to different niches within landfills, are poorly developed. In the aerobic regions around landfills methane oxidation is known to occur (13) and could be exploited to restrict migration of methane and so reduce the risk of explosion. Any reduction of methane yield due to anaerobic methane oxidation is unknown.

We report below on some initial studies of landfill microbiology. We have isolated and characterised methanogenic bacteria from a variety of landfill sites and have compared methods for enumeration of methanogens in landfill samples. Enumeration of methanogens by conventional micriobiological methods is prone to error because of the particulate nature of refuse, the tendency of microorganisms to adhere to solid matter, and because methanogens commonly exist in clumps or filaments. Assay of coenzyme F_{420} is a method widely used as an indirect guide to methanogenic biomass in other ecosystems (14,15). We have assessed the applicability of assaying coenzyme F_{420} and have devised an assay for use with landfill samples. An analytical method for estimating methanogenic biomass would find use in laboratory studies of refuse degradation and in the monitoring of reactors or defined regions of a landfill.

MATERIALS AND METHODS

Landfill samples

Samples were supplied by the Landfill Research and Management Section of Harwell Laboratory, U.K. Samples were transported after being packed into anaerobic jars (Oxoid) in an anaerobic environment (Gaspak:Oxoid).

Non-methanogenic bacteria

The presence of aerobic and anaerobic methane oxidising bacteria was investigated. Landfill samples were obtained from Calvert landfill (U.K.) at various depths. The refuse was estimated to be between 2 and 12 months old. 0.5g wet wt. landfill was introduced to 4.5ml medium (13) and serial dilutions then made. Each sample at each dilution was examined in quadruplicate. The head space in the tubes was replaced by CH_4/O_2 (7/3, by vol.) at 1.4atm. pressure. Tubes were incubated at 22°C statically for 53 days. Removal of CH_4 (estimated by gas chromatography) was taken as evidence of methane oxidation. For detection of anaerobic methane oxidation, 50g samples of landfill were added to 400ml medium (16) supplemented with sodium lactate, sodium acetate or yeast extract (each 0.1g/l) as additional carbon sources and Na_2SO_4, $NaNO_3$, $NaNO_2$, $MnCl_2$ or $FeCl_3$ (each 1g/l) as potential electron acceptors. Cultures were overgassed with CH_4 and bromoethane sulfonate (1mmol/l) was added to some cultures to inhibit methanogenesis. After approximately 4 weeks incubation at 37°C samples were serially diluted into tubes containing 4.5ml of the same media and incubated for 64 weeks at 37°C. Removal of CH_4 was estimated by gas chromatography.

Sulfate-reducing bacteria were isolated from Calvert landfill using methods described by Postgate (17). Repeated transfers of isolates from broth to agar - solidified media were required to ensure purity of isolates.

Methanogenic bacteria

The methods used in the isolation and characterisation of methanogenic bacteria from a variety of landfills have been described elsewhere (18).

Assay of coenzyme F_{420}

The extraction of coenzyme F_{420} from cultured methanogenic species and analysis by reversed-phase high performance liquid chromatography (HPLC) was based on the method described previously (19). The elution procedure for the HPLC required modification for assay of the F0 fragment of coenzyme F_{420} and a gradient, rather than an isocratic, solvent elution was employed (20). Purified standards of coenzyme F_{420} analogues were kindly provided by Professor G. Vogels, University of Nijmegen, The Netherlands. Prior to injection into the HPLC extracted samples from either cultured methanogens or landfills were cleaned by passage through C_{18} sep-pak columns (Waters Assoc. Ltd., U.K.). Extracted samples were applied to the Sep-pak column either when the pH had been adjusted to 4.0 by addition of HCl or, alternatively, were resuspended in 100mM sodium acetate (pH 4.0). A wash with glass distilled water eluted material more polar than coenzyme F_{420}, which otherwise interfered with the HPLC separation and detection of coenzyme F_{420}. Coenzyme F_{420} was subsequently eluted from the sep-pak columns in methanol.

The applicability of assaying coenzyme F_{420} as a measure of methanogenic biomass was examined in pure cultures of methanogenic bacteria grown in Met 3 medium (21). <u>Methanosarcina barkeri</u> strain MS and <u>Ms. mazei</u> strain S-6 were studied in detail because these species contain all known analogues of coenzyme F_{420}. <u>Methanobacterium bryantii</u> strain FR2 was used as a representative non-acetotrophic methanogen. The quantification of methanogenic biomass by assay of coenzyme F_{420} was examined in mixtures of known amounts of pure species and in landfill samples spiked with known amounts of cultured methanogens. The assay of coenzyme F_{420} was compared with the enzyme-linked immunosorbent assay (ELISA) of methanogens (22).

Protein was estimated using the Folin reagent (23).

RESULTS

Non-methanogenic bacteria

The total numbers of anaerobic bacteria in a variety of landfill samples, based on dilution series methods, have been reported elsewhere (24). In this paper we report on attempts to detect methane oxidising activity and to isolate sulfate-reducing species.

Aerobic methane oxidation, judged by removal of methane at 99% certainty, was found in 5 out of 16 tubes at the lowest dilution examined (Table 1).

TABLE 1

Aerobic methane oxidation in samples from Calvert landfill (U.K.)

Depth in landfill (m)	Number of tubes showing removal of methane (Number of replicates in parenthesis)
2-3	0 (4)
4-5	0 (4)
7-8	3 (4)
9	2 (4)

There was no evidence for methane oxidation under anaerobic conditions despite the use of a variety of alternative electron acceptors to oxygen, and the presence of several different supplemental carbon sources.

Sulfate-reducing bacteria were obtained in pure culture using the diagnostic media of Postgate (17). As the purpose of this work was to examine landfill samples for the presence of sulfate-reducing bacteria, rather than carrying out a systematic study, detailed characterisation was not performed. All isolates stained Gram-ve and blackened diagnostic media due to sulfate-reduction. Vibrios and rod-shaped organisms were obtained and all the rod-shaped isolates survived heating at 70°C for 10 min.

During the initial procedures for isolating anaerobic methane-oxidising bacteria an obligately anaerobic, non-sulfate-reducing, agar-degrading bacterium was isolated. The original sample of landfill was from Calvert (U.K.) at 9m depth.

The organism grew well in a mineral salts medium containing agar (Ionogar No. 2:Oxoid) as the sole carbon source. The isolate survived 80°C, 10 min. and contained terminal/sub-terminal spores which give rise to swollen ends of the, otherwise, rod-shaped cells. The Gram staining was variable. The isolate is tentatively described as a Clostridium sp. and is shown in Fig. 1.

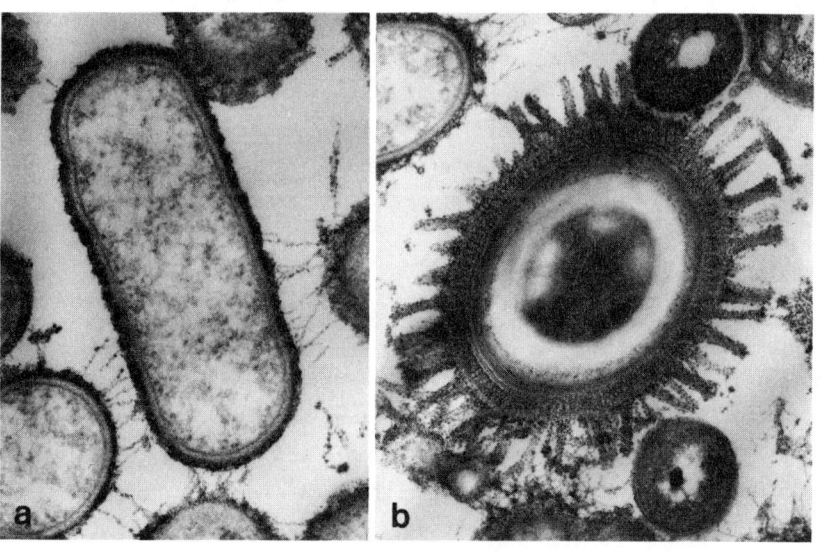

Fig. 1. Agar-degrading isolate : (a) cell, (b) spore.

Methanogenic bacteria

The approximate enumeration of methanogens in landfill samples (24,25) and isolation of methanogens (18) have been documented previously. Other isolates and methanogenic enrichments have since been obtained. Species obtained have been similar to those reported (18) except that enrichments containing organisms resembling Methanothrix have been produced. Species obtained in pure culture and characterised by conventional methods (26) including serology are listed in Table 2.

TABLE 2

Methanogenic bacteria isolated from landfills.
Adapted from Fielding et al. (18).

Isolate	Source landfill (U.K.)	Designation
EF1	Aveley, Essex	Methanobacterium formicicum
EF2	Aveley, Essex	coccoid methanogen
EF3	Stangate, Kent	Methanobacterium bryantii
EF4	Laboratory enrichment	Methanobacterium bryantii
EF5	Aveley, Essex	Methanosarcina barkeri
EF6	Enderby, Leicestershire	Methanobacterium sp.
EF7	Blue Circle, Kent	Methanobacterium sp.

Assay of coenzyme F_{420}

The structures of coenzyme F_{420} and various degradation fragments or analogues are illustrated in Fig. 2. The structure of coenzyme F_{420} extracted from Methanobacterium bryantii was elucidated by Eirich et al. (27). This molecule contains 2 glutamic acid residues in the side chain (hence coenzyme F_{420}-2). Since then, coenzyme F_{420} molecules with different numbers of glutamic acid residues (coenzymes F_{420}-3, F_{420}-4, F_{420}-5) have been identified (28).

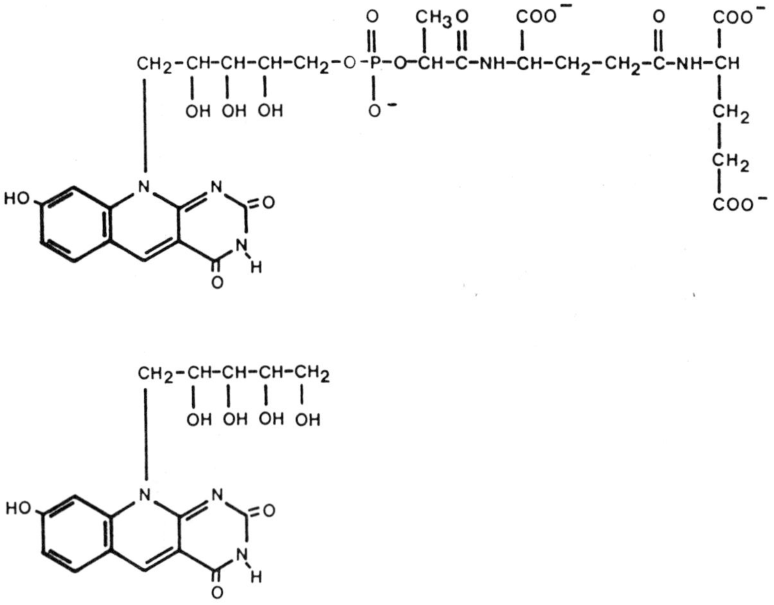

Fig. 2 Structure of coenzyme F_{420}-2 (top) and F0 (bottom). From Eirich et al. (35).

The successful result of the clean-up procedure using C_{18} sep-pak columns is illustrated in Fig. 3. UV absorbing and fluorescent material which elutes early in the run, and which obscures the fluorescence due to coenzymes F_{420}-4 and F_{420}-5, is removed from extracts of pure methanogenic cultures (Fig. 3) and landfills (not shown). During exponential growth of Ms.mazei or Ms. barkeri the total coenzyme F_{420} content (per wt. cells) remained constant (Table 3).

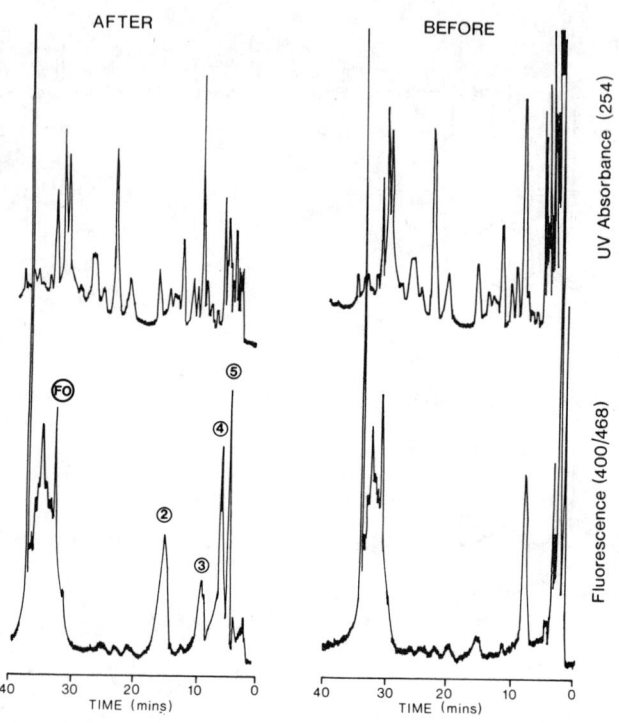

Fig. 3. UV absorbance (245nm) and fluorescence (400/468nm) from an HPLC chromatograph of coenzymes F_{420} from Ms. mazei S-6 before and after clean up on a C_{18} sep-pak column. A small amount of purified enzymes F_{420}-4 and F_{420}-5 was added to the extract prior to clean-up. During exponential growth of Ms.mazei or Ms.barkeri the total coenzyme F_{420} content (per wt. cells) remained constant (Table 3).

TABLE 3

Total coenzyme F_{420} content of Methanosarcina spp.

Species	Substrate	Coenzyme F_{420} nmol/g protein
Ms. barkeri strain MS	methanol	530
Ms. barkeri strain MS	$H_2 + CO_2$	474
Ms. mazei strain S-6	methanol	230

Cell lysis which ensued in stationary phase released coenzyme F_{420} into the medium. Results are presented for Ms. mazei S-6 grown with methanol (Fig. 4).

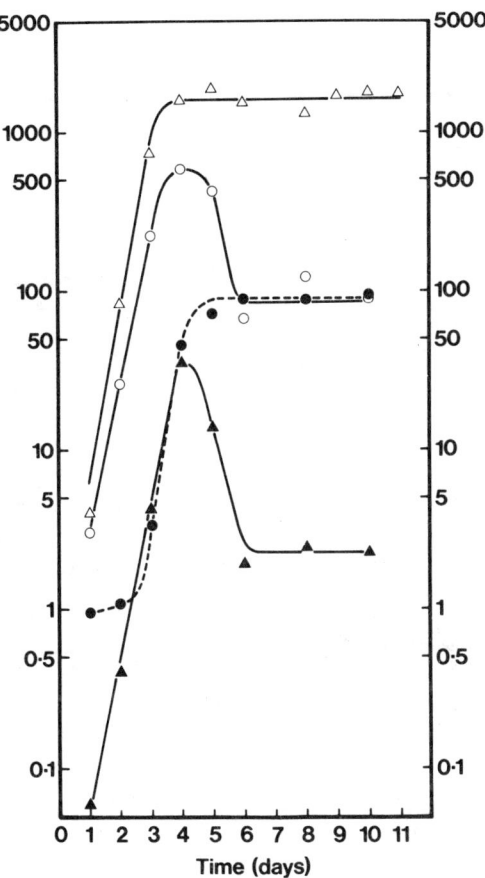

Fig. 4. Growth and coenzyme F_{420} content of Ms. mazei S-6 grown in 400ml cultures with methanol as substrate. △, accumulated methane (mmol per culture x 100); ○, cell protein (mg per culture x 10), ▲, intracellular total coenzyme F_{420} (nmol per culture); ●, extracellular total coenzyme F_{420} (nmol per culture).

Although the total cellular coenzyme F_{420} content remained constant during growth the relative abundance of analogues varied (Fig. 5). The relative abundance of FO increased during stationary phase and it finally represented 28% of the total cellular coenzyme F_{420}. This was despite most of the FO (>97%) always being extracellular.

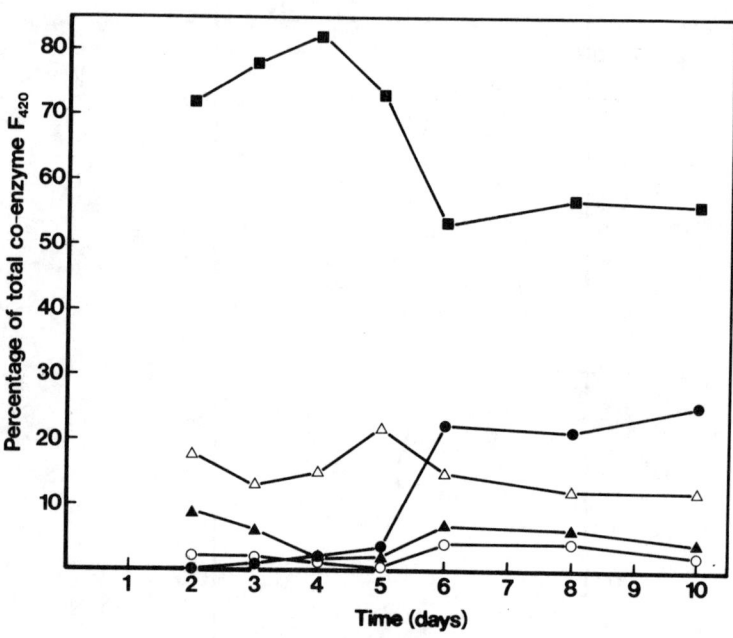

Fig. 5 Relative abundance of the intracellular coenzyme F_{420} analogues of <u>Ms.mazei</u> S-6 grown on methanol in batch cultures. ■, coenzyme F_{420}-2; Δ, coenzyme F_{420}-3; ▲, coenzyme F_{420}-4; ○, F_{420}-5; ●, FO fragment.

These results and others (not shown) provide confidence that the measurement of total coenzyme F_{420}, if not individual analogues, can be a guide to the total methanogenic biomass. Different authors, however, have published widely differing values for the specific content of total coenzyme F_{420} in methanogens.

This reflects different techniques in extracting and quantifying coenzyme F_{420} but also reflects the dependence of coenzyme F_{420} content on cultural conditions. In this study, a relatively small difference in the total content of coenzyme F_{420}, and content of individual analogues, was noted when Ms.barkeri strain MS was cultivated on methanol or $H_2 + CO_2$. Others have reported similar results (28,29). With this in mind, assay of coenzyme F_{420} might have value in estimating to an order of magnitude the methanogenic biomass in samples, such as those from landfills, which are not easily amenable to other analyses.

Known quantities of Ms.mazei S-6 and Mb.bryantii FR2 were added either singly or together to landfill samples. FO was added as internal standard. Samples were extracted and assayed for coenzyme F_{420} content. Coenzyme F_{420}-3 was taken as a measure of Ms. mazei because Mb. bryantii lacks this analogue. A detailed description of the experimental procedure will be presented elesewhere. Recovery of the added FO was 43.0 ± 6.5 (mean ± s.d.)% and recoveries of expected levels of added coenzyme F_{420} analogues was lower. Improvements in the method are anticipated so that much better recoveries can be expected in line with that experienced in similar work with samples from anaerobic digesters (29). Recoveries estimated by ELISA were less than 10%. Such a low recovery is not surprising with such a technique designed for use with homogeneous, well dispersed samples.

DISCUSSION

Environmental conditions within landfills vary widely and it should not be assumed that the biology of waste degradation in landfills will be entirely analogous to other better studied environments. The high solids content and extremes of available nutrients, pH, Eh, and, to a lesser extent, temperature may encourage a flora peculiar to landfills. The codisposal of toxic chemicals may introduce local perturbations to the flora. The obligate anaerobes are probably introduced to the landfill with the refuse which is largely aerobic in origin. The size of the original inoculum of anaerobes is unknown but may be small and its composition may be affected by the oxygen tolerance of the organisms, although anaerobes in association with facultative organisms can survive in apparently aerobic environments.

Microbiological studies of landfills have been mainly concerned with upper aerobic regions (30). In the anaerobic zone little work has been done although microbial activity has been assessed through associated enzyme assays (31). Although cellulolytic bacteria have been isolated from landfills (32) there has not been an extensive study. Landfills may lack the anaerobic cellulolytic fungi and protozoa found in the rumen although rumen protozoa have been exploited in acidogenesis from refuse (11). We have demonstrated the presence of sulfate-reducing species and methanogenic species in landfills. Aerobic methane-oxidizing species have been reported previously (13) and although we have confirmed their finding we could not detect anaerobic methane oxidation. This may reflect the difficulties in assessment rather than indicating a true lack of anaerobic methane oxidation and more thorough investigation is required.

Hydrogen sulfide is not normally a significant component of landfill gas (1). Sulfate-reducing bacteria might still be expected to be active, however, when levels of sulfate encourage their growth or, otherwise, as acetogens. Precipitation of insoluble metal sulfides within a landfill is likely to restrict the egress of gaseous hydrogen sulfide. The methanogenic bacteria isolated indicate that landfills house a number of different species which include hydrogenotrophic and acetotrophic organisms (18,33). In that respect landfills are similar to many other heterogeneous anaerobic environments. The peculiarities of the landfill envirionment may well encourage the proliferation of particular species or may favour microbial associations not necessarily common in other environments. If so, the impact on refuse degradation could have consequences for landfill management.

Assay of coenzyme F_{420} has been commonly used as an indicator of methanogenic biomass in anaerobic digesters. Many of the assays, used, however, do not specifically assay coenzyme F_{420} and, at best, coenzyme F_{420} can only be a guide to the methanogenic biomass.

In conjunction with assay of other methanogen-specific cofactors the assay of coenzyme F_{420} becomes a more powerful approach (28,29,34) and one which can be performed in analytical laboratories not equipped to culture methanogens. Environmental constraints and the attractions of realising the energy potential of landfills necessitate stricter control over refuse degradation. Closer monitoring of the process will be required and assay of methanogenic biomass can then be useful. Notwithstanding the problems of obtaining representative samples, the high solids content of landfill samples renders assay of methanogens by conventional microbiological or serological (22) techniques prone to error. Developments of the technique for assay of coenzyme F_{420} described in this paper may provide the means for the simple assay of methanogenic biomass in landfill samples.

CONCLUSIONS

1. Landfills are heterogeneous environments of high solids content. They are characterised by extremes of nutrient content, pH, Eh and, to a lesser extent, temperature. A single sample for analysis does not represent a landfill.

2. Landfills harbour a diversity of aerobic and anaerobic species of bacteria. Among the anaerobes are cellulolytic species, sulfate-reducing species and methanogens.

3. The quantification of methanogenic bacteria by assay of coenzyme F_{420} may be appropriate with landfill samples because of their high solids content. Quantification of methanogens in landfills by conventional microbiological techniques is inaccurate.

4. Assay of coenzyme F_{420} requires separation of the analogues from other fluorescent material which otherwise interferes in the assay. However, the cellular content of coenzyme F_{420} analogues varies with the state of growth of the methanogen although the specific content of total coenzyme F_{420} remains relatively constant during exponential growth.

5. In studies of pure cultures of methanogens the FO fragment of coenzyme F_{420} is released into the medium and is therefore unsuitable as an internal control. In mixed cultures the turnover of FO may prevent its accumulation.

ACKNOWLEDGEMENTS

We thank the U.K. Department of the Environment and AERE Harwell for financial support. We thank D.J.V. Campbell and J.R. Emberton for providing landfill samples and for many discussions. We are grateful to G. Vogels for providing authentic samples of coenzyme F_{420} and S.Clark and M. Parker for electron microscopy.

REFERENCES

1. Richards, K.M., Landfill gas exploitation - demonstration schemes in the U.K.. In Energy from Landfill Gas, eds, J.R. Emberton and R.F. Emberton, A.E.R.E Harwell, U.K., 1986, pp 267-274.
2. Senior, E. and Balba, M.T.M., Landfill biotechnology. In Bioenvironmental Systems, vol. 2, ed. D.L. Wise, CRC Press, Boca Raton, USA, 1987, pp.17-65.
3. Bell, J.M., Characteristics of municipal refuse. Proceedings of the National Conference on Solid Waste Research, American Public Works Association, 1963, Special Report No. 29.
4. Sumner, J., The storage and collection of refuse - methods, practices, technical developments and trends - an international survey. In Reports of the 9th International Conference of INTAPAC (Paris) 1967, 1-65.
5. Rees, J.F., Optimisation of methane production and refuse decomposition in landfills by temperature control. J. Chem. Tech. Biotechnol. 1980, 30,458-465.
6. Anderson, R.J., The public health aspect of solid waste disposal. Public Health Report 1964,79, 93-96.
7. Hanks, T.G., Solid waste/disease relationships; a literature survey. Public Health Service Publication No. 999-UIH-6, U.S. Govt. Printing Office, 1967.
8. Pfeffer, J.T., Evolution of the RefCOM system, In Proceedings of the 1st Symposium on Biotechnological Advances in Processing Municipal Wastes for Fuels and Chemicals, ed. A.A. Antonopoulos, Argonne National Laboratory, U.S.A., 1985, pp. 331-341.
9. DeBaere, L. and Verstraete, W., Anaerobic fermentation of semi-solid and solid substrates, Commission of European Communities Report EUR9347, 1984, pp. 195-208.
10. VALORGA, Z.I. de Vendargues, F-34740, France.

11. Gijzen, H.J., Zwart, K.B., van Gelder, P.T. and Vogels, G.D., Continuous cultivation of rumen microorganisms; a system with possible application to the anaerobic degradation of lignocellulosic waste material. Appl. Microbiol. Biotechnol. 1986, **25**, 155-163.
12. Gijzen, H.J., Zwart, K.B., Verhagen, F.J.M. and Vogels G.D., High-rate two-phase process for the anaerobic degradation of cellulose, employing rumen microorganisms for an efficient acidogenesis. Biotech. Bioeng. 1988, **31**, 418-425.
13. Mancinelli, R.L., and McKay, L.P., Methane-oxidising bacteria in sanitary landfills, In Proceedings of the 1st Symposium on Biotechnological Advances in Processing Municipal Wastes for Fuels and Chemicals, ed. A.A. Antonopoulos, Argonne National Laboratory, U.S.A., 1985, pp.437-450.
14. Delafontaine, M.J., Naveau, H.P. and Nyns, E.J., Fluorimetric monitoring of methanogenesis in anaerobic digesters. Biotech. Lett. 1979, **1**, 71-74.
15. Van Beelen, P., Dijkstra, A.C. and Vogels, G.D., Quantitation of coenzyme F_{420} in methanogenic sludge by the use of reversed-phase high performance liquid chromatography and a fluorescence detector. Eur. J. Appl. Microbiol. Biotechnol. 1983, **18**, 67-69.
16. Panganiban, A.T., Patt, T.E., Hart, W. and Hanson, R.S., Oxidation of methane in the absence of oxygen in lake water samples. Appl. Environ. Microbiol. 1979, **37**, 303-309.
17. Postgate, J.R., The Sulphate-Reducing Bacteria, 2nd ed., Cambridge University Press, Cambridge, U.K., 1984.
18. Fielding, E.R., Archer, D.B., Conway de Macario, E. and Macario, A.J.L., Isolation and characterization of methanogenic bacteria from landfills. Appl. Environ. Microbiol. 1988, **54**, 835-836.
19. Peck, M.W. and Archer, D.B., Improved assay of coenzyme F_{420} analogues from methanogenic bacteria. Biotechnol. Tech. 1987, **1**, 279-284.
20. Peck. M.W., in preparation.
21. Davis, R.P. and Harris, J.E., Spontaneous protoplast formation by Methanosarcina barkeri. J. Gen. Microbiol. 1985, **131**, 1481-1486.
22. Kemp, H.A., Archer, D.B. and Morgan, M.R.A., Enzyme-linked immunosorbent assays for specific and sensitive quantification of Methanosarcina mazei and Methanobacterium bryantii. Appl. Environ. Microbiol. 1988, **54**, 1003-1008.
23. Lowry, O.H., Rosebrough, N.J., Farr, A.L. and Randall,R.J., Protein measurement with the Folin phenol reagent. J. Biol. Chem. 1951, **193**, 265-275.
24. Fielding, E.R. and Archer, D.B., Microbiology of landfill. Identification of methanogenic bacteria and their enumeration. Inst. Chem. Eng. Symp. Ser. 96, 1986, 331-341.
25. Campbell, D.J.V., Fielding, E.R. and Archer D.B., Understanding refuse decomposition processes to improve landfill gas energy potential. In Proceedings of 3rd E.C. Conference Energy from Biomass, eds. W Palz, J. Coombs, and D.O. Hall, Elsevier Applied Science Publishers, Amsterdam, The Netherlands, 1985, pp. 1151-1155.
26. Boone, D.R. and Whitman, W.B., Proposal of minimal standards for describing new taxa of methanogenic bacteria. Int. J. Syst. Bacteriol. 1988, **38**, 212-219.

27. Eirich, L.D., Vogels, G.D. and Wolfe, R.S., Proposed structure for coenzyme F_{420} from Methanobacterium. Biochemistry 1978, **17**, 4583-4593.
28. Gorris, L.G.M. and van der Drift, C., Methanogenic cofactors in pure cultures of methanogens in relation to substrate utilization. In Biology of Anaerobic Bacteria, eds. H.C. Dubourguier, G. Albagnac, J. Montreuil, C. Ramond, P. Sautiere and J. Guillaume, Elsevier Applied Science Publishers, Amsterdam, The Netherlands, 1986, pp. 144-150.
29. Gorris, L.G.M., Kemp, H.A. and Archer, D.B., Quantification of methanogenic biomass by enzyme-linked immunosorbent assay and by analysis of specific methanogenic cofactors. Biomass, 1987, **14**, 195-208.
30. Archer, D.B. and Robertson, J.A., The fundamentals of landfill microbiology, In Energy from Landfill Gas, eds. J.R. Emberton and R.F. Emberton, A.E.R.E. Harwell, U.K. 1986, pp.116-122.
31. Jones, K.L., Rees, J.F. and Grainger, J.M., Methane generation and microbial activity in a domestic refuse landfill site, Eur. J. Appl. Microbiol. Biotechnol. 1983, **18**, 242-245.
32. Bagnara, C., Toci, R., Gaudin, C. and Belaich, J.P., Isolation and characterization of a cellulolytic microorganism, Cellulomonas fermentans sp. nov., Int. J. Syst. Bacteriol. 1985, **35**, 502-507.
33. Mouton, C., Beckelynck, J., Albagnac, G. and Dubourguier, H.C., Production et récupération de biogaz produit par les ordures ménagères enfouies en décharge. T.S.M. L'eau, 1985, **9**, 391-404.
34. Gorris, L.G.M., de Kok, T.M., Kroon, B.M., van der Drift, C. and Vogels, G.D., Relationships between methanogenic cofactor content and maximum specific activity of anaerobic granular sludge. Appl. Environ. Microbiol, 1988, **54**, 1126-1130.
35. Eirich, L.D., Vogels, G.D. and Wolfe, R.S., Distribution of coenzyme F_{420} and properties of its hydrolytic fragments. J. Bacteriol. 1979, **140**, 20-27.

METHANOGENESIS AND REDUCTIVE DECHLORINATIONS IN AN ALKALINE, HYPERSALINE SEDIMENTS AND GROUNDWATER

DAVID R. BOONE[1], RICHARD L. JOHNSON[1],
DEBORA C. CHEN[1], INDRA M. MATHRANI[2], AND ROBERT A. MAH[2]
Environmental Science and Engineering, Oregon Graduate Center,
Beaverton, OR 97006-1999, USA[1], and School of Public Health,
University of California, Los Angeles, CA 90024, USA[2]

ABSTRACT

Sediments from an alkaline, hypersaline lake (Alkali Lake, Oregon, USA) and nearby groundwaters were examined for the presence and activity of methanogenic, sulphate-reducing, and acetigenic bacteria. Like the methanogen *Methanohalophilus zhilinae*, lake sediments produced CH_4 from trimethylamine, but not from acetate or H_2. *Methanohalophilus zhilinae*, a methanogen isolated from a similar environment (Wadi el Natrun, Egypt), was unable to use H_2 in the presence of CO_2, methanol, and trimethylamine; activities of enrichment cultures derived from West Alkali Lake indicate that physiologically similar methanogens may exist there. H_2 disappeared from enrichment cultures when sulphate was present, indicating the presence of sulphate-reducing bacteria. A chemical disposal site near Alkali Lake adds chlorophenols and chlorophenoxyphenols to the groundwater. These compounds are readily detectable in the upper, oxic groundwater, but in the anoxic zone the concentration drops more rapidly than their diffusion and dispersion would predict. This suggests that the chlorinated compounds may be biologically degraded in the anoxic subsurface. The partial pressure of H_2 in samples from the anoxic zone is high, and may be related to the rapid degradation of chlorinated phenolic compounds.

INTRODUCTION

Catabolic pathways in alkaline, hypersaline lakes and groundwaters have not been systematically examined. These lakes often have high concentrations of soluble organic matter, indicating that degradative reactions may be inhibited [1]. From these and similar environments, methanogens have been enriched which catabolize methyl-containing substrates such as methanol, trimethylamine, dimethylsulfide, and methane thiol [2-4], but not H_2, formate, or acetate. The lack of halophilic methanogens able to degrade non-methyl-containing substrates may be due to competition with sulphate-reducers, which generally have access to excess quantities of sulphate in hypersaline environments [5]. Thus, the non-competitive use of methyl-containing compounds by methanogens [5,6] may be the source of methane in those ecosystems.

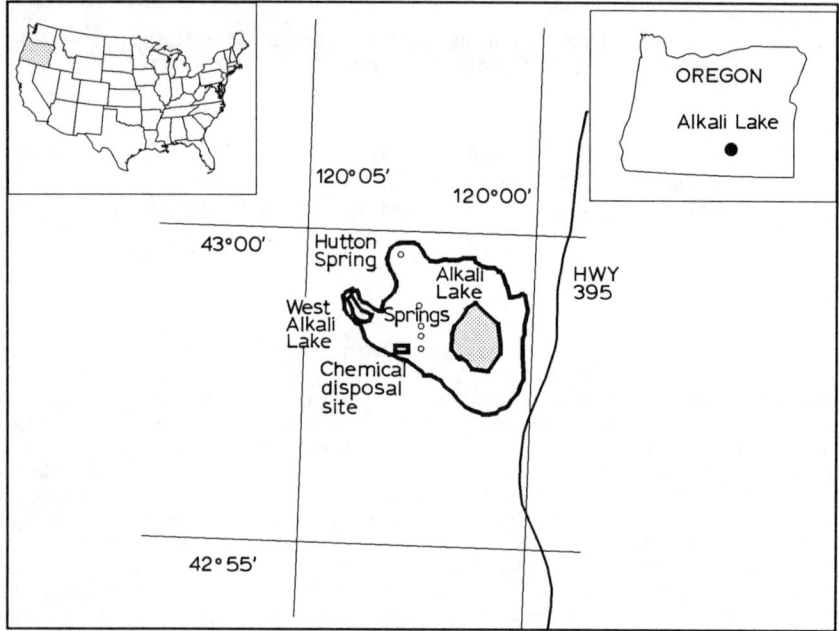

Figure 1. Location of Alkali Valley, Oregon, USA

Site Description

Alkali Lake and West Alkali Lake are hypersaline, alkaline lakes in the high desert of south central Oregon, USA (fig. 1). In former times Alkali Lake and West Alkali Lake were a single lake, but are now separated by a narrow, shallow valley in which the water table is typically within 1 m of the ground surface. The source of some of the near-surface water is a group of artesian springs located near the east end of the connecting valley. These springs deliver fresh water to both Alkali Lake and West Alkali Lake through groundwater flow.

The lakes are high in Na_2CO_3; they have a salinity of 90 to 120 g l^{-1} and pH of 9.7 to 10.1. The lakes and groundwater contain large amounts of the dissolved cations sodium (5.1 M) and potassium (0.23 M), with carbonate and bicarbonate (1.6 M, total concentration), chloride (1.3 M), and sulphate (0.48 M) being the major anions [7].

Between 1969 and 1971 more than 25,000 55-gallon drums (206 l) containing chlorophenols (CP), chlorphenoxyphenols (CPP), and other herbicide manufacturing wastes were deposited in the shallow valley just west of the springs (fig. 2). In 1976, the drums were placed in shallow trenches, crushed, and covered with gravel. The contents contaminated the groundwater, creating a plume which currently extends approximately 0.5 km west of the chemical disposal site, moving westward at approximately 10 cm day^{-1} [8].

MATERIALS AND METHODS

Collection of Samples

Sediment samples from West Alkali Lake were collected from within 2 m of the shore, indicated in figure 2. We filled the sample bottles with lake water, and then added sediments obtained between 3 and 6 cm deep (from the top of

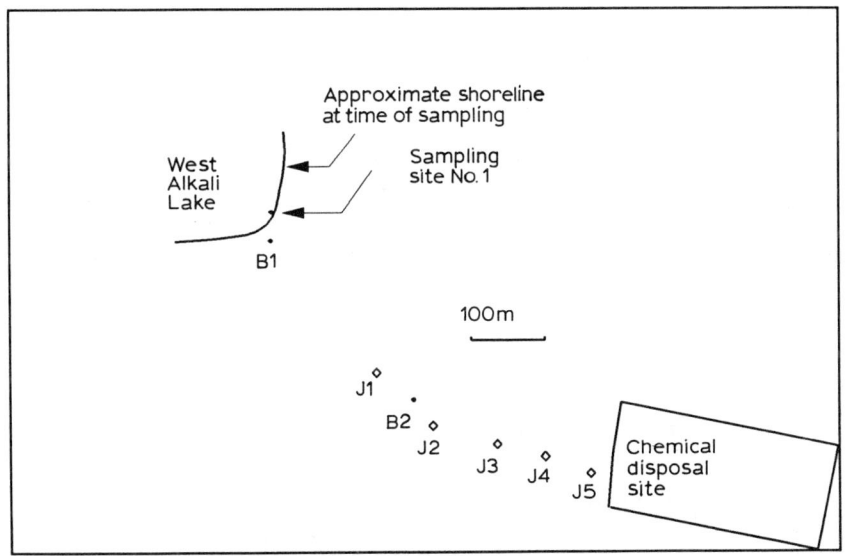

Figure 2. Location of sampling sites.

the sediment). The bottles were completely filled with water and sealed with butyl rubber stoppers. The sediment sample collected on July 15, 1988, had a temperature of 29°C.

For microbiological analysis, O_2-free slurries of subsurface solids and groundwater were collected on the same day. We hand-augured a 100-mm diameter well approximately 2.8 m deep and installed a casing of polyvinyl chloride pipe without a screen and open at the bottom. The well was emptied of water, and we waited 30 min for the water level in the well to recover. To collect the sample, we recirculated water from the bottom of the well to a point about 1 m above the bottom, thereby creating an O_2-free slurry of groundwater and subsurface solids. The temperature of this recirculating stream was 21°C at each of the two wells where samples were collected. Serum bottles (161-ml internal volume) were filled with slurry. We continued adding the slurry after the bottles were full, thereby displacing the original suspension to ensure elimination of O_2. Some gas bubbles formed in the suspension, probably because decompression of the samples released dissolved gas. The serum bottles were sealed with butyl rubber stoppers held in place with aluminum crimp seals. After settling, the solids occupied about one half of the volume in the collection vials.

Samples of sediments and subsurface slurries were held at room temperature and returned to the laboratory.

Subsurface cores were collected by hand-auguring and analyzed for CPs and CPPs in September, 1982. The soil cores were extracted with a mixture of methanol and water (1:1) followed by a sequential liquid-liquid extraction into methylene chloride. Internal standards were added to track the extraction efficiency. These cores were taken 4, 25, and 50 m from the west (down-gradient) boundary of the chemical disposal site.

Subsurface water samples were obtained from multi-level wells installed along the down-gradient path from the chemical disposal site (fig. 2).

Culture Techniques and Media
Culture techniques included syringe modifications [9] and serum-tube modifications [10] of the Hungate technique [11]. Culture medium was the same as that described previously [12] except that $NaHCO_3$ and $NaCl$ concentrations were modified in the following way. For MSH medium, the $NaHCO_3$

concentration was decreased to 83 mM and NaCl was increased to 0.5 M. For MS medium, $NaHCO_3$ was 50 mM and NaCl was 1.5 M. The gas phase was N_2 and the final pH was about 8.3. Catabolic substrates were added as sterile, O_2-free solutions or gases were added with over-pressure after inoculation.

Analytical Methods
Methane was determined by gas chromatography with thermal conductivity detection. H_2 was determined with a reduction gas analyzer (Trace Analytical, Palo Alto, Calif.); this method separates gases by gas chromatography and detects reducing gases such as H_2 by their reduction of heated mercuric oxide; reduced mercury (vapor) is determined by atomic absorption. Dissolved O_2 was determined by an O_2-monitor (Yellow Springs Instruments model 57, Yellow Springs, Ohio). Mono- and di-chlorophenols were determined by high-resolution, capillary-column gas chromatography with split injection and flame-ionization detection. CPPs were determined by gas chromatography with on-column injection and mass spectroscopy detection [8,13]. A Zeiss Universal Research microscope was used for epi-fluorescence microscopy with an O2 filter set (excitation spectrum with a peak at 365 nm and cutoff at 395 nm, and a 420 nm long-pass barrier filter).

RESULTS

H_2-using and Trimethylamine-using Bacteria
We examined an O_2-free slurries of alkaline, hypersaline waters and solids for the presence of H_2-using bacteria. Slurries from the lake sediment and from well B1 were tested. These slurries were re-suspended by inverting the vials and then diluted (1:10) in sterile culture medium. Each diluted suspension was used to inoculate various enrichment media (0.1 ml of the dilution, equivalent to 0.01 ml of the original suspension). Enrichment media contained (1) H_2, (2) H_2 and sulphate, (3) trimethylamine, and (4) trimethylamine and H_2. All cultures (inoculated with B1 or lake-sediment slurry) in media with trimethylamine produced methane, and no methane was found after incubation of the cultures without trimethylamine. This indicated the presence of trimethylamine-fermenting methanogens and demonstrated the lack of H_2-oxidizing methanogens. H_2 disappeared only in cultures which also had sulphate, indicating the presence of H_2-using sulphate-reducers.

The the enrichment cultures with trimethylamine as substrate were serially diluted and inoculated into roll-tube medium with trimethylamine as substrate. Thus far, colonies have only developed in media from the lake-sediments enrichment culture. Small (< 1 mm in diameter) epi-fluorescent colonies developed after 25 days incubation at 37°C. These were picked, diluted serially, and re-inoculated into roll-tube media for purification, which is continuing. Microscopic examination of the colony contents revealed very small (< 1 μm in diameter) coccoid cells. No motility was evident. The behaviour of enrichment cultures is consistent with the presence of bacteria similar to *Methanohalophilus zhilinae*, although the cells seen here were much smaller. *M. zhilinae*, like the methanogens present in our enrichment cultures, are unable to catabolize H_2, and can only grow by using methylamines (or perhaps methylmercaptans) as catabolic substrate.

H_2 Partial-pressure in Subsurface Slurries

H_2 was measured after 1 day equilibration of a small amount of headspace gas in 8 l of subsurface slurry from well B1. The partial pressure of H_2 was approximately 11 Pa.

Vertical Distribution of O_2 in the Groundwater Between the Lakes

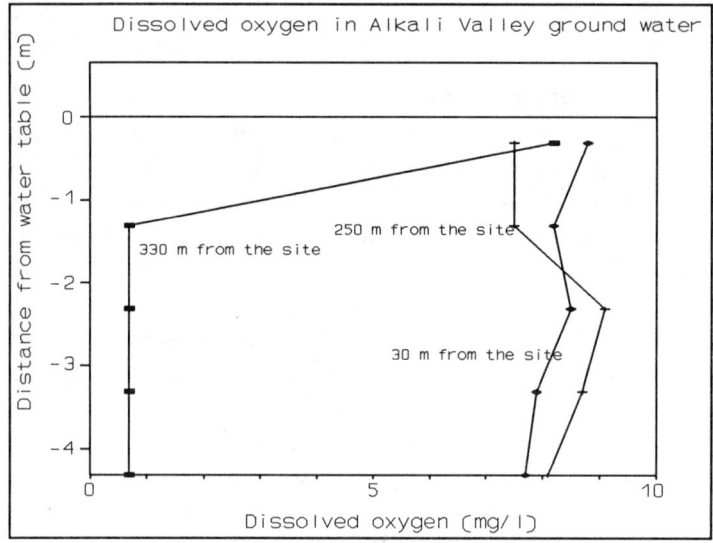

Figure 3. Vertical distribution of O_2 in ground water 30 m, 250 m, and 330 m from the disposal site.

Figure 3 shows the distribution of O_2 in the subsurface water at wells J5, J2, and J1 (30, 250 m and 330 m from the disposal site, respectively). In the more distant well, O_2 declined with depth; samples below 2 m had less than 1 mg of O_2 per litre, and were probably completely anoxic. However, in the wells nearer to the disposal site, O_2 penetrated more deeply, probably because toxic compounds present in the waste inhibited microbial activity and O_2 utilization. CPPs are known to be toxic to microbial activity, and the only commercially produced CPP, Irgasan DP300, is marketed as a bactericide.

Vertical Distribution of CPs and CPPs in Groundwater Between the Lakes

Cores were taken 4 m, 25 m, and 50 m west (down-gradient) from the western boundary of the chemical disposal site in 1982. Figure 4 shows the depth profile of 2,4-dichlorophenol and 2,4,6-trichlorophenol at well 37 (4 m from the site boundary). The maximum concentration of each of these compounds was found between 1 and 1.6 m below the water table. The profile of these same contaminants 25 m from the site is shown in figure 5. This profile

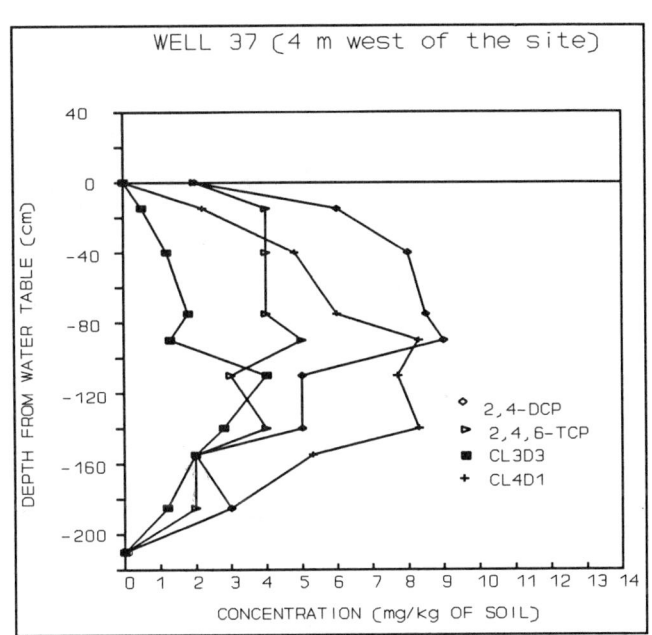

Figure 4. Vertical distribution of contaminants 4 m west of the site.

shows a distinct decrease in concentration of CPP at depths below about 1 m. This decrease was not expected based on diffusion and dispersion of the contaminants, and is consistent with microbial biodegradation in the anoxic zone. Figure 6 also shows decreases in concentration in the anoxic zones which may be accounted for by anaerobic biodegradation. This possibility is being investigated in experiments with subsurface slurries from well B1. These slurries are being incubated in serum bottles under anoxic conditions with 2,4-dichlorophenol or Irgasan DP300 added. We are monitoring the disappearance of these compounds and the appearance of expected biodegradation products.

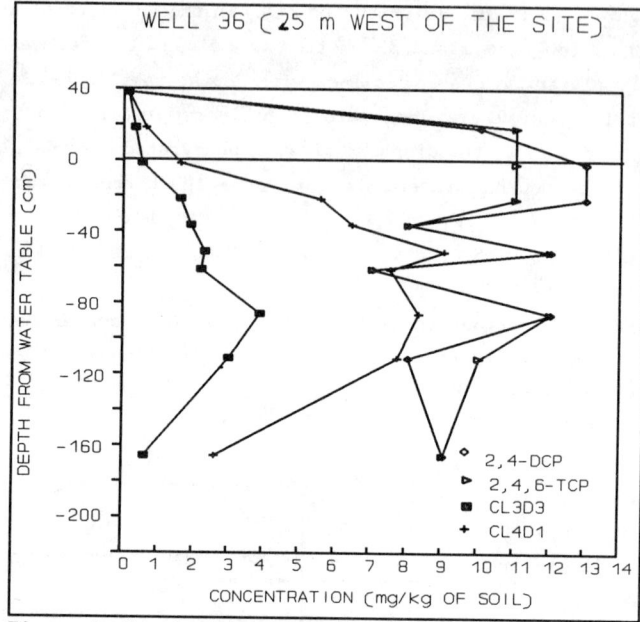

Figure 5. Vertical profile of contaminants at well 36, 25 m west of the site.

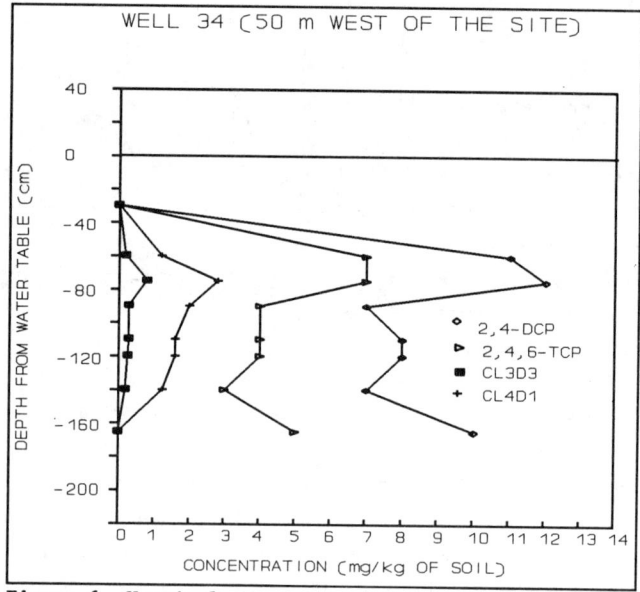

Figure 6. Vertical distribution of contaminants 50 m west of the site.

DISCUSSION

The dechlorination of CPs and the depolymerization of CPPs are reductive reactions. The ability of microbes to carry out such reactions may depend on the availability of reducing equivalents in the form of reduced pyridine nucleotides. When electron acceptors such as O_2 and nitrate are available, bacteria can obtain a great deal of energy by reducing them, so in oxic environments reduced pyridine nucleotides or other electron donors may be scarce. Anoxic environments lack powerful electron acceptors such as O_2, resulting in a low redox potential and probably more abundant electron donors. For example, H_2 is a central intermediate in many anoxic environments, and its partial pressure has been correlated with the redox couple of the electron-accepting reaction [14,15]. Thus, the partial pressure of H_2 is very low in oxic or nitrate-containing environments, and progressively higher in sulphate-reducing environments and methanogenic environments. This higher pressure of H_2 may be the reason why some compounds such as halogenated aliphatic hydrocarbons are more rapidly biodegraded in anoxic environments than oxic ones [16]. Further, the uncontaminated anoxic subsurface environment in the Alkali Lake basin has a higher partial pressure of H_2 than other sulphate-containing environments, perhaps because the extremity in pH and salinity inhibits the bacteria present or limits microbial diversity. The elevated H_2 partial-pressure may be responsible for accelerated depolymerization and dechlorination in the anoxic zone of the subsurface.

ACKNOWLEDGEMENTS

We thank Susan M. Brillante, Chee Choy and Darryl Coleman for technical assistance.

This work was supported by grants 14-08-0001-G1636 and 14-08-0001-A0510 from the U. S. Geological Survey.

REFERENCES

1. Zehr, J. P., Harvey, R. W., Oremland, R. S., Cloern, J. E., George, L. H., and Lane, J. L., Big Soda Lake (Nevada). 1. Pelagic bacterial heterotrophy and biomass. *Limnol. Oceanogr.*, 1987, **32**, 781-93.

2. Oremland, R. S., Marsh, L. M., and DesMarais, D. J., Methanogenesis in Big Soda Lake, Nevada: an alkaline, moderately hypersaline desert lake. *Appl. Environ. Microbiol.*, 1982, **43**, 140-3.

3. Boone, D. R., Worakit, S., Mathrani, I. M., and Mah, R. A., Alkaliphilic methanogens from high-pH lake sediments. *System. Appl. Microbiol.*, 1986, **7**, 230-4.

4. Keine, R. P., Oremland, R. S., Catena, A., Miller, L. G., and Capone, D. G., Metabolism of reduced methylated sulfur compounds in anaerobic sediments and by a pure culture of an estuarine methanogen. *Appl. Environ. Microbiol.*, 1986, **52**, 1037-45.

5. Oremland, R. S., Marsh, L. M., and Polcin, S., Methane production and simultaneous sulphate reduction in anoxic, salt marsh sediments. *Nature*, 1982, **296**, 143-5.

6. Oremland, R. S., and Polcin, S., Methanogenesis and sulfate reduction: competitive and noncompetitive substrates in estuarine sediments. *Appl. Environ. Microbiol.*, 1982, **44**, 1270-6.

7. Jones, B. F., Geochemical evolution of closed basin water in the western Great Basin. In *Second Symposium on Salt, Northern Ohio Geological Society*, ed., J. L. Rau, 1966, **1**, 181-200.

8. Pankow, J. F., Johnson, R. L., Houck, J. E., Brillante, S. M., and Bryan, W. J., Migration of chlorophenolic compounds at the chemical waste disposal site at Alkali Lake, Oregon--1. Site description and ground-water flow. *Ground Water*, 1984, **22**, 593-601.

9. Smith, P. H., Pure culture studies of methanogenic bacteria. *Proceedings 20th Purdue Industrial Waste Conference*, 1966, 583-8.

10. Balch, W. E., and Wolfe, R. S., New approach to the cultivation of methanogenic bacteria: 2-mercaptoethanesulfonic acid (HS-CoM)-dependent growth of *Methanobacterium ruminantium* in a pressurized atmosphere. *Appl. Environ. Microbiol.*, 1976, **32**, 781-91.

11. Hungate, R. E., A roll tube method for cultivation of strict anaerobes. In *Methods in microbiology*, ed., J. R. Norris and D. W. Ribbons, 1969, **3B**, 117-32.

12. Worakit, S., Boone, D. R., Mah, R. A., Abdel-Samie, M.-E., and El-Halwagi, M. M., *Methanobacterium alcaliphilum* sp. nov., an H_2-utilizing methanogen that grows at high pH values. *Int. J. System. Bacteriol.*, 1986, **36**, 380-2.

13. Johnson, R. L., Brillante, S. M., Isabelle, L. M., Huck, J. E., and Pankow, J. F., Migration of chlorophenolic compounds at the chemical waste disposal site at Alkali Lake, Oregon--2. Contaminant distributions, transport, and retardation. *Ground Water*, 1985, **23**, 652-66.

14. Lovley, D. R., Minimum threshold for hydrogen metabolism in methanogenic bacteria. *Appl. Environ. Microbiol.*, 1985, **49**, 1530-1.

15. Cord-Ruwisch, R., Seitz, H.-J., and Conrad, R., The capacity of hydrogenotrophic anaerobic bacteria to compete for traces of hydrogen depends on the redox potential of the terminal electron acceptor. *Arch. Microbiol.*, 1988, **149**, 350-7.

16. McCarty, P. L., Rittmann, B. E., and Bouwer, E. J., Microbiological processes affecting chemical transformations in groundwater. In *Groundwater pollution microbiology*, ed., G. Bitton and C. P. Gerba, 1984, John Wiley & Sons, New York, pp. 89-115.

PROPERTIES OF (NIFE) AND (NIFESE) HYDROGENASES FROM METHANOGENIC BACTERIA

GUY FAUQUE
Section Enzymologie et Biochimie Bactérienne
A.R.B.S., C.E.N. Cadarache
13108 Saint-Paul-Lez-Durance Cedex, France

ABSTRACT

Hydrogenase is a constitutive enzyme in methanogenic bacteria which possess multiple forms of nickel-containing hydrogenases. Two groups of hydrogenases are present in methanogens : a high-molecular weight, coenzyme F_{420}-reducing hydrogenase and a methyl viologen-dependent hydrogenase of lower molecular mass. Another classification in two types of hydrogenases can be proposed for the F_{420}-reducing enzymes of methanogenic bacteria, like in the sulfate-reducing microorganisms : the non-heme iron-nickel-containing hydrogenases (NiFe hydrogenases) and the nickel-(iron-sulfur)-selenium-containing hydrogenases (NiFeSe hydrogenases). The catalytic and physico-chemical properties of two examples of each class of hydrogenases will be compared : the (NiFe) hydrogenases from Methanosarcina barkeri (DSM 800) and from a thermophilic species of Methanosarcina (DSM 2905) and the (NiFeSe) hydrogenases from two species of Methanococcus : Methanococcus voltae (DSM 1537) and Methanococcus vannielii (DSM 1224). A comparison will be made with the homologous hydrogenases found in the sulfate-reducing bacteria of the genus Desulfovibrio.

NOTATION LIST

F_{420}, 7,8-didemethyl-8-hydroxy-5-deazariboflavin 5'-phosphoryllactylglutamylglutamate; EPR, electron paramagnetic resonance; MV, methyl viologen; BV, benzyl viologen; FMN, flavin mononucleotide; FAD, flavin adenine dinucleotide; SDS, sodium dodecyl sulfate; DSM, Deutsche Sammlung von Mikroorganismen (German Collection of Microorganisms).

INTRODUCTION

Hydrogenases (hydrogen : oxidoreductase E.C. 1.12.) play a central role in the energy metabolism of a wide variety of microorganisms (1-9) and may also have applications in the in vitro evolution of hydrogen (10-14). Hydrogenases constitute a class of redox proteins that are very diversified in their structure and active center composition and catalyze the reversible oxidation of the dihydrogen molecule to protons and electrons (1,5-7, 15).

Most methanogenic bacteria can grow on hydrogen plus carbon dioxide as sole energy source (16-18), except some obligatory methylotrophic or acetotrophic species belonging to the genera Methanothrix (19, 20), Methanosarcina (21, 22), Methanococcoides (23), Methanolobus (24) and Methanosphaera (25). Hydrogenases play an essential role in the process of interspecies hydrogen transfer which occurs in complex fermentations such as cellulose degradation leading to methane production (26). Interspecies hydrogen transfer can be simply described as the transfer of molecular hydrogen from a H_2-evolving bacterium to a H_2-utilizing microorganism in mixed or co-cultures with the maintenance of a low H_2 partial pressure (27-32). Hydrogenase is a constitutive enzyme in methanogens (33, 34) ; similar hydrogenase activity was found in Methanosarcina (Ms.) barkeri strain 227 grown on acetate, methanol or on H_2 and CO_2 (35). Hydrogenase is also present in obligatory acetotrophic and methylotrophic methanogenic bacteria (20, 36-39).

Multiple forms of nickel-containing hydrogenases have been reported in several genera of methanogens (33, 34, 40-42). These hydrogenases can be divided into two groups according to their reactivities with the cofactor F_{420}, their molecular weights and their subunit structures (34, 40-42). One group of hydrogenases reduces the coenzyme F_{420} and methyl viologen (F_{420}-reducing hydrogenases) and the other group reduces methyl viologen but not F_{420} (MV-reducing hydrogenases). These two groups of hydrogenases contain non-heme iron and nickel but the F_{420}-reducing hydrogenases are high molecular weight proteins with three subunits and possess FMN or FAD as prosthetic group (40-43). The presence of selenium has also been reported in two F_{420}-reducing hydrogenases purified from the genus Methanococcus (44-48). The properties of F_{420}-reducing hydrogenases (43-58) and MV-reducing hydrogenases (59-65) from methanogenic bacteria are reported in Tables 1 and 2, respectively. Two classes of nickel-containing hydrogenases, (NiFe) and (NiFeSe) hydrogenases, are also present in the sulfate-reducing bacteria of the genus Desulfovibrio and can be distinguished by their catalytic properties (66-69), metal contents (15, 70-75) and immunological cross-reactivities (76, 77).

I report here, the purification, characterization and catalytic properties of two soluble F_{420}-reducing (NiFe) hydrogenases from a thermophilic strain of Methanosarcina

TABLE 1

Properties of coenzyme F_{420}-reducing hydrogenases from methanogenic bacteria [a]

Hydrogenase	Localization	Molecular mass (kDa)	Subunits (kDa)	Metal and cofactor content [b]
Msp. hungatei	membrane	720	50.7, 30.7	6-7 Ni
Ms. barkeri MS	soluble	800	60	0.6-0.8 Ni, 8-10 Fe
Mc. vannielii	soluble	340	42, 35, 27	2 Ni, 3.8 Se, 18 Fe, 2 FAD
Mc. voltae	membrane	745	45, 37, 27	0.63 Ni, 0.66 Se, 4.5 Fe, 1 FAD
Mb. bryantii MOH	soluble	680	N.R.[*]	Fe, Ni +
Mb. formicicum MF	soluble	600	42.6, 34, 23.5	3 Ni, 20 Fe, FAD +
Mb. thermoautotrophicum ΔH	soluble	800	47, 31, 26	0.6-0.7 Ni, 13-14 Fe, 0.8-0.9 FAD, Cu, Zn +

[*] N.R. = Not reported ; + = Present.

[a] Adapted from references 34, 40, 42, and 46.

[b] Per minimum molecular mass except for Msp. hungatei, Mc. vannielii and Mb. formicicum MF (per native hydrogenase).

TABLE 2
Properties of methyl viologen-reducing hydrogenases from methanogenic bacteria [a]

Hydrogenase	Localization	Molecular mass (kDa)	Subunits (kDa)	Metal and cofactor content
Mb. formicicum MF	soluble	70	48, 38	0.49 Ni, 9.4 Fe [b] 1.6 Cu, 0.8 Zn
Mb. strain G2R	membrane	900	80, 50.7, 38.5	N.R.*
Mb. bryantii MOH	soluble	70	N.R.*	Fe, Ni +
Mb. thermoautotrophicum ΔH	soluble	N.R.*	57, 45, 42, 33	Ni, Fe, Cu, Zn +
Mb. thermoautotrophicum Marburg	soluble	60	N.R.*	0.6-0.8 Ni

* N.R. = Not reported ; + = Present.

[a] Adapted from references 34, 40 and 42.

[b] Presence of a (3Fe-xS) cluster.

(DSM 2905) and Ms. barkeri (DSM 800) and of two soluble F_{420}-reducing (NiFeSe) hydrogenases from Methanococcus (Mc.) voltae (DSM 1537) and Mc. vannielii (DSM 1224).

MATERIALS AND METHODS

Organisms
Methanosarcina barkeri strain MS (DSM 800) was obtained from the German Collection of Microorganisms, Göttingen, F.R.G., the thermophilic Methanosarcina strain MST-A$_1$ (DSM 2905) was kindly provided by Dr. J.P. Touzel (INRA, Lille) and Mc. voltae strain PS (DSM 1537) was a kind gift from Dr. W.B. Whitman (University of Georgia, Athens).

Medium and Growth Conditions
Methanosarcina barkeri was grown at 37°C in a methanol-containing medium as previously described (78). The thermophilic species of Methanosarcina was grown at 55°C on 100 mM methanol (79) and Mc. voltae was grown at 37°C on a complex medium containing 5 mM leucine, 5 mM isoleucine and 100 mM sodium formate (80).

Preparation of Crude Extracts
The cells were harvested by anaerobic centrifugation at the late logarithmic phase of growth and stored at - 30°C under hydrogen. The crude extract of Ms. barkeri was obtained from 1.9 kg of cells (wet weight) as previously reported (50). The crude extracts of Mc. voltae and Methanosarcina (DSM 2905) were prepared from 200 g of cells (wet weight) which were suspended in 200 ml of 20 mM potassium phosphate buffer (pH 6.8) containing 1 mM $MgCl_2$, 5 mM dithiothreitol, 40 mM NaCl and 0.5 mM phenylmethylsulfonyl fluoride (47). A few crystals of DNAse and streptomycin sulfate (0.5 g/g of protein) were added and the cell suspensions were ruptured twice through a French press at 15,000 p.s.i. under a nitrogen/hydrogen atmosphere (95/5, v/v). The crude extracts were obtained after centrifugation at 120,000 x g for 50 min. The entire purification procedures were carried out at 4-6°C in an anaerobic glove box (Coy Instruments, Ann Arbor, MI) and all buffers, phosphate and Tris-(hydroxy metyl amino methane hydrochloride) (Tris - HCl), were boiled and gassed with argon before use.

Assays of Hydrogenase Activity
Hydrogenase activity was determined by three assays. Hydrogen production was followed by gas chromatography in the presence of 15 mM sodium dithionite as electron donor and 1 mM methyl viologen as mediator (81). Hydrogen uptake was determined spectrophotometrically by the reduction of 15 mM benzyl viologen or F_{420} (47). Specific hydrogenase activity is expressed in micromoles of hydrogen produced or consumed per minute per mg of protein. The proton-deuterium exchange reaction catalyzed by hydrogenase was performed in a reaction vessel connected via a teflon membrane - inlet to the ion source

of a mass-spectrometer (model MM 8-80, V.G. Instruments) monitored by an Apple II data acquisition system (82). The buffered medium was saturated with the oxygen-free N_2-D_2 mixture (20 % D_2 in N_2) before hydrogenase injection. The exchange reaction was then followed directly on the dissolved gas phase by the appearance of deuterium hydride (HD) and dihydrogen (H_2). Specific hydrogenase exchange activity is expressed in micromoles of HD plus H_2 evolved per minute per mg of protein.

Analyses
Protein concentrations were determined by the Lowry (83) or Bradford (84) method using bovine serum albumin as standard. The purity of hydrogenases was established by polyacrylamide gel electrophoresis under non-denaturing conditions (85) and the molecular mass of the native hydrogenases was estimated by gel filtration on a Biogel A 1.5 M column. Analytical SDS-polyacrylamide gel electrophoresis was used to determine the molecular weights of subunits (86). The flavin content of hydrogenases was determined by thin-layer chromatography on cellulose and silica plates (50) and the flavin chromophore extracted quantitated by the D-amino acid oxidase assay (87). Metals were screened and quantitated by plasma emission spectroscopy (88) using a Mark II Jarrell-Ash model 965 Atom Comp.

Spectroscopic Instrumentation
The UV-visible optical absorption spectra were recorded on a Beckman model DU-6 spectrophotometer with a kinetic console. Electron paramagnetic resonance (EPR) spectroscopy was carried out on a Varian E-109 spectrometer interfaced with a Hewlett-Packard HP-85 microcomputer.

RESULTS

Properties of the F_{420}-Reducing (NiFe) Hydrogenase from Ms. barkeri (DSM 800)
The hydrogenase activity in Ms. barkeri (DSM 800) is present both in the soluble and membrane fractions but only the soluble form has been purified (50). The soluble F_{420}-reducing hydrogenase from Ms. barkeri was the first hydrogenase isolated from an acetoclastic methanogen (50-52). The physico-chemical and catalytic properties of this enzyme are summarized in Table 3. This hydrogenase (Mr = 800 kDa), which contains only one subunit of 60 kDa, is the unique F_{420}-reducing hydrogenase possessing FMN instead of FAD as prosthetic flavin group. This hydrogenase is rather stable at high temperatures (70 % activity remaining after one hour of incubation at 65°C) and at exposure to air at 4°C (50). This F_{420}-reducing hydrogenase shows an extremely broad reactivity since it can reduce under hydrogen several natural electron acceptors such as ferredoxins, mono-and plurihemic c-type cytochromes from Desulfovibrio species and Desulfu-

romonas acetoxidans (51). This enzyme can also reduce monoelectronic acceptors such as methyl and benzyl viologens but is unable to catalyze the reduction of physiological two-electron acceptors such as NAD or NADP (52). The Ms. barkeri hydrogenase is more active in the H_2 uptake than in H_2 evolution and possesses a very low specific activity in the D_2-H^+ exchange reaction (Table 3). This hydrogenase has a H_2 to HD ratio lower than one in the D_2-H^+ exchange assay (H_2/HD ratio = 0.40 at pH 7.6) (52). This enzyme is particularly resistant to inhibition by CO (only 50 % inhibition with 57 μM CO in the H_2 production and 50 μM CO in the D_2-H^+ exchange reaction) (52).

TABLE 3
Properties of the F_{420}-reducing (NiFe) hydrogenase from Methanosarcina barkeri (DSM 800) grown on methanol [a]

Molecular mass (kDa)	800
Subunit (kDa)	60
Nickel/Subunit	0.6-0.8
Iron/Subunit	8-10
FMN/Subunit	1
Specific activity [b]	
H_2 evolution (MV)	270
H_2 uptake (BV)	960
D_2-H^+ exchange (20 % D_2)	35
Thermal stability	70 % at 65°C (1 hour)
Inhibition by CO [c]	50 % with 50 μM

[a] Adapted from references 50 and 52.

[b] Specific activity is expressed in micromoles of H_2 produced (H_2 evolution) or consumed (H_2 uptake) and in micromoles of HD plus H_2 evolved (D_2-H^+ exchange) per minute per mg of protein.

[c] In the D_2-H^+ exchange reaction.

Purification and Characterization of the F_{420}-Reducing (NiFe) Hydrogenase from Methanosarcina (DSM 2905)

The hydrogenase from the thermophilic species of Methanosarcina strain MST-A1 (DSM 2905) has been purified from the soluble extract in four chromatographic steps (2 DEAE Biogel A, 1 Biogel A 1.5 M and 1 Amicon Blue A columns) with a yield of 12 % and a 103.6-fold purification (18 mg of pure hydrogenase have been obtained from 200 g of cells) (47, 48). This hydrogenase has a golden-brown color and presents an absorption spectrum characteristic of iron-sulfur-containing proteins with two maxima at 379 and 278.5 nm and a weak shoulder

at 318 nm. The purity index (A 379 nm/A 278.5 nm = 0.33) of
this enzyme is in agreement with values reported for other-
nickel-containing hydrogenases. This hydrogenase is able to
reduce under an H_2 atmosphere the two-electron acceptor F_{420}
as well as methyl and benzyl viologens, methylene blue, FMN
and FAD (47). This hydrogenase has a high molecular weight
of 760 kDa and contains 3 non identical subunits with respective molecular weights of 45, 33 and 31 kDa. Based on a minimum molecular mass of 109 kDa ($\alpha \beta \gamma$ subunit arrangement),
0.43 FAD mole, 0.31 nickel atom and 7.50 iron atoms were
found. The aerobically isolated hydrogenase exhibits EPR
nickel signals with g-values at 2.30, 2.14 and 2.02 (47).
The catalytic properties of this (NiFe) thermophilic hydrogenase are reported in Table 4. This enzyme has a specific
activity of 168.5 in the D_2-H^+ exchange reaction, 220.6 in
the H_2 uptake (with benzyl viologen as electron acceptor)
and 490 in the H_2 evolution. The pH optimum is around 6.8
for the D_2-H^+ exchange and the reduction of cofactor F_{420}.
In the D_2-H^+ exchange reaction, the H_2 to HD ratio is lower
than one and pH-independent. This hydrogenase is relatively
resistant to inhibition by carbon monoxide (50 % inhibition
obtained with 10 μM CO at 60°C) (48).

**Properties of the F_{420}-Reducing (NiFeSe) Hydrogenase from
Mc. vannielii (DSM 1224)**
The first selenium-containing hydrogenase has been purified
and characterized from the soluble extract of Mc. vannielii
grown on formate in the presence of tungstate and (^{75}Se)
selenite (44, 45). The physico-chemical characteristics of
this hydrogenase are reported in Table 5. This enzyme with
a molecular weight of 340 kDa tends to aggregate in a
tetrameric form of 1,300 kDa ; both molecular forms are able
to reduce the cofactor F_{420} and 2,3,5-triphenyltetrazolium
chloride with molecular hydrogen (44). This hydrogenase
contains three subunits of Mr = 42, 35 and 27 kDa and 4 gram
atoms of selenium, 2 gram atoms of nickel, 18-20 gram atoms
of iron and 2 equivalents FAD are found per mole of native
enzyme (340 kDa)(45a). The selenium is present exclusively
in the 42 kDa subunit in the chemical form of selenocysteine.
This (NiFeSe) hydrogenase is extremely oxygen-sensitive and
can be reactivated by incubation in the presence of dithiothreitol and hydrogen (44).

**Purification and Characterization of the Mc. voltae F_{420}-
Reducing (NiFeSe) Hydrogenase**
The hydrogenase from Mc. voltae has been purified 201.1-fold
from the soluble fraction in three chromatographic steps
(DEAE-cellulose 52, hydroxylapatite-Ultrogel and Ultrogel
ACA 22) with a yield of 23.2 % (15 mg of pure enzyme obtained
from 200 g of cells). This hydrogenase presents an optical
spectrum typical of non-heme iron proteins with maxima at
380 and 275.5 nm and a prominent shoulder at 320 nm.The native
hydrogenase has a molecular weight of 660 kDa with 3 subunits
of 46, 35 and 27 kDa. This enzyme contains 0.81 FAD molecule,
0.41 nickel atom, 8.50 iron atoms and 0.40 mole of selenium

TABLE 4

Catalytic properties of (NiFe) hydrogenases from Ms. barkeri and Methanosarcina (DSM 2905) and (NiFeSe) hydrogenase from Mc. voltae [a]

Properties	Ms. barkeri (DSM 800)	Methanosarcina (DSM 2905)	Mc. voltae (DSM 1537)
Specific activity [b]			
H_2 uptake (BV)	960	220.6	350
H_2 evolution (reduced MV)	270	490	N.D.*
D_2-H^+ exchange (20 % D_2)	35	168.5 [d]	2,800
H_2/HD (at pH optimum) [c]	0.40	0.50	1.12
CO inhibition (50 %) [c]	50 μM	10 μM [d]	1.5 μM

* N.D. = Not determined.

[a] Adapted from references 48, 50 and 52.

[b] Specific activity is expressed in micromoles of H_2 (consumed or produced) or of HD plus H_2 evolved per minute per mg of protein.

[c] In the D_2-H^+ exchange reaction.

[d] At 60°C.

per minimum molecular mass unit (108 kDa). In the presence of hydrogen, this hydrogenase is able to reduce the coenzyme F_{420} as well as FMN, FAD, methylene blue, benzyl and methyl viologens (47, 48). The anaerobically isolated (NiFeSe) hydrogenase presents an EPR spectrum with g-values at 2.20, 2.15 and 2.01 characteristic of the Ni (III)-hydride intermediate (Ni-signal C) observed with other (NiFe) hydrogenases (15,75). This nickel signal disappears by reduction under an H_2 atmosphere (47). The catalytic activity of the Mc. voltae hydrogenase was tested in the H_2 uptake and the D_2-H^+ exchange reaction where the specific activities are 350 and 2,800, respectively (Table 4). This hydrogenase has a H_2/HD ratio above one at pH higher than 6.0 in the D_2-H^+ exchange reaction where 50 % inhibition is obtained with 1.5 μM CO and 0.1 mM NO_2^-.

TABLE 5
Properties of the F_{420}-reducing (NiFeSe) hydrogenase from Methanococcus vannielii [a]

Molecular mass (kDa)	340
	1,300 (aggregation)
Subunits (kDa)	42 (selenocysteine)
	35
	27
Nickel [b]	2
Iron [b]	18-20
Selenium [b]	4
FAD [b]	2

[a] Adapted from references 44 and 45a.

[b] Per mole of native protein (340 kDa).

DISCUSSION

Hydrogen is an important metabolite in the degradation of organic compounds in anaerobic ecosystems (26, 28). Methanogenic bacteria are generally considered as hydrogen consumers rather than producers (89) ; however, when grown on methanol, CO or acetate, methanogens can evolve molecular hydrogen (32, 90). Although the presence of hydrogenase activity in crude extracts of methanogenic bacteria has been reported since the first studies of in vitro methane formation (91), the purification and the characterization of hydrogenases has only began in the last decade. Hydrogenases play a central role in most methanogens in which H_2 oxidation is utilized to drive reduction of CO_2 and generate energy for the cell (18, 33, 34, 40, 42, 89).

All the hydrogenases so far isolated from methanogenic bacteria are non-heme iron-sulfur proteins containing nickel (33, 34, 40, 42). The F_{420}-reducing hydrogenases have been purified and characterized from four genera of methanogens (Methanospirillum, Methanosarcina, Methanobacterium and Methanococcus) (Table 1) and the MV-reducing hydrogenases have been isolated from five Methanobacterium species (Table 2) and from the thermophilic Methanosarcina strain MST-A_1 (47). A common characteristic of coenzyme F_{420}-dependent methanogenic hydrogenases seems to be the presence of flavin groups, either FMN or FAD, which function both in one - and two - electron transfer reactions. These flavins could mediate electron flow between monoelectronic iron-sulfur clusters and the obligate two-electron acceptor coenzyme F_{420} (43). Some F_{420}-reducing hydrogenases from Methanobacterium species have been reported to be unstable because of the loss of FAD from the enzymes (56, 60). The MV-reducing hydrogenases from methanogens are generally smaller molecules showing molecular weights comparable to the (NiFe) hydrogenases from other groups of microorganisms (1, 5, 7, 15). They are present in lower amounts than the F_{420}-dependent hydrogenases (47, 55). The function of MV-dependent methanogenic hydrogenases is still unknown and no natural electron carrier has been reported to interact with these enzymes. The MV-reducing hydrogenases may be involved in the synthesis of ATP or in the generation of membrane potentials (43). Most hydrogenases from methanogens are considered to be soluble proteins and their true localization seems to be on the cytoplasmic side of the cell membrane (34).

Two types of F_{420}-reducing hydrogenases, (NiFe) and (NiFeSe), have been found in methanogenic bacteria; the (NiFeSe) type seems to be characteristic of methanogens belonging to the genus Methanococcus. The physico-chemical and catalytic properties of these two classes of methanogenic hydrogenases present similarities with those of the homologous (NiFe) and (NiFeSe) hydrogenases isolated from the sulfate-reducing bacteria of the genus Desulfovibrio (3, 15, 66-77). This genus also contains a third class of hydrogenases with only iron-sulfur centers [(Fe) hydrogenase]. The periplasmic (Fe) hydrogenase found in Desulfovibrio (D.) vulgaris Hildenborough is the most active of all the three types of hydrogenases in the three reactions (D_2-H^+ exchange, H_2 uptake and H_2 evolution). In addition, the ratio between the initial H_2 and HD production is lower than one, as it is the case with the (NiFe) hydrogenases from D. gigas and D. multispirans (Table 6). In contrast, the (NiFeSe) hydrogenases from D. baculatus and D. salexigens have a H_2 to HD ratio higher than one in the D_2-H^+ exchange reaction. The (NiFeSe) Desulfovibrio hydrogenases are also more sensitive than the (NiFe) hydrogenases to inhibitors such as CO and NO_2^- (69). Thus, the classification proposed for the nickel-containing hydrogenases from Desulfovibrio species seems adequate for methanogenic bacteria.

TABLE 6

Catalytic properties of (Fe), (NiFe) and (NiFeSe) hydrogenases from _Desulfovibrio_ [a]

Hydrogenases	Specific Activity [b]			
	H_2 evolution	H_2 uptake	D_2-H^+ exchange	H_2/HD
D. vulgaris (Fe)	4,800	50,000	2,700	0.4-0.6
D. gigas (NiFe)	440	1,500	267	0.25-0.4
D. multispirans (NiFe)	790	590	600	0.3-0.4
D. baculatus (NiFeSe)	466	120	350	1.1-1.5
D. salexigens (NiFeSe)	1,830	1,300	900	> 1

[a] Adapted from reference 68.

[b] Specific activity is expressed in micromoles of H_2 (produced or consumed) or of HD + H_2 evolved per minute per mg of protein.

ACKNOWLEDGEMENTS

I would like to thank my following colleagues for experimental contributions and valuable discussions : Drs. Y. Berlier, M. Czechowski, D.V. DerVartanian, J. LeGall, P.A. Lespinat, I. Moura, J.J.G. Moura, H.D. Peck Jr., M. Teixeira, S.B. Woo and A.V. Xavier. The skillful secretarial assistance of Claude Beugnon is gratefully acknowledged. This work was supported in part by the Commission des Communautés Européennes under contract BAP-0269-F.

REFERENCES

1. Adams, M.W.W., Mortenson, L.E. and Chen, J.S., Hydrogenases. Biochim. Biophys. Acta, 1981, **594**, 105-76.

2. Odom, J.M. and Peck, H.D.Jr., Hydrogenase, electron transfer proteins and energy coupling in the sulfate-reducing bacteria, Desulfovibrio. Ann. Rev. Microbiol., 1984, **38**, 551-92.

3. LeGall, J. and Fauque, G., Dissimilatory reduction of sulfur compounds. In Biology of Anaerobic Microorganisms, ed., A.J.B. Zehnder, Chapter 11, John Wiley and Sons, Inc., New York, 1988, pp. 587-639.

4. Gest, H., Oxidation and evolution of molecular hydrogen by microorganisms. Bacteriol. Rev., 1954, **18**, 43-73.

5. Cammack, R., Fernandez, V.M. and Schneider, K., Nickel in hydrogenases from sulfate-reducing, photosynthetic and hydrogen-oxidizing bacteria. In Bioinorganic Chemistry of Nickel, ed., J.R. Lancaster, Jr., VCH Publishers, Deerfield Beach, Florida, 1988 (in press).

6. Evans, H.J., Harker, A.R., Papen, H., Russel, S.A., Hanus, F.J. and Zuber, M., Physiology, biochemistry, and genetics of the uptake hydrogenase in Rhizobia. Ann. Rev. Microbiol., 1987, **41**, 335-61.

7. Vignais, P.M., Colbeau, A.,Willison, J.C. and Jouanneau, Y., Hydrogenase, nitrogenase and hydrogen metabolism in the photosynthetic bacteria. Adv. Microb. Physiol., 1985, **26**, 155-234.

8. LeGall, J. and Peck, H.D. Jr., Hydrogenases : physiology, location and relevance for sulfate-reducing and methane-forming bacteria. In The Biological Chemistry of Iron, ed., H.B. Dunford, D. Dolphin, K. Raymond and L. Sieker, NATO Advanced Studies Institutes Series, D. Reidel Publishing Company, Edmonton, 1982, pp. 207-22.

9. Kondratieva, E.N. and Gogotov, I.N., Production of molecular hydrogen in microorganisms. In Advances in Bio-

chemical Engineering / Biotechnology, Vol. 28, ed., A. Fiechter, Springer Verlag, Berlin and New York, 1983, pp. 139-91.

10. Weaver, P.F., Lien, S. and Seibert, M., Photobiological production of hydrogen. Solar Energy, 1980, **24**, 3-45.

11. Cammack, R., Hall, D.O. and Rao, K.K., Hydrogenases : structure and applications in hydrogen production. In Microbial Gas Metabolism : Mechanistic, Metabolic and Biotechnological Aspects, ed., R.K. Poole and C.S. Dow, Society for General Microbiology, Academic Press Inc., London, 1985, pp. 75-102.

12. Klibanov, M., Biotechnological potential of the enzyme hydrogenase. Process Biochem., 1983, **18**, 13-23.

13. Krasna, A.I., Hydrogenase : properties and applications. Enzyme Microb. Technol., 1979, **1**, 165-72.

14. Okura, I., Hydrogenase and its application for photo-induced hydrogen evolution. Coordination Chemistry Reviews, 1985, **68**, 53-99.

15. Moura, J.J.G., Moura, I., Teixeira, M., Xavier, A.V., Fauque, G.D. and LeGall, J., Nickel-containing hydrogenases. In Metal Ions in Biological Systems, Vol. 23, ed., H. Sigel, Marcel Dekker, Inc., New York and Basel, 1988, pp. 285-314.

16. Balch, W.E., Fox, G.E., Magrum, L.J., Woese, C.R. and Wolfe, R.S., Methanogens : reevaluation of a unique biological group. Microbiol. Rev., 1979, **43**, 260-96.

17. Whitman, W.B., Methanogenic bacteria. In The Bacteria. A Treatise on Structure and Function, Vol. VIII Archaebacteria, ed., J.R. Sokatch, L.N. Ornston, C.R. Woese and R.S. Wolfe, Academic Press, Inc., New York, 1985, pp. 3-84.

18. Jones, W.J., Nagle, D.P. and Whitman, W.B., Methanogens and the diversity of archaebacteria. Microbiol. Rev., 1987, **51**, 135-77.

19. Huser, B.A., Wuhrmann, K. and Zehnder, A.J.B., Methanothrix soehngenii gen. nov. sp. nov., a new acetotrophic non-hydrogen-oxidizing methane bacterium. Arch. Microbiol., 1982, **132**, 1-9.

20. Patel, G.B., Characterization and nutritional properties of Methanothrix concilii sp. nov., a mesophilic, aceticlastic methanogen. Can. J. Microbiol., 1984, **30**, 1383-96.

21. Zinder, S.H. and Mah, R.A., Isolation and characterization of a thermophilic strain of Methanosarcina unable

to use H_2-CO_2 for methanogenesis. Appl. Environ. Microbiol., 1979, **38**, 996-1008.

22. Sowers, K.R., Baron, S.F. and Ferry, J.G., Methanosarcina acetivorans sp. nov., an acetotrophic methane-producing bacterium isolated from marine sediments. Appl. Environ. Microbiol., 1984, **47**, 971-78.

23. Sowers, K.R. and Ferry, J.G., Isolation and characterization of a methylotrophic marine methanogen, Methanococcoides methylutens gen. nov., sp. nov. Appl. Environ. Microbiol., 1983, **45**, 684-90.

24. König, H. and Stetter, K.O., Isolation and characterization of Methanolobus tindarius, sp. nov., a coccoid methanogen growing only on methanol and methylamines. Zentralb. Bakteriol. Parasitenkd. Infektionskr. Hyg. Abt.1 Orig., 1982, **C3**, 478-90.

25. Miller, T.L. and Wolin, M.J., Methanosphaera stadtmaniae gen. nov., sp. nov. : a species that forms methane by reducing methanol with hydrogen. Arch. Microbiol., 1985, **141**, 116-22.

26. Peck, H.D. Jr. and Odom, M., Anaerobic fermentations of cellulose to methane. In Trends in the Biology of Fermentations for Fuels and Chemicals, ed., A. Hollaender, Plenum Press, New York and London, 1981, pp. 375-95.

27. Bryant, M.P., Wolin, E.A., Wolin, M.J. and Wolfe, R.S., Methanobacillus omelianskii, a symbiotic association of two species of bacteria. Arch. Mikrobiol., 1967, **59**, 20-31.

28. Wolin, M.J., Interaction between H_2-producing and methane-producing species. In Symposium on Microbial Production and Utilization of Gases (H_2, CH_4, CO_2), ed., H. Schlegel, G. Gottschalk and N. Pfennig, K. Goltze, Göttingen, 1976, pp. 141-50.

29. Bryant, M.P., Campbell, L.L., Reddy, C.A. and Crabill, M.R., Growth of Desulfovibrio in lactate or ethanol media low in sulfate in association with H_2-utilizing methanogenic bacteria. Appl. Environ. Microbiol., 1977, **33**, 1162-69.

30. Wolin, M.J., Hydrogen transfer in microbial communities. In Microbial Interactions and Communities, Vol. 1, ed., A.T. Bull and J.H. Slater, Academic Press, London, 1982, pp. 323-56.

31. Wolin, M.J. and Miller, T.L., Interspecies hydrogen transfer : 15 years later. ASM News, 1982, **48**, 561-65.

32. Phelps, T.J., Conrad, R. and Zeikus, J.G., Sulfate-

dependent interspecies H_2 transfer between Methanosarcina barkeri and Desulfovibrio vulgaris during the coculture metabolism of acetate and methanol. Appl. Environ. Microbiol., 1985, **50**, 589-94.

33. Daniels, L., Sparling, R. and Sprott, G.D., The bioenergetics of methanogenesis. Biochim. Biophys. Acta, 1984, **768**, 113-63.

34. Keltjens, J.T. and Van der Drift, C., Electron transfer reactions in methanogens. FEMS Microbiol. Rev., 1986, **39**, 259-303.

35. Baresi, L. and Wolfe, R.S., Levels of coenzyme F_{420}, coenzyme M, hydrogenase, and methyl coenzyme M methylreductase in acetate-grown Methanosarcina. Appl. Environ. Microbiol., 1981, **41**, 388-91.

36. Zeikus, J.G., Metabolism of one-carbon compounds by chemotrophic anaerobes. Adv. Microbiol. Physiol., 1983, **24**, 215-99.

37. Nelson, M.J.K. and Ferry, J.G., Carbon monoxide-dependent methyl coenzyme M methylreductase in acetotrophic Methanosarcina spp. J. Bacteriol., 1984, **160**, 526-32.

38. Bhatnagar, L., Krzycki, J.A. and Zeikus, J.G., Analysis of hydrogen metabolism in Methanosarcina barkeri : regulation of hydrogenase and role of CO-dehydrogenase in H_2 production. FEMS Microbiol. Lett., 1987, **41**, 337-43.

39. Lovley, D.R. and Ferry, J.G., Production and consumption of H_2 during growth of Methanosarcina spp. on acetate. Appl. Environ. Microbiol., 1985, **49**, 247-49.

40. Hausinger, R.P., Nickel utilization by microorganisms. Microbiol. Rev., 1987, **51**, 22-42.

41. LeGall, J., Moura, J.J.G., Peck, H.D.Jr. and Xavier, A.V., Hydrogenase and other iron-sulfur proteins from the sulfate-reducing and methane-forming bacteria. In Iron-Sulfur Proteins, ed., T.G. Spiro, John Wiley and Sons, New York, 1982, pp. 177-248.

42. Vogels, G.D., Keltjens, J.T. and Van der Drift, C., Biochemistry of methane production. In Biology of Anaerobic Microorganisms, ed., A.J.B. Zehnder, Chapter 13, John Wiley and Sons, Inc., New York, 1988, pp. 707-70.

43. Fox, J.A., Livingston, D.J., Orme-Johnson, W.H. and Walsh, C.T., 8-hydroxy-5-deazaflavin-reducing hydrogenase from Methanobacterium thermoautotrophicum : 1. Purification and characterization. Biochemistry, 1987, **26**, 4219-27.

44. Yamazaki, S., A selenium-containing hydrogenase from Methanococcus vannielii. Identification of the selenium

moiety as a selenocysteine residue. J. Biol. Chem., 1982, **257**, 7926-29.

45. Yamazaki, S., Tsai, L., Stadtman, T.C., Teshima, T., Nakaji, A. and Shiba, T., Stereochemical studies of a selenium-containing hydrogenase from Methanococcus vannielii : Determination of the absolute configuration of C-5 chirally labeled dihydro-8-hydroxy-5-deazaflavin cofactor. Proc. Natl. Acad. Sci. USA, 1985, **82**, 1364-66.

45a. Stadtman, T.C., Specific occurence of selenium in certain enzymes and amino acid transfer ribonucleic acids. Phosphorus and Sulfur, 1985, **24**, 199-216.

46. Muth, E., Mörschel, E. and Klein, A., Purification and characterization of an 8-hydroxy-5-deazaflavin-reducing hydrogenase from the archaebacterium Methanococcus voltae. Eur. J. Biochem., 1987, **169**, 571-77.

47. Woo, S.B., Hydrogen metabolism and energy conservation in methanogenic bacteria. Ph.D. Thesis, University of Georgia, Athens, U.S.A., 1988, 261 pages.

47a. Muth, E., Localization of the F_{420}-reducing hydrogenase in Methanococcus voltae cells by immuno-gold technique. Arch. Microbiol., 1988, **150**, 205-07.

48. Fauque, G., Woo, S.B., Berlier, Y., Lespinat, P.A., Peck, H. D. Jr. and LeGall, J., Purification, characterization and catalytic properties of hydrogenases from Methanococcus voltae and from a thermophilic species of Methanosarcina. In Fifth International Symposium on Anaerobic Digestion, ed., A. Tilche and A. Rozzi, Monduzzi Editore S.p.A., Bologna, 1988, pp. 35-38.

49. Sprott, G.D., Shaw, K.M. and Beveridge, T.J., Properties of the particulate enzyme F_{420}-reducing hydrogenase isolated from Methanospirillum hungatei. Can. J. Microbiol., 1987, **33**, 896-904.

50. Fauque, G., Teixeira, M., Moura, I., Lespinat, P.A., Xavier, A.V., DerVartanian, D.V., Peck, H.D. Jr., LeGall, J. and Moura, J.J.G., Purification, characterization and redox properties of hydrogenase from Methanosarcina barkeri (DSM 800). Eur. J. Biochem., 1984, **142**, 21-28.

51. Fauque, G., Teixeira, M., Moura, I., Lino, A.R., Lespinat, P.A., Xavier, A.V., DerVartanian, D.V., Peck, H.D. Jr., LeGall, J. and Moura, J.J.G., A link between hydrogen and sulfur metabolisms in Methanosarcina barkeri (DSM 800). In Biotechnological Advances in Processing Municipal Wastes for Fuels and Chemicals, ed., A.A. Antonopoulos, Energy and Environmental Systems Division, Argonne National Laboratory, 1985, pp. 97-111.

52. Fauque, G., Relations structure-fonction de différentes

protéines d'oxydo-réduction isolées de bactéries sulfato-réductrices et methanigènes. <u>Doctorat</u> <u>d'Etat</u> <u>Thesis,</u> Université de Technologie de Compiègne, France, 1985, 222 pages.

53. Lancaster, J.R. Jr., Soluble and membrane-bound paramagnetic centers in <u>Methanobacterium</u> <u>bryantii</u>. <u>FEBS</u> <u>Lett.</u>, 1980, **115**, 285-88.

54. Lancaster, J.R. Jr., New biological paramagnetic center : Octahedrally coordinated nickel (III) in the methanogenic bacteria. <u>Science,</u> 1982, **216**, 1324-25.

55. Jin, S.-L., Blanchard, D.K. and Chen, J.-S., Two hydrogenases with distinct electron-carrier specificity and subunit composition in <u>Methanobacterium</u> <u>formicicum</u>. <u>Biochim. Biophys. Acta,</u> 1983, **748**, 8-20.

56. Nelson, M.J.K., Brown, D.P. and Ferry, J.G., FAD requirement for the reduction of coenzyme F_{420} by hydrogenase from <u>Methanobacterium</u> <u>formicicum</u>. <u>Biochem. Biophys. Res. Commun.,</u> 1984, **120**, 775-81.

57. Jacobson, F.S., Daniels, L., Fox, J.A., Walsh, C.T. and Orme-Johnson, W.H., Purification and properties of an 8-hydroxy-5-deazaflavin-reducing hydrogenase from <u>Methanobacterium</u> <u>thermoautotrophicum</u>. <u>J. Biol. Chem.,</u> 1982, **257**, 3385-88.

58. Wackett, L.P., Hartwieg, E.A., King, J.A., Orme-Johnson, W.H. and Walsh, C.T., Electron microscopy of nickel-containing methanogenic enzymes : methyl reductase and F_{420}-reducing hydrogenase. <u>J. Bacteriol.,</u> 1987, **169**, 718-27.

59. Adams, M.W.W., Jin, S.-L.C., Chen, J.-S. and Mortenson, L.E., The redox properties and activation of the F_{420}-non-reactive hydrogenase of <u>Methanobacterium</u> <u>formicicum.</u> <u>Biochim. Biophys. Acta,</u> 1986, **869**, 37-47.

60. Mckellar, R.C. and Sprott, G.D., Solubilization and properties of a particulate hydrogenase from <u>Methanobacterium</u> strain G2R. <u>J. Bacteriol.,</u> 1979, **139**, 231-38.

61. Kojima, N., Fox, J.A., Hausinger, R.P., Daniels, L., Orme-Johnson, W.H. and Walsh, C.T., Paramagnetic centers in the nickel-containing, deazaflavin-reducing hydrogenase from <u>Methanobacterium</u> <u>thermoautotrophicum</u>. <u>Proc. Natl. Acad. Sci. USA,</u> 1983, **80**, 378-82.

62. Walsh, C.T. and Orme-Johnson, W.H., Perspectives in biochemistry. Nickel enzymes. <u>Biochemistry,</u> 1987, **26**, 4901-06.

63. Fuchs, G., Moll, J., Scherer, P. and Thauer, R., Activity, acceptor specificity and function of hydrogenase in

Methanobacterium thermoautotrophicum. In Hydrogenases : Their Catalytic Activity, Structure and Function, ed., H.G. Schlegel and K. Schneider, K. Goltze, Göttingen, 1978, pp. 83-92.

64. Graf, E.-G. and Thauer, R.K., Hydrogenase from Methanobacterium thermoautotrophicum, a nickel-containing enzyme. FEBS Lett., 1981, **136**, 165-69.

65. Albracht, S.P.J., Graf, E.-G. and Thauer, R.K., The EPR properties of nickel in hydrogenase from Methanobacterium thermoautotrophicum. FEBS Lett., 1982, **140**, 311-13.

66. Lespinat, P.A., Berlier, Y., Fauque, G., Czechowski, M., Dimon, B. and LeGall, J., The pH dependence of proton-deuterium exchange, hydrogen production and uptake catalyzed by hydrogenases from sulfate-reducing bacteria. Biochimie, 1986, **68**, 55-61.

67. Fauque, G.D., Berlier, Y.M., Czechowski, M.H., Dimon, B., Lespinat, P.A. and LeGall, J., A proton-deuterium exchange study of three types of Desulfovibrio hydrogenases. J. Ind. Microbiol., 1987, **2**, 15-23.

68. Fauque, G., Berlier, Y., Choi, E.S, Peck, H.D.Jr., LeGall, J. and Lespinat, P.A., The carbon monoxide inhibition of the proton-deuterium exchange activity of iron, nickel-iron and nickel-iron-selenium hydrogenases from Desulfovibrio vulgaris Hildenborough. Biochem. Soc. Trans., 1987, **15**, 1050-51.

69. Berlier, Y., Fauque, G., LeGall, J., Choi, E.S., Peck, H.D.Jr. and Lespinat, P.A., Inhibition studies of three classes of Desulfovibrio hydrogenase : Application to the further characterization of the multiple hydrogenases found in Desulfovibrio vulgaris Hildenborough. Biochem. Biophys. Res. Commun., 1987, **146**, 147-53.

70. LeGall, J., Ljungdahl, P.O., Moura, I., Peck, H.D..Jr., Xavier, A.V., Moura, J.J.G., Teixeira, M., Huynh, B.H. and DerVartanian, D.V., The presence of redox-sensitive nickel in the periplasmic hydrogenase from Desulfovibrio gigas. Biochem. Biophys. Res. Commun., 1982, **106**, 610-16.

71. Rieder, R., Cammack, R. and Hall, D.O., Purification and properties of the soluble hydrogenase from Desulfovibrio desulfuricans (strain Norway 4). Eur. J. Biochem., 1984, **145**, 637-43.

72. Teixeira, M., Moura, I., Fauque, G., Czechowski, M., Berlier, Y., Lespinat, P.A., LeGall, J., Xavier, A.V. and Moura, J.J.G., Redox properties and activity studies on a nickel-containing hydrogenase isolated from a halophilic sulfate reducer, Desulfovibrio salexigens. Biochimie, 1986, **68**, 75-84.

73. Czechowski,M., Fauque, G., Galliano, N., Dimon, B., Moura, I., Moura, J.J.G., Xavier, A.V., Barata, B.A.S., Lino, A.R. and LeGall, J., Purification and characterization of three proteins from a halophilic sulfate-reducing bacterium, Desulfovibrio salexigens. J. Ind. Microbiol., 1986, 1, 139-47.

74. Teixeira, M., Fauque, G., Moura, I., Lespinat, P.A., Berlier, Y., Prickril, B., Peck, H.D.Jr., Xavier, A.V., LeGall, J. and Moura, J.J.G., Nickel-(iron-sulfur)-selenium-containing hydrogenases from Desulfovibrio baculatus (DSM 1743). Redox centers and catalytic properties. Eur. J. Biochem., 1987, 167, 47-58.

75. Moura, J.J.G., Teixeira, M., Moura, I. and LeGall, J., (NiFe) hydrogenases from sulfate-reducing bacteria : Nickel catalytic and regulatory roles. In Bioinorganic Chemistry of Nickel, ed., J.R. Lancaster, Jr., VCH Publishers, Deerfield Beach, Florida, 1988 (in press).

76. Lissolo, T., Choi, E.S., LeGall, J. and Peck, H.D.Jr., The presence of multiple intrinsic membrane nickel-containing hydrogenases in Desulfovibrio vulgaris (Hildenborough). Biochem. Biophys. Res. Commun., 1986, 139, 701-08.

77. Prickril, B.C., He, S.H., Li, C., Menon, N., Choi, E.S., Przybyla, A.E., DerVartanian, D.V., Peck, H.D.Jr., Fauque, G., LeGall, J., Teixeira, M., Moura, I., Moura, J.J.G., Patil, D. and Huynh, B.H., Identification of three classes of hydrogenase in the genus, Desulfovibrio. Biochem. Biophys. Res. Commun., 1987, 149, 369-77.

78. Blaylock, B.A. and Stadtman, T.C., Methane biosynthesis by Methanosarcina barkeri. Properties of the soluble enzyme system. Arch. Biochem. Biophys., 1966, 116, 138-52.

79. Touzel, J.P. and Albagnac, G., Acetoclastic methanogens in anaerobic digesters. In Biotechnological Advances in Processing Municipal Wastes for Fuels and Chemicals, ed., A.A. Antonopoulos, Energy and Environmental Systems Division, Argonne National Laboratory, 1985, pp. 35-39.

80. Whitman, W.B., Ankwanda, E. and Wolfe, R.S., Nutrition and carbon metabolism of Methanococcus voltae. J. Bacteriol., 1982, 149, 852-63.

81. Peck, H.D. Jr. and Gest, H., A new procedure for assay of bacterial hydrogenase. J. Bacteriol., 1956, 71, 70-80.

82. Berlier, Y.M., Dimon, B., Fauque, G. and Lespinat, P.A., Direct mass-spectrometric monitoring of the metabolism and isotope exchange in enzymic and microbiological investigations. In Gas Enzymology, ed., H. Degn, R.P. Cox and H. Toftlund, D. Reidel Publishing Company, Dordrecht, 1985, pp. 17-35.

83. Lowry, O.H., Rosebrough, N.J., Farr, A.L. and Randall, R.J., Protein measurement with the folin phenol reagent. J. Biol. Chem., 1951, **193**, 265-75.

84. Bradford, M., A rapid and sensitive method for the quantitation of microgram quantities of protein utilizing the principle of protein-dye binding. Anal. Biochem., 1976, **72**, 248-54.

85. Laemmli, U.K., Cleavage of structural proteins during the assembly of the head of bacteriophage T_4. Nature, 1970, **227**, 680-85.

86. Weber, K. and Osborn, M., The reliability of molecular weight determination of dodecyl sulfate polyacrylamide gel electrophoresis. J. Biol. Chem., 1969, **244**, 4406-12.

87. Fonda, M.L. and Anderson, B.M., D-amino acid oxidase. I. Spectrophotometric studies. J. Biol. Chem., 1967, **242**, 3957-62.

88. Jones, J.B. Jr., Elemental analysis of soil extracts and plant tissue ash by plasma emission spectroscopy. Commun. Soil Sci. Plant Anal., 1977, **8**, 349-65.

89. Zeikus, J.G., The biology of methanogenic bacteria. Bacteriol. Rev., 1977, **41**, 514-41.

90. O'Brien, J.M., Wolkin, R.H., Moench, T.T., Morgan, J.B. and Zeikus, J.G., Association of hydrogen metabolism with unitrophic or mixotrophic growth of Methanosarcina barkeri on carbon monoxide. J. Bacteriol., 1984, **158**, 373-75.

91. Wolin, E.A., Wolin, M.J. and Wolfe, R.S., Formation of methane by bacterial extracts. J. Biol. Chem., 1963, **238**, 2882-86.

METHANOGENESIS IN ARTIFICIALLY CREATED EXTREME ENVIRONMENTS*

ANN C. WILKIE and PAUL H. SMITH
Department of Microbiology and Cell Science,
University of Florida - IFAS,
Gainesville, Florida 32611, USA.

ABSTRACT

Perturbations of the methanogenic fermentation may result in the creation of extreme environments for all anaerobic bacterial life involved. Quantitative measurement of perturbations induced by increased organic loading rates and reduced hydraulic retention times, in laboratory-scale anaerobic digesters, are reported. The initiation of the perturbation was the result of something other than the inhibition of methane production from hydrogen and carbon dioxide, from organic acid conversion, or by pH. Perturbation of methane formation observed due to increased propionic acid additions was pH independent and was unrelated to either the concentration of the propionic acid in the liquid fraction of the fermenting material or the hydrogen concentration in the evolved biogas. On the basis of our initial results, we consider it to be true that the bacteria which metabolize propionic acid produce some factor which is inhibitory to the hydrolysis of large polymers during the fermentation of plant material to methane.

INTRODUCTION

The earth's atmosphere of gases and small particles has, for millenia, filtered incoming radiant energy from the sun and removed most of the short wavelengths having energy levels too great to be compatible with the stability of carbon-to-carbon bonds [1]. In prehistoric times, carbon-to-carbon bonds became arranged in configurations able to absorb radiant energy and to couple the absorbed energy to a wide variety of chemical reactions [2]. The chemical and physical environmental conditions of the primitive earth became well suited for the growth of plants, which then synthesized massive amounts of carbon compounds utilizing atmospheric carbon dioxide. The formation of great quantities of oxygen accompanied this synthesis. Dead plant material was then oxidized, replenishing the atmosphere with carbon dioxide and other components essential for plant

*Florida Agricultural Experiment Station Journal Series No. 9565.

growth. A cyclying of carbon was initiated which continuously produced adequate quantities of carbon dioxide for the growth of plants, and which has continued to the present time.

Microorganisms metabolized the dead plant material via a diversity of metabolic pathways. Material exposed to the oxygen of the atmosphere was metabolized utilizing molecular oxygen as a terminal electron acceptor. Carbon dioxide and the methyl group of acetic acid were, most likely, the predominant terminal electron acceptors in the anaerobic metabolism of plant material, with methane and carbon dioxide as end products. These gases were returned to the atmosphere and the cycling of carbon continued. Recalcitrant and incompletely metabolized plant material became deposited. The fossil fuels of our generation constitute a fraction of those deposits. Had plant material been completely metabolized, either aerobically or anaerobically, deposition of fossil fuels would not have occurred and the course of human history would have been greatly altered.

The methanogenic bacteria are ancient forms of life [3]. They are normally found wherever organic matter decays under anaerobic conditions. Methanogenic bacteria and the related Archaebacteria are among the most hardy of bacterial species. Since these organisms are among the most ancient of bacterial species, it follows that they have excellent characteristics for survival and geographic distribution. Early conclusions that methanogenic bacteria were fragile were based more on the difficulties encountered in culturing them [4] than on experiments testing their ability to withstand abuse. Found with these organisms is a unique group of bacteria which oxidizes propionic and butyric acids [5,6], along with microorganisms which produce these acids. Many additional bacterial species are also found in anaerobic environments, such as those which metabolize a variety of organic monomers and polymers.

Many factors undoubtedly contributed to the incomplete metabolism which led to the deposition of vast amounts of plant residue during the growth of the great coal forests. This phenomenon can be explained by the assumptions that: (1) the range of microbial species prevalent at that time had inadequate biochemical capacity to efficiently catalyze complete anaerobic oxidation of the diversity of complex organic molecules produced by the many emerging plant species, and (2) aerobic processes were limited by ineffective oxygen transport within these vast deposits of plant material. Thus, ancient microbial conversions were insufficient to recycle the large amounts of biomass formed. Such a cycle would require anaerobic conversion of biomass to methane and carbon dioxide by microbial activities similar to those occurring in present-day anaerobic digesters. Incomplete anaerobic oxidations during this period could have been caused by organic acid accumulation in amounts inhibitory to the methane fermentation, as occurs during digester failure, due to the absence of a sufficiently efficient hydrogenogenic microflora at that time [7]. Prominant among these acids would be acetic acid, propionic acid and butyric acid. Under conditions of rapid biodegradation of organic matter, the methanogenic bacteria create a low partial pressure of hydrogen, a condition which favours the metabolic activities of bacteria which metabolize propionic and butyric acids [8-11]. However, when there is a perturbation of the fermentation the rate of methanogenesis decreases and may ultimately reach zero. Under these conditions, organic acids accumulate and eventually anaerobic bioconversion ceases.

For the purpose of this discussion, an extreme environment is considered to be an environment which is not compatible with the normal course of biological processes. These processes have been fine-tuned by evolution to guarantee successful continuity of the carbon cycle. Failure of the methanogenic bacteria and the organic acid-oxidizing bacteria to function in harmony may create such an extreme environment for all anaerobic bacterial life. Extreme environments created by this mechanism were likely contributing factors in the deposition of organic matter on and within the surface of the earth from prehistoric time to the present, and impact continually on the carbon cycle.

Combustion of fossil fuels for industrial and domestic use has added a new dimension to the carbon cycle which is yet to be clearly understood. Currently, man-made carbon dioxide accounts for approximately 50% of worldwide atmospheric emissions for all gases, including methane and carbon dioxide [12]. Worldwide emissions of carbon dioxide more than tripled between 1950 and 1980, and the effective doubling of atmospheric carbon dioxide is expected within 50 to 100 years [12]. The possible effects on the earth's environment of the continual changes in the carbon dioxide concentration of the atmosphere are the subject of wide speculation, with scientists unable to predict the outcome.

Biotechnology is the application of biological principles to the solution of practical problems. Given the ability to biologically regulate and control the methanogenic process, it would conceivably be possible to alter the carbon cycle in a small way, at least sufficient to accommodate carbon emissions which have resulted from the utilization of fossil fuels. The ratio of methane to non-digested biomass could be decreased in a methane-from-biomass system. The methane could be utilized as an energy source to replace fossil fuels for industrial and domestic use, and the non-digested biomass stored. Over a period of centuries, large amounts of carbon could be removed from the atmosphere by purpose-grown biomass. Such a system would, by necessity, be a worldwide endeavour and would lead to biological production of non-fossil fuel (methane) and deposition of large amounts of a stored carbon reserve. Such an undertaking would test the limits of biotechnology.

Any biotechnologically engineered shifts in the rates of methanogenesis must not, however, create a potential for perturbation of natural biological processes. Genetic manipulation of mesophilic bacteria, for example, could produce an organism capable of direct and rapid conversion of carbohydrates or organic acids to methane. Such an organism could become established in the digestive tract of ruminants and cause their starvation by rapidly metabolizing the organic acids on which they depend for energy, with catastrophic effects on world food supply.

The research and conclusions presented here were developed over the period of time from 1959 to the present and include the conclusions of the authors' most recent investigations. In particular, this paper addresses the quantitative measurement of perturbations of the methanogenic process, in laboratory-scale anaerobic digesters, caused by increased organic loading rates, reduced hydraulic retention times, and increased rates of organic acid addition. It has been speculated that short-chain organic acids inhibit the conversion of organic matter to methane by some mechanism which is pH independent [13]. The nature of this mechanism is obscure. High concentrations of propionic acid appear to be more toxic for anaerobic digestion than other acids [14]. Indeed, our observation has been that the metabolism of this acid is the most difficult parameter to control during the start-up of Napiergrass-fed digesters (Wilkie, unpublished).

RESULTS AND DISCUSSION

Biological formation of propionic acid is well understood relative to the understanding of propionic acid dissimilation under anaerobic conditions. Propionic acid formation has been a research area since the initial observation by Redtenbacher in 1846 that propionic acid is produced during the fermentation of glycerol [15]. Propionic acid-producing bacteria were first described by van Niel, in 1928 [16]. A major contribution to the understanding of propionic acid fermentation was the demonstration of heterotrophic CO_2 fixation by Wood and Werkman, in 1936 [17]. Since that time, pathways have been established for the formation of propionic acid via the acrylic acid pathway [18] and the methylmalonyl pathway [19]. These pathways appear to provide an adequate background for evaluation of the formation of this molecule in natural environments. Enzymes needed to catalyze the methylmalonyl reactions, including the phosphorylation enzymes, have been demonstrated to exist within bacteria in sufficient quantities to account for observed product formation and cellular growth [20].

In contrast, the understanding of the bacteria which utilize propionic acid is among the most primitive areas in the field of microbiology. It is currently somewhat below the level of understanding of the methane fermentation at the time of Schnellen's isolation of the first, indisputably pure methanogenic culture, in 1938 [21,22]. Bacteria capable of utilizing propionic acid as an energy source for growth have yet to be obtained in pure culture. The initially reported isolation of a pure culture of a propionate-utilizing methanogen [23] has not been repeated.

In earlier years one of the authors (P.H. Smith) spent considerable time trying to isolate a pure culture of a bacterial species capable of converting propionic acid to methane. Results were unsuccessful. In 1966, Smith reported the isolation of an unusual spirillum with a unique colony structure (Plate 1) from propionic acid enrichments of sewage sludge [24], which was subsequently described in 1974 by Ferry et al. [25]. Colonies would become quite large when the organism was grown in medium with added propionic acid and hydrogen. Transfer of large colonies into liquid medium with added hydrogen and propionic acid resulted in rapid uptake of hydrogen, but propionic acid utilization could not be initially detected. After three months of further incubation, propionic acid was utilized. The experiments were repeatable. However, after the colonies had been repeatedly subcultured and thoroughly purified, propionic acid metabolism could not be detected, even after four months of incubation. After many repetitions it was concluded that this interesting organism was methanogenic but incapable of metabolizing propionic acid. This organism was named Methanospirillum hungatii, in honor of Dr. R.E. Hungate's many contributions to the field of anaerobic microbiology [25].

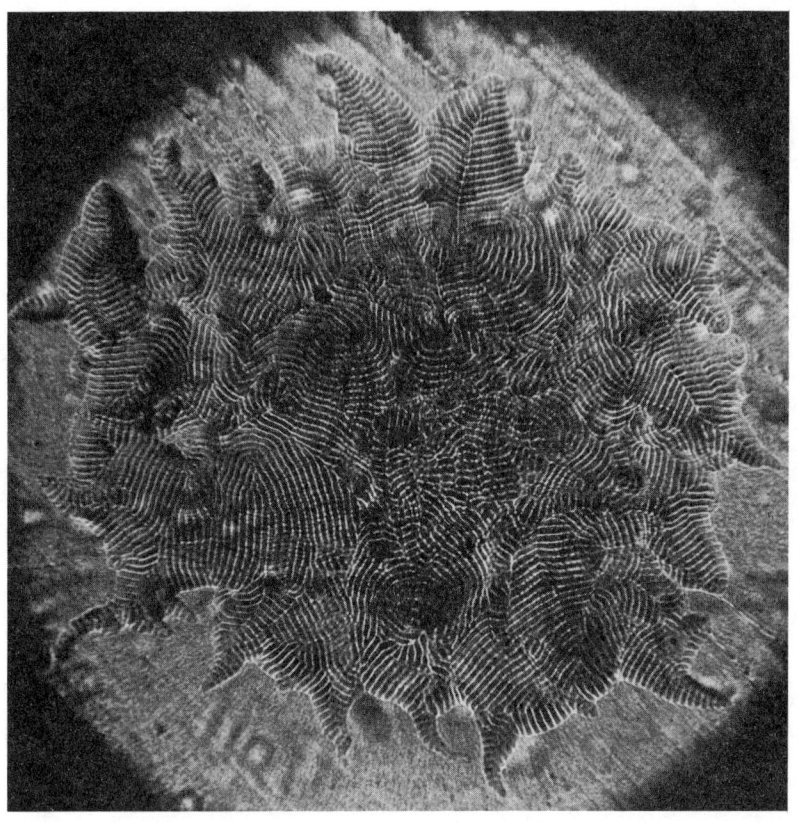

PLATE 1. Surface colony of M. hungatii sp.nov. exhibiting ridges two cell lengths (16 µm) apart [25].

   The fact that M. hungatii was repeatedly isolated in medium having high concentrations of propionic acid and the fact that, when subcultured, this organism tended to remain associated with bacteria which metabolize propionic acid suggests that M. hungatii may have a unique affinity for propionic acid-utilizing bacteria. This experience is reminiscent of the experience of Bryant et al. with ethanol [8], except that pure cultures of the propionate-utilizing organism could not be obtained. The observed close relationship of M. hungatii with propionate metabolism suggests that this organism would be attractive as a biotechnological tool in the development of a rapid fermentation of propionic acid, which could have potential in the treatment of "stuck" digesters or in the enhancement of anaerobic metabolism to assist in bioremediation of waste sites.

Prior to the pioneering work of Boone [26], no conclusive evidence existed which excluded the formation of methane directly from propionic acid. Speculation developed that propionate-utilizing organisms were not methanogenic, based on the negative evidence that these organisms could not be isolated on a hydrogen energy source (P.H. Smith, unpublished), that ethanol metabolism produced hydrogen rather than methane [8], and that molecular hydrogen inhibited propionic acid formation in domestic sludge [9,10]. Boone [26] observed that hydrogen was produced during the metabolism of propionic acid by propionic acid enrichments. Oxygen removal techniques adequate for the culturing of methanogenic bacteria were inadequate for quantitative product recoveries with Boone's enrichment system. Use of the culture technique developed by Zehnder and Wuhrmann [27], which utilizes the titanous ion to remove oxygen, permitted the reactions to progress in the absence of inhibition and quantitative recoveries were obtained. Similar experiments using reduced copper for oxygen removal were unsuccessful. These experiments demonstrated that propionic acid could be quantitatively converted to molecular hydrogen, acetic acid and carbon dioxide by a propionic acid enrichment [11,26]. Boone and Bryant [5] first grew propionic acid metabolizing bacteria in biculture and described Syntrophobacter wolinii. Progress towards obtaining pure cultures of these bacteria has not been reported.

Repeated unsuccessful efforts to isolate a pure culture of a bacterium which utilizes propionic acid stimulated a change in the direction of the research effort, from isolation studies to investigations of rates of propionate formation, propionate conversions during anaerobic digestion, and the interactive effects of propionic acid metabolism during conversion of complex organic matter to methane.

Propionic acid appears to be an important intermediate in the rapid metabolism of complex organic matter [28,29]. The rate of dissimilation of an intermediate is the product of pool size times its turnover rate [29]. In Smith's laboratory, numerous turnover experiments were conducted [30] utilizing the procedures developed to evaluate the contribution of acetic acid to methane formation in domestic sludge [31]. Results summarized in Table 1 show propionic acid as an important intermediate in all cases examined. The subtotal from acetic acid shows that a substantial amount of acetic acid is derived from propionic and butyric acid (Table 1). The location of the propionic acid in the digesting material was determined by separating propionic acid in the liquid phase (extracellular) from propionic acid in the solids phase (intracellular), using the procedure described for acetate [31]. These experiments demonstrated that propionic acid formation results in the deposition of the acid in the liquid fraction of the digesting material and that the microflora which metabolize these molecules would be associated with the liquid fraction and possibly the surface of the solids fraction [30]. The formation of this molecule may be significant in the evaluation of natural processes because propionic acid formation may have regulatory impacts on overall anaerobic digestion processes, above and beyond pH effects created by acid accumulation. The accumulation of this acid may create extreme environments in microniches where it may be formed and lead to inhibition of other biological conversions. Propionic acid may also create conditions in the liquid fraction of fermenting organic material which could be inhibitory to both the methanogenic and the non-methanogenic microflora.

TABLE 1

Organic acid precursors to methane formed during the anaerobic digestion of domestic waste [30].

Organic acid	Percentage contribution to total methane formed
Total from acetic acid[a]	70±5%
Total from propionic acid[a]	23 – 30
Total from butyric acid[a]	10 – 40
Subtotal from acetic acid[b]	21 – 49

[a] Total rates of formation of methane were measured directly. Total rates of formation of individual acids were experimentally determined under steady-state conditions by measuring the concentrations of the individual acids and measuring a rate constant for their formation by isotope dilution procedures.
[b] Calculated by adding the stoichiometric amounts of acetic acid which should be derived from propionic and butyric acid, based on their rates of formation.

Within the frame of reference initially established in this paper, an environment is extreme when it will no longer support the normal course of existing biological processes. When anaerobic dissimilation of complex organic matter to methane is inhibited, an extreme environment is created. This applies equally to tropical rain forests such as existed during the Carboniferous period and to anaerobic digesters which have become inhibited. The former case is no longer subject to direct experimentation since the biological materials involved have not been viable for millions of years. The latter case is amenable to experimentation which could also elucidate fundamental biological mechanisms prevalent during the Carboniferous period.

In laboratory-scale anaerobic digesters, we have experimentally examined three different forms of organic feedstock: domestic sewage sludge, plant biomass and organic acids. These experiments were conducted to investigate fundamental features of the inhibition of anaerobic digestion created by artificially imposed imbalance. With domestic sewage sludge, the two means chosen to create imbalance, to the point of digester failure, were: (1) to gradually overload the fermentation with excess organic substrate, and (2) to gradually decrease the hydraulic retention time. Parameters monitored included methane production rates, hydrogen uptake rates, organic acid concentrations, and pH. Hydrogen uptake rates were measured as a possible predictive parameter for identifying the onset of digester failure. The rationale of the hydrogen uptake experiments was that if the artificially created imbalance were the sole result of effects on the hydrogen-utilizing methanogens, it would then follow, a priori, that decreases in methane production from externally supplied hydrogen should precede decreases in the capacity of the digester microflora to produce methane from other metabolic sources. Similarly, if the imbalance were the sole result of effects on the organic acid-utilizing bacteria, it would then follow, a priori, that organic acid accumulation should precede decreases in the methane production rates.

Fresh domestic sludge was fed to two mesophilic laboratory-scale digesters (35°C, 4 liter culture volume). The hydraulic retention time was varied with one digester (A) and the amount of substrate fed was varied with the second digester (B) (Fig. 1). Changes in grams solids fed were correlated with hydraulic retention time (HRT) to give approximately the same rates of change in methane production in the two digesters. Methane production increased initially with decreases in the hydraulic retention time (digester A) and with increases in the amount of substrate fed (digester B) (Fig. 2).

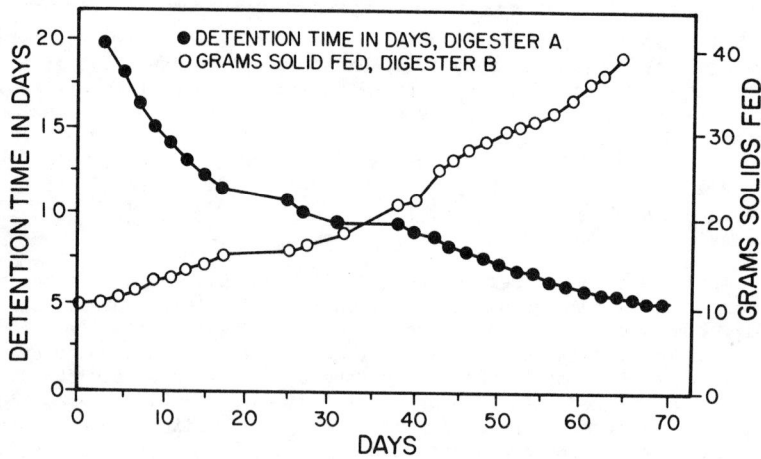

FIGURE 1. HRT and amount of substrate fed in digesters being induced to fail.

Hydrogen uptake capacity was measured by removing a 50 ml sample and measuring the total amount of methane produced when the sample was incubated under an atmosphere of 70% hydrogen and 30% carbon dioxide for 30 minutes. A similar sample was incubated under an atmosphere of 70% nitrogen and 30% carbon dioxide (Fig. 2). Methane production from externally supplied hydrogen (Fig. 3) was then calculated by subtracting the methane production rate under nitrogen and carbon dioxide (Fig. 2) from the methane production rate under hydrogen and carbon dioxide. The results show that the digesting sludge always had a capacity to convert excess hydrogen to methane (Fig.3). This capacity was equal, in methane equivalents, to approximately 70% of total methane formation in the absence of added hydrogen. Under the conditions imposed, the capacity of sludge to produce methane from hydrogen and carbon dioxide reflects the capacity of sludge to produce methane from other metabolic sources. During failure, decrease in methane production from hydrogen did not precede a decrease in methane formation from other sources.

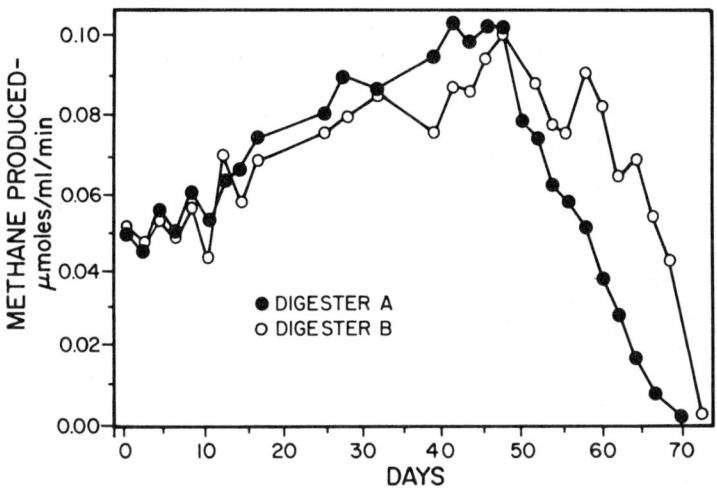

FIGURE 2. Methane production rates in digesters being induced to fail.

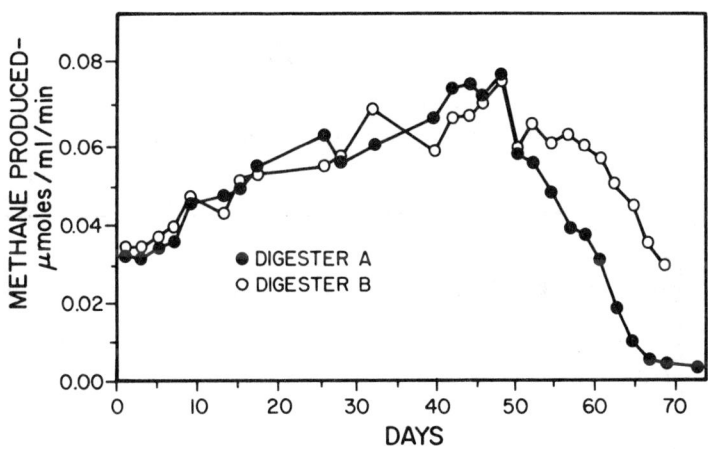

FIGURE 3. Methane production rates from externally supplied H_2 in digesters being induced to fail.

The concentration of organic acids in the mixed liquor was measured over the same time period (Figs. 4-8). The pH of the digestion was maintained at approximately pH 7 during the course of the experiments. In all cases, the concentration of organic acids varied inversely to the rate of methane production. As with the measurements of hydrogen uptake capacity, changes in these parameters did not precede decreases in the rate of methane formation during digester failure.

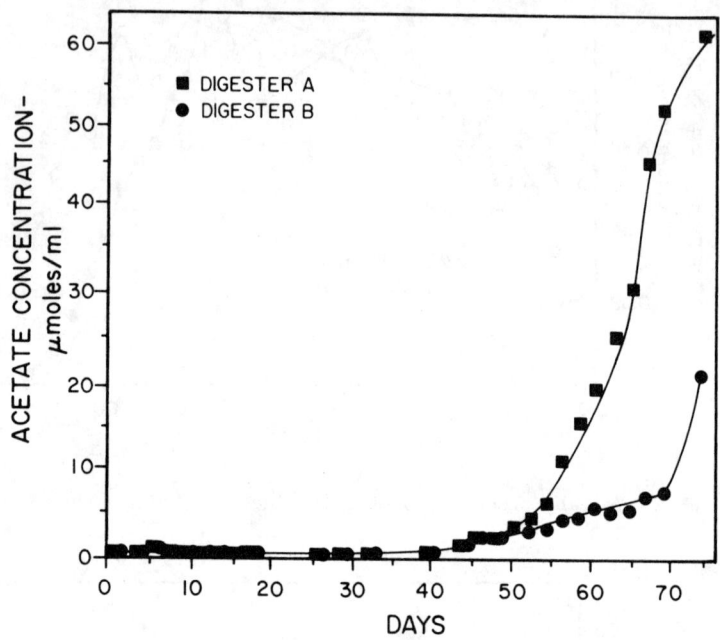

FIGURE 4. Acetate concentration.

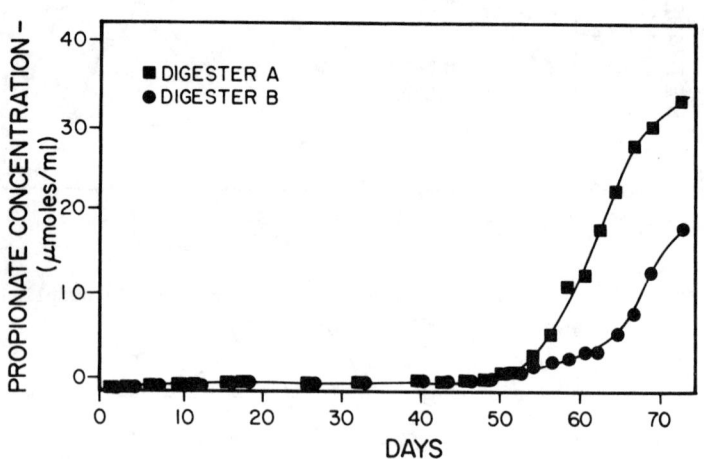

FIGURE 5. Propionate concentration.

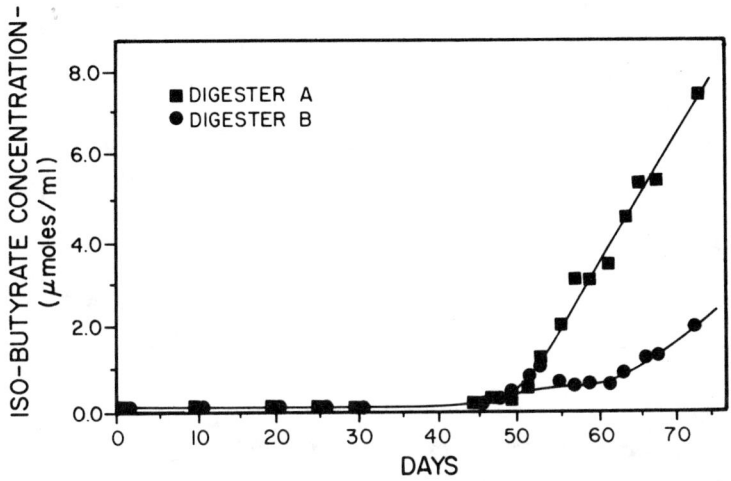

FIGURE 6. iso-Butyrate concentration.

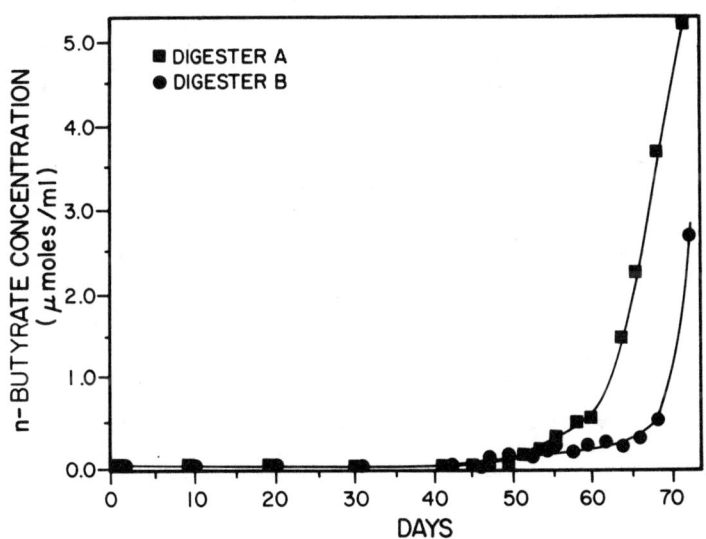

FIGURE 7. n-Butyrate concentration.

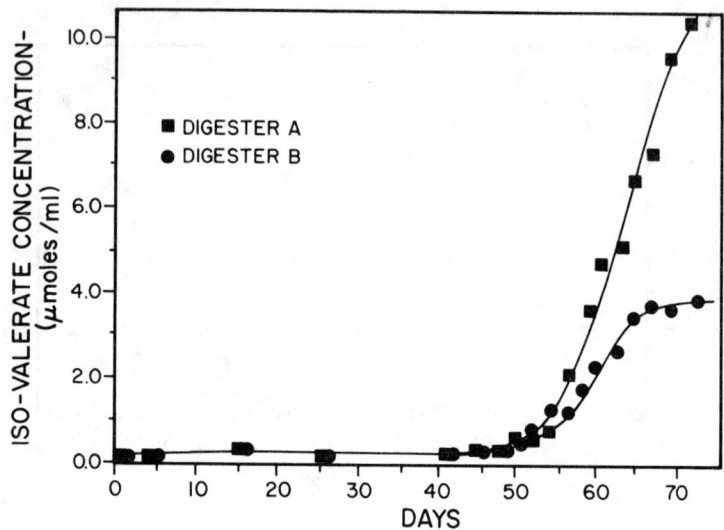

FIGURE 8. iso-Valerate concentration.

These results show that both the hydrogen-utilizing methanogenic bacteria and the bacteria which utilize organic acids were equally sensitive to the cause of the perturbation of the fermentation. The continued increase in the concentrations of organic acids suggests that the bacteria which produce these acids were not inhibited. The cause of the effects observed is obscure. We can conclude that the initiation of the perturbation was the result of something other than the inhibition of methane production from hydrogen and carbon dioxide, from organic acid conversion, or by pH.

Theoretical modeling has been used to characterize the effects of organic acids on methane formation [32,33], but no data are available to evaluate the effects of the rates of formation or the concentration of these acids on the rate and stability of anaerobic digestion of biomass in a single-stage process. There has been no procedure available which permits the assessment of these phenomena, because batch addition of the acids does not take into account possible shifts in the microbial population. However, the question could be addressed if it were possible to alter the acid formation rates in isolation. This cannot be done by increasing the rate of substrate addition because this also increases the concentration of many other molecules.

Experiments were conducted, by infusing the sodium salts of organic acids into thermophilic digesters (55°C, 3.6 liter culture volume), to determine the net effect of organic acid formation rate and concentration on overall stability of biomass-fed digesters [7,34]. Following the establishment of new steady-state organic acid concentrations and methane production rates, the rates of infusion were increased. This was repeated until digester failure occurred. These experiments showed that thermophilic digesters could operate at higher external pool sizes of propionic and butyric acid than usually observed in CSTR reactors. Infusion of propionic or butyric acid salts into thermophilic digesters at rates of 25 µmol/ml of mixed liquor/day caused digester failure [34]. The inhibition observed could be attributed to either the accumulation of organic acids or the accumulation of sodium ion.

Wilkie modified the previously mentioned experiments and devised infusion procedures in which organic acids, not their sodium salts, could be continuously infused into mesophilic anaerobic digesters (35°C, 3.6 liter culture volume) being batch-fed Napiergrass on a daily basis. Since microbial activity is dependent upon optimal supply of nutrients, the Napiergrass fed to both the infused and control digesters was supplemented with specific micronutrients. These micronutrients have been shown to have a stimulatory influence on the metabolism of organic acids in Napiergrass digesters [35]. In four separate experiments, the concentrations of acetic, propionic and butyric acids decreased below the detection limits of 0.1 mM for all three acids, after micronutrient addition [35].

Organic acids were infused into stably operating, Napiergrass-fed digesters. The infusion experiments were designed to impose an extreme condition of organic acid formation rate on a normal fermentation of plant biomass. The rationale of these experiments is that the bacterial population in the digester cannot distinguish between metabolically produced organic acids and acids infused from an external source. This experimental procedure is designed to permit the growth of those bacteria which utilize the organic acids and to maximize the effects of these bacteria on other aspects of the fermentation. Under such circumstances the effects of the acid should become recognizable by comparison of the resulting fermentation balance of the infused digesters with the fermentation balance of control digesters to which no acid was added. Infusion of propionic acid created an inhibition. We have, therefore, artificially produced what can be characterized as an extreme environment.

The theoretical methane yield of infused digesters is calculated by adding the actual CH_4 yield from control digesters to the stoichiometric CH_4 yield expected from complete conversion of the particular organic acid infused. The experimental methane yield in the propionic acid infused digester was less than theoretically expected, at all infusion rates tested (Table 2). No similar inhibition has been observed with acetic acid, up to an infusion rate of 20 µmol/ml of mixed liquor/day (Table 2). With propionic acid, there is approximately a 13% inhibition of methane production at an infusion rate of 12.5 µmol/ml/day (Table 2). The extracellular concentration of propionic acid did not increase or decrease, ranging from < 0.1 to 0.2 mM, so there is no inhibition of propionate metabolism. The evolved hydrogen concentration in the gas phase remained the same as in controls with no propionic acid added, averaging 40-50 ppm, so there is no inhibition of hydrogen utilization. On the basis of our initial results, we consider it to be true that the bacteria which metabolize propionic acid produce some factor which is inhibitory to the hydrolysis of large polymers during the fermentation of plant material to methane.

TABLE 2

Continuous infusion of metabolic intermediates into Napiergrass-fed digesters.

Metabolic intermediate infused	Infusion rate (μmol/ml/day)	Acetic acid equivalents generated[a]	% Methane yield[b]
Acetic acid	20	20	98
Propionic acid	5	5	95
Propionic acid	7.5	7.5	92
Propionic acid	12.5	12.5	87

[a] μmol/ml/day of acetic acid based on either the direct amount of acetic acid infused or the stoichiometric amount of acetic acid which could be theoretically obtained from the amount of propionic acid infused.

[b]
$$\text{\% Methane yield} = \frac{\text{experimental methane yield}}{\text{theoretical methane yield}} \times 100$$

$$[\text{theoretical CH}_4 \text{ yield} = \text{control digester CH}_4 \text{ yield} + \text{stoichiometric CH}_4 \text{ yield from infused acid}]$$

CONCLUSIONS

When anaerobic dissimilation of complex organic matter to methane is inhibited, an extreme environment is created. Experiments were conducted to elucidate fundamental features of anaerobic digestion inhibition/failure created by artificially imposed imbalance. Hydrogen uptake capacity was measured as a possible predictive parameter for identifying the onset of digester failure. However, the initiation of the perturbation was shown to result from something other than the inhibition of methane production from hydrogen and carbon dioxide, from organic acid conversion, or by pH. Therefore, the assay of hydrogen conversion to methane is not a promising assay for predicting digester failure. Perturbation of methane formation observed due to increased propionic acid additions was pH independent and was unrelated to either the concentration of the propionic acid in the liquid fraction of the fermenting material or the hydrogen concentration in the evolved biogas. Until proven otherwise, we conclude that the bacteria which metabolize propionic acid produce some factor which is inhibitory to the initial phases of the conversion of complex organic material to methane. This inhibition, accompanied by a multiplicity of other factors relating to high concentrations of organic acids, such as low pH, could account for the incomplete metabolism of plant material which led to the formation of fossil fuels.

ACKNOWLEDGEMENTS

This investigation was supported by the U.S. Environmental Protection Agency, Grant No. 17070-DJV, and the Methane from Biomass and Wastes program sponsored by the Institute of Food and Agricultural Sciences (IFAS) of the University of Florida and the Gas Research Institute (GRI) of Chicago.

REFERENCES

1. Wald, G., Life and light. *Scientific American*, 1959, 201, 92-108.
2. Oparin, A.I., *The Origin of Life*, 2nd edition, Dover Publications, Inc., New York, 1953, 270pp.
3. Woese, C.R. and Fox, G.E., Phylogenetic structure of the prokaryotic domain: The primary kingdoms. *Proceedings of the National Academy of Sciences USA*, 1977, 74, 5088-90.
4. Smith, P.H. and Hungate, R.E., Isolation and characterization of Methanobacterium ruminantium n.sp. *Journal of Bacteriology*, 1958, 75, 713-8.
5. Boone, D.R. and Bryant, M.P., Propionate-degrading bacterium, Syntrophobacter wolinii sp.nov.gen.nov., from methanogenic ecosystems. *Applied and Environmental Microbiology*, 1980, 40, 626-32.
6. McInerney, M.J., Bryant, M.P., Hespell, R.B. and Costerton, J.W., Syntrophomonas wolfei gen.nov.sp.nov., an anaerobic, syntrophic, fatty acid-oxidizing bacterium. *Applied and Environmental Microbiology*, 1981, 41, 1029-39.
7. Smith, P.H., Bordeaux, F.M., Goto, M., Shiralipour, A., Wilkie, A., Andrews, J.F., Ide, S. and Barnett, M.W., Biological production of methane from biomass. In *Methane from Biomass: A Systems Approach*, eds., W.H. Smith and J.R. Frank, Elsevier Applied Science Publishers, London, 1988, pp. 291-334.
8. Bryant, M.P., Wolin, E.A., Wolin, M.J. and Wolfe, R.S., Methanobacillus omelianskii, a symbiotic association of two species of bacteria. *Archiv fur Mikrobiologie*, 1967, 59, 20-31.
9. Smith, P.H., Bordeaux, F.M. and Shuba, P.J., Methanogenesis in sludge digestion. *Abstracts of the 159th Meeting, American Chemical Society*, 1970, WATR 49.
10. Smith P.H. and Shuba, P.J., Terminal anaerobic dissimilation of organic molecules. In *Proceedings of the Bioconversion Energy Research Conference*, Institute for Man and His Environment, University of Massachusetts, Amherst, 1973, pp. 8-12.
11. Boone, D.R. and Smith, P.H., Hydrogen formation from volatile organic acids by methanogenic enrichments. *Abstracts of the Annual Meeting, American Society for Microbiology*, 1978, Q82, pp. 208.
12. Shepard, M., The politics of climate. *EPRI Journal*, 1988, 13(4), 4-15.
13. Buswell, A.M. and Hatfield, W.D., Anaerobic fermentations. *Illinois State Water Survey Bulletin*, 1936, 32, 193pp.
14. McCarty, P.L. and Brosseau, M.H., Effect of high concentrations of individual volatile acids on anaerobic treatment. In *Proceedings of the 18th Industrial Waste Conference*, Purdue University, Engineering Extension Series, 1964, 115, 283-96.
15. Redtenbacher, J., Notiz uber eine neue entstehungsweise der metacetonsaure. *Annalen der Chemie und Pharmacie*, 1846, 57, 174-7.
16. Van Niel, C.B., *The propionic acid bacteria*. Ph.D. thesis, Delft University of Technology, 1928 (N.V. Uitgeverszaak J.W. Boissevain & Co., Haarlem, publisher).
17. Wood, H.G. and Werkman, C.H., The utilisation of CO_2 in the dissimilation of glycerol by the propionic acid bacteria. *The Biochemical Journal*, 1936, 30, 48-53.
18. Johns, A.T., The mechanism of propionic acid formation by *Clostridium propionicum*. *Journal of General Microbiology*, 1952, 6, 123-7.

19. Swick, R.W. and Wood, H.G., The role of transcarboxylation in propionic acid fermentation. Proceedings of the National Academy of Sciences USA, 1960, 46, 28-41.
20. Wood, H.G. and Goss, N.H., Phosphorylation enzymes of the propionic acid bacteria and the roles of ATP, inorganic pyrophosphate, and polyphosphates. Proceedings of the National Academy of Sciences USA, 1985, 82, 312-5.
21. Schnellen, C.G.T.P., Onderzoekingen over de methaangisting. Ph.D. thesis, Delft University of Technology, 1947 (De Maasstad, Rotterdam, publisher).
22. Hungate, R.E., Development of ideas on the nature and agents of biomethanogenesis. In Biotechnological Advances in Processing Municipal Wastes for Fuels and Chemicals, ed., A.A. Antonopoulos, ANL/CNSV-TM-167, Argonne National Laboratory, Argonne, Illinois, 1985, pp. 1-14.
23. Stadtman, T.C. and Barker, H.A., Studies on the methane fermentation: VIII. Tracer experiments on fatty acid oxidation by methane bacteria. Journal of Bacteriology, 1951, 61, 67-80.
24. Smith, P.H., The microbial ecology of sludge methanogenesis. Developments in Industrial Microbiology, 1966, 7, 156-61.
25. Ferry, J.G., Smith, P.H. and Wolfe, R.S., Methanospirillum, a new genus of methanogenic bacteria, and characterization of Methanospirillum hungatii sp. nov. International Journal of Systematic Bacteriology, 1974, 24, 465-9.
26. Boone, D.R., Mechanism of the dissimilation of volatile organic acids by methanogenic enrichments. Ph.D. dissertation, University of Florida, Gainesville, 1977.
27. Zehnder, A.J.B. and Wuhrmann, K., Titanium (III) citrate as a nontoxic oxidation-reduction buffering system for the culture of obligate anaerobes. Science, 1976, 194, 1165-6.
28. Tarvin D. and Buswell, A.M., The methane fermentation of organic acids and carbohydrates. Journal of the American Chemical Society, 1934, 56, 1751-5.
29. Hungate, R.E., The Rumen and Its Microbes, Academic Press, New York, 1966, 533pp.
30. Smith, P.H., Studies of methanogenic bacteria in sludge. EPA-600/2-80-093, U.S. Environmental Protection Agency, Cincinnati, Ohio, 1980, 100pp.
31. Smith, P.H. and Mah, R.A., Kinetics of acetate metabolism during sludge digestion. Applied Microbiology, 1966, 14, 368-71.
32. Andrews, J.F. and Graef, S.P., Dynamic modeling and simulation of the anaerobic digestion process. In Anaerobic Biological Treatment Processes, Advances in Chemistry Series No. 105, American Chemical Society, Washington, D.C., 1971, pp. 126-62.
33. Moletta, R., Verrier, D. and Albagnac, G., Dynamic modelling of anaerobic digestion. Water Research, 1986, 20, 427-34.
34. Henson, J.M., Bordeaux, F.M., Rivard, C.J. and Smith, P.H., Quantitative influences of butyrate or propionate on thermophilic production of methane from biomass. Applied and Environmental Microbiology, 1986, 51, 288-92.
35. Wilkie, A., Goto, M., Bordeaux, F.M. and Smith, P.H., Enhancement of anaerobic methanogenesis from Napiergrass by addition of micronutrients. Biomass, 1986, 11, 135-46.

APPROACHES TO THE DEVELOPMENT OF GENE TRANSFER SYSTEMS IN
Methanobacterium thermoautotrophicum

LEO MEILE, THOMAS RECHSTEINER, URS JENAL, MARTIN JORDAN, THOMAS LEISINGER Mikrobiologisches Institut ETH, ETH-Zentrum
CH-8092 Zürich, Switzerland

INTRODUCTION

The concept of archaebacteria and progress in understanding the biochemistry of methanogenesis have led to an interest in the development of gene transfer and cloning systems for methanogenic bacteria. Besides serving as tools in research, such systems may, in the long run, lead to applications in the construction of methanogens utilizing more complex growth substrates, resistant to toxic chemicals in organic waste or able to perform desirable chemical transformations. Experiments to establish methods for the transfer of genes within methanogens and between methanogens and other bacteria have concentrated on methanococci, particularly on *Methanococcus voltae*, and on *Methanobacterium* sp. A number of auxotrophic and drug-resistant mutants have been isolated and characterized in *M. voltae*, and virus-like particles as well as a system for the transformation of this organism with linear DNA fragments have been reported [1, 2]. The genetic tools for *Mb. thermoautotrophicum* Marburg that have been developed in our laboratory, are listed in Table 1. They include a number of auxotrophic and drug-resistant mutants, the cloned tryptophan biosynthetic genes from the organism, the cryptic plasmid pME2001 and bacteriophage ΨM1. In addition, mutants resistant to base analogues and, most importantly, a method for the transformation of this organism by fragments of chromosomal DNA have been described by Nagle and co-workers [1, 8].
In the following we report on some properties of the two known extrachromosomal genetic elements, plasmid pME2001 and bacteriophage ΨM1, of *Mb. thermoautotrophicum* Marburg.

RESULTS

TABLE 1
Elements of gene transfer systems for *Mb. thermoautotrophicum* Marburg

Element	Description	Reference
Mutants	Trp⁻, Leu⁻, Thi⁻, Ade⁻	3
	Pseudomonic acid resistance	4
Selective markers	Cloned genes of tryptophan biosynthesis	5
Potential cloning vector	Plasmid pME2001	6
Generalized transducing phage	Bacteriophage ΨM1	7

Plasmid pME2001

The 4.5 kb cryptic plasmid pME2001 of *Mb. thermoautotrophicum* has been used for the construction of potential shuttle vectors capable of replication in eubacteria and yeast [9]. It was partially sequenced and shown to encode a 0.6 kb transcript whose 5'- and 3'-ends have been mapped [10]. Here we describe a plasmid-free mutant of *Mb. thermoautotrophicum* and an experiment to determine the copy number of pME2001.

Strain MBT9, the plasmid-free mutant, and strain MBT10, its pseudomonic acid-resistant derivative, grew at slightly reduced rates but to the same extent as the wild type strain. This indicates that plasmid pME2001 is dispensable for its host. The cured strain was originally isolated as a leucine auxotroph from a culture which had been treated with the alkylating agent N-methyl-N'-nitro-N-nitrosoguanidine, a mutagen causing closely spaced multiple mutations. The plasmid-free mutant strains may be of use as recipients in transformation experiments involving cloning vectors derived from pME2001. However, it cannot be decided whether the cured strains have lost pME2001 due to mutation(s) in the plasmid or whether they carry lesions in one or several host functions responsible for plasmid maintenance.

The experiment to estimate the copy number of plasmid pME2001 reported in Fig. 1 was dependent on a method for efficient labelling of DNA. Extracellular adenosine proved to be a good precursor of DNA in *Mb. thermoautotrophicum*. A gentle method for cell lysis that avoids shearing of plasmid DNA was another prerequisite for copy number estimation. It was afforded by the enzymatic lysis of *Mb. thermoautotrophicum* using pseudomurein endopeptidase [11]. The data presented in pseudomurein endopeptidase [11]. The data presented in Fig. 1 and those obtained in a second experiment with independently labelled DNA were used to estimate the copy number of the 4.5 kb plasmid pME2001. Based on a total genome size of 1725 kb for *Mb. thermoautotrophicum* [12, 13] copy numbers of 23.6 and 20.5 were calculated from the two experiments. These numbers indicate that pME2001 is present at 15-30 copies per genome equivalent.

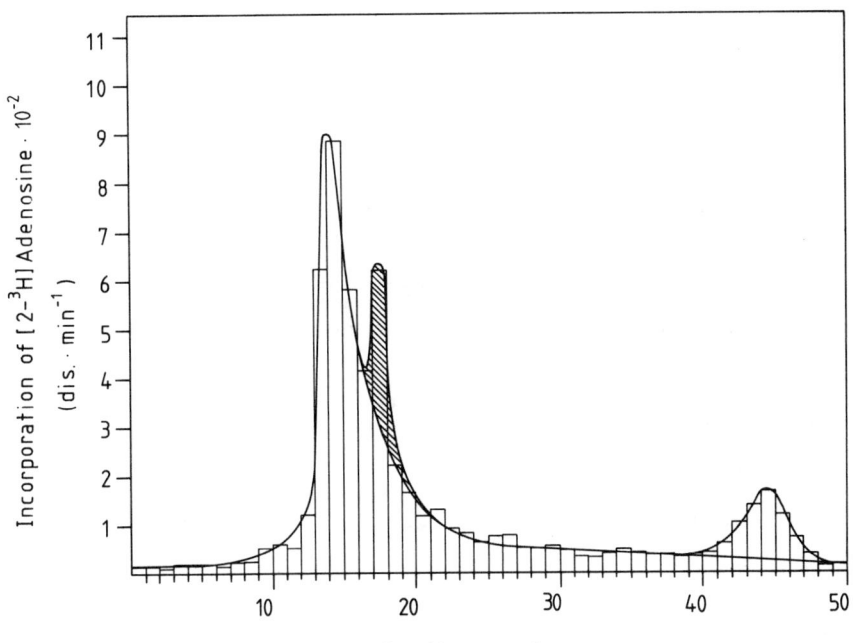

Figure 1. Estimation of pME2001 copy number.
Radioactivity profile of a 1% low melting agarose gel run with 30 µg of RNase-treated *Sma*I-digested, ^{3}H-labelled total DNA of strain MBT1. Labelling was done by growth for 2.5 generations in minimal medium containing 360 µM [2-^{3}H] adenosine. The start of the gel corresponds to fraction 1 on the figure. The first peak (fraction 15) of the profile was due to fragments of chromosomal DNA. The position of the second peak (fraction 18) corresponded to the position of linearized plasmid DNA in an ethidium bromide stained gel run under the same conditions. The hatched area indicates the radioactivity contributed by plasmid pME2001. The third peak (fraction 45) was due to undigested RNA and was therefore not taken into consideration for calculating the copy number.

Bacteriophage ΨM1
Bacteriophage ΨM1 is a virulent, oxygen resistent phage of *Mb. thermoautotrophicum*. Its genome consists of linear double-stranded DNA with a size of 30.4 ± 1.0 kb. Restriction and hybridization analysis of DNA extracted from phage particles revealed two types of linear molecules with the size of the phage genome. About 85 percent of the DNA molecules in such preparations consisted of ΨM1 DNA, whereas approximately 5 percent were multimers of plasmid pME2001. Phage ΨM mediated generalized transduction. The three biosynthetic markers Leu$^+$, Trp$^+$ and Ade$^+$ as well as resistance to pseudomonic acid were transduced at frequencies of 10^{-4} to 10^{-5} per PFU.

CONCLUSIONS

Methods for the exchange of genetic information within *Mb. thermoautotrophicum* by transformation with linear chromosomal DNA and by generalized transduction have become available. Also available are some of the elements of a cloning system, namely the *Mb. thermoautotrophicum* replicon pME2001 and cloned methanogen genes suitable as selective markers. To use these elements cloning vectors have to be constructed and a protocol for plasmid transformation of the Marburg strain needs to be developed.

Acknowledgements
This work was supported by the Swiss National Foundation for Scientific Research (project No. 3.193-0.85).

REFERENCES

1. Nagle, D.P., Development of genetic systems in methanogenic archaebacteria. Dev. Ind. Microbiol., 1988, in press.
2. Bertani, J. and Baresi, L., Genetic transformation in the methanogen *Methanococcus voltae* PS. J. Bacteriol., 1987, **69**, 2730-2738.
3. Kiener, A., Holliger, C. and Leisinger, T., Analogue-resistant and auxotrophic mutants of *Methanobacterium thermoautotrophicum*. Arch. Microbiol., 1984, **139**, 87-90.
4. Kiener, A., Rechsteiner, T. and Leisinger, T., Mutation to pseudomonic acid resistance of *Methanobacterium thermoautotrophicum* leads to an altered isoleucyl-tRNA synthetase. FEMS Microbiol. Lett., 1986, **33**, 15-18.
5. Meile, L., Kotik, M. and Leisinger, T., 1988, unpublished.

6. Meile, L., Kiener, A. and Leisinger, T., A plasmid in the archaebacterium *Methanobacterium thermoautotrophicum*. <u>Mol. Gen. Genet.</u>, 1983, **191**, 480-484.
7. Meile, L., Jenal, U., Studer, D., Jordan, M. and Leisinger, T., Characterization of ΨM1, a virulent phage of *Methanobacterium thermoautotrophicum*, 1988, submitted.
8. Worrell, V.E., Nagle, D.P., McCarty, D. and Eisenbraun, A., Genetic transformation system in the archaebacterium *Methanobacterium thermoautotrophicum* Marburg. <u>J. Bacteriol.</u>, 1988, **170**, 653-656.
9. Meile, L. and Reeve, J.R., Potential shuttle vectors based on the methanogen plasmid pME2001, <u>Biotechnology</u>, 1985, **January**, 69-72.
10. Meile, L., Madon, J. and Leisinger, T., Identification of a transcript and its promoter region on the archaebacterial plasmid pME2001. <u>J. Bacteriol.</u>, 1988, **170**, 478-481.
11. Kiener, A., König, H., Winter, J. and Leisinger, T., Purification and use of *Methanobacterium wolfei* pseudomurein endopeptidase for lysis of *Methanobacterium thermoautotrophicum*. <u>J. Bacteriol.</u>, 1987, **169**, 1010-1016.
12. Mitchell, R.M., Loeblich, L.A., Klotz, L.C. and Loeblich, A.R., DNA Organization of *Methanobacterium thermoautotrophicum*. <u>Science</u> 1979, **204**, 1982-1084.
13. Brandis, A., Thauer, R.K. and Stetter, K.O., Relatedness of strains ΔH and Marburg of *Methanobacterium thermoautotrophicum*. <u>Zbl. Bakt. Hyg. I Abt. Orig.</u>, 1987, **C2**, 311-317.

IN VIVO NMR STUDIES OF THE METABOLISM AND BIOENERGETICS OF METHANOSARCINA BARKERI

H. Santos, P. Fareleira, R. Toci*, Y. Berlier*,
J. LeGall[+],*, Harry, D. Peck, Jr.[+] and A. V. Xavier

Centro de Química Estrutural, Complexo I, Av. Rovisco Pais, 1096 Lisboa Codex, Portugal
* Section d'Enzymologie Bactérienne A.R.B.S., CEN Cadarache, 13108 Saint Paul Lez Durance Cedex, France
[+] Department of Biochemistry, University of Georgia, School of Chemical Sciences, Athens, GA 30602 U.S.A.

Methanogenic bacteria are strictly anaerobic archaebacteria capable of coupling the exergonic process of methanogenesis with the synthesis of ATP [1]. Most species produce methane by reduction of CO_2 by H_2. *Methanosarcina barkeri* is the metabolically most versatile methanogen; besides H_2 + CO_2, it can utilize methanol, acetate, methanol + H_2 and methylamines. Methanol is disproportionated to carbon dioxide and methane by resting cells of *M.barkeri* according to the formula: $4CH_3OH \rightarrow 3CH_4 + CO_2$.

The process of methanogenesis, involving the reduction of CO_2 to CH_4, is accomplished in a series of intermediate steps where the C_1 units at the various stages of reduction are carried by several unusual coenzymes [2]. It is now well established that reduction of methyl-coenzyme M (CH_3S-CoM) is the terminal step in methane formation by all methanogens [2]. Furthermore, recent studies have shown that N-7-mercaptoheptanoylthreonine phosphate (HS-HTP), is the electron donor for the reduction of CH_3S-CoM to CH_4 [3, 4]. Although it is well known that the formation of methane is coupled to ATP synthesis, the mechanism by which ATP synthesis is coupled to CH_3SCoM reduction, remains a matter of controversy [1].

Nuclear Magnetic Resonance (NMR) has already proved to be a very powerful technique to investigate cellular processes in a non-invasive way [5]. *In vivo* ^{31}P NMR provides information about intracellular pH, energetic charge, rates of enzymatic reactions, transport of phosphorylated metabolites, etc. In vivo ^{13}C NMR can provide information on the variation of metabolite levels and fluxes through biochemical pathways. We have used NMR and mass-spectrometry to investigate the metabolism of methanol and pyruvate by resting cells of M.barkeri; NMR giving information on the energetic state of the cells and on the concentrations of substrates and products in the liquid phase and mass--spectrometry allowing to monitor the gas phase composition.

Although thick cell suspensions are required for the NMR experiments (20-30 mg of protein /ml), the rate of consumption of [^{13}C] methanol by M.barkeri is found to be similar to the rates reported in the literature for cell suspensions about ten times less concentrated [6]. Upon addition of [^{13}C] methanol, ^{13}C resonances due to labeled methane, CO_2 and HCO_3, appear and increase with time until the methanol is exhausted. Methanogenesis is associated with an increase in the intracellular ATP content, as monitored by ^{31}P NMR. Internal concentrations of 4 to 5 mM of ATP are determined, considering an internal cell volume of 3.2 μl/mg of protein [7]. Upon energization with methanol, an acidification of both the intracellular and the extracellular spaces is observed and a gradient of 0.5 pH units across the cytoplasmic membrane is determined from the ^{31}P NMR data. Addition of methyl viologen (5 mM) causes an immediate stop of the methanol consumption and the resonances due to ATP rapidly disappear. The same result is observed upon addition of the protonophore tetrachlorosalicylanilide (TCS) or chloroform. The inhibitor of the ATP synthase, N, N' - dicyclohexylcarbodiimide (DCCD) is ineffective in lowering the intracellular ATP levels; high concentrations of DCCD (corresponding to 300 nmol mg protein $^{-1}$) were required to decrease the ATP content by ~60% and under these conditions, formation of acetyl phosphate is detected [8]. In the last conditions the consumption rate of methanol is not affected by DCCD, and acetate is detected as a final product by ^{13}C NMR.

When sodium pyruvate (100 mM) is administered to a starving cell suspension of M.barkeri, methane formation and ATP synthesis are observed, showing that pyruvate can also be utilized by whole cells of M.barkeri as a substrate for methanogenesis. Pyruvate was consumed at a rate of 12 nmol min $^{-1}$ mg protein $^{-1}$. Similar experiments monitored by mass-spectrometry, using a membrane-inlet vessel, showed that H_2 and CO_2 are also produced in addition to methane. From the proton NMR spectra, acetate and acetone were identified as the other main products of pyruvate metabolism. In order to investigate the mechanism of ATP synthesis, the effect of methyl viologen and TCS, on the ATP levels were tested. Methyl viologen causes a transitory decrease of the intracellular ATP content with accumulation of acetyl phosphate, which is subsequently consumed as the ATP level recovers. Methanogenesis is inhibited, but the rates of CO_2 and H_2 production are stimulated. Addition of TCS causes only a small decrease in the ATP content. The results obtained with pyruvate demonstrate the activity of pyruvate dehydrogenase in whole cells of *M.barkeri* and are considered as strong evidence for the presence of a mechanism of substrate level phosphorylation for the ATP formation.

The detection of an energy-rich phosphocompound (acetyl phosphate), when *M.barkeri* is energized with methanol is rather interesting, providing additional information for the controversial question about the mechanism of energy conservation in methanogenic bacteria. In fact, although in a different context, evidence has been obtained in recent years in favour for a chemiosmotic type mechanism for the synthesis of ATP in *M.barkeri*, from a variety of substrates [9]. In contrast to this, synthesis of ATP has been observed in whole cells of *Methanobacterium thermoautotrophicum* [10] and *Methanococcus voltae* [11], apparently in the absence of an electrochemical proton gradient. These results were interpreted by Lancaster [12] as indicating that ATP is synthesized by the mechanism of substrate level phosphorylation. Further support for this hypothesis has been recently obtained by Keltjens *et al* [13], who have reported inorganic pyrophosphate synthesis during methanogenesis from methylcoenzyme M by cell-free extracts of *Methanobacterium thermoautotrophicum*.

Our own results on the detection of acetyl phosphate in *M.barkeri* suggest a mechanism by which ATP is produced at the expense of acetyl phosphate. However it is not yet clear whether acetyl phosphate is also formed in the absence of DCCD, since the failure to detect a resonance from acetyl phosphate under these conditions, might be due to low steady state concentration of this compound in the undisturbed cells.

(This work was supported by Junta Nacional de Investigação Científica e Tecnológica (JNICT), Grant nr. 832.86.178).

References

[1] Daniels, L., Sparling, R. and Sprott, G. D., Biochim. Biophys. Acta, 1984, **768**, 113-163.
[2] Wolfe, R. S. Trends Biochem. Sci., 1985, **10**, 396-399.
[3] Ellerman, J., Hedderich, R., Böcher, R. and Thauer, R. K., Eur. J. Biochem., 1988, **172**, 669-677.
[4] Bobik, T. A. and Wolfe, R. S., Proc. Natl. Acad. Sci. USA, 1988, **85**, 60-63.
[5] Shulman, R. G., Trends Biochem. Sci., 1988, **13**, 37-39.
[6] Eikmanns, B. and Thauer, R. K., Arch. Microbiol., 1984, **138**, 365-370.
[7] Blaut, M. and Gottschalk, G., Eur. J. Biochem., 1984, **41**, 217-222.

[8] Santos, H., Fareleira, P., Toci, R., LeGall, J., Peck, H. D. Jr. and Xavier, A. V., Eur. J. Biochem., 1988, submitted.
[9] Blaut, M. and Gottschalk, G., Trends Biochem. Sci., 1985, **10**, 486-489.
[10] Schonheit, P. and Beimborn, D. B., Eur. J. Biochem., 1985 **148**, 545-550.
[11] Crider, B. P., Carper, S. W. and Lancaster, J. R. Jr., Proc. Natl. Acad. Sci. USA, 1985, **82**, 6793-6796.
[12] Lancaster, J. R. Jr., FEBS Lett.,1986, **199**, 12-18.
[13] Keltjens, J. T., van ERP, R., Mooijaart, R. J., van der Drift, C. and Volgels, G. D., Eur. J. Biochem., 1988, **172**, 471-476.

TAXONOMY OF HALOPHILIC BACTERIA

ANTONIO VENTOSA
Department of Microbiology,
Faculty of Pharmacy, University of Sevilla,
Sevilla, Spain

ABSTRACT

Hypersaline habitats are typical extreme environments that include saline lakes, salterns and saline soils. Halophilic bacteria able to grow at high salt concentrations are included in two main physiological groups: moderately and extremely halophilic bacteria. The moderately halophilic bacteria are a very heterogeneous taxonomic group and include species that belong to very different genera. With the exception of three recently described species included in the archaebacterial genera Methanohalophilus and Halomethanococcus, they are eubacteria included in the genera: Vibrio, Flavobacterium, Deleya, Halomonas, Haloanaerobium, Halobacteroides, Sporohalobacter, Spirochaeta, Micrococcus, Sporosarcina, Marinococcus, Paracoccus, Ectothiorhodospira and Rhodospirillum. Systematic studies of hypersaline natural environments could originate the recognition of new taxa in the future.

Extremely halophilic bacteria are, not exclusively but essentially, members of the family Halobacteriaceae. The genera currently accepted in this family are: Halobacterium, Haloarcula, Haloferax, Halococcus, Natronobacterium and Natronococcus. However, some new groups may exist on the basis of polar lipid analyses and nucleic acid hybridization studies. The current taxonomic distribution and characteristics of these bacteria are reviewed.

INTRODUCTION

Hypersaline environments include water and soil habitats. The former are divided into thalassohaline and athalassohaline (i.e. the relative proportions of individual salts are similar or dissimilar to those found in seawater, respectively)[1]. The most extensively studied habitats are natural lakes (Great Salt Lake, in U.S.A., the Dead Sea, in Israel, Wadi Natrun in Egypt and Lake Magadi in Kenya) and solar salterns [1-6]. These environments have some typical characteristics which imply that the microorganisms that live there are very well adapted; these characteristics are besides the high salinity, high temperature, low oxygen availability, high nutrient availability, high light intensity and pH near neutrality or, in some cases, extremely alkaline [1].

The response of the microorganisms to salt can be divided into different categories. Kushner [7] proposed a classification that includes: i) non-halophilic microorganisms, if they grow best in media containing less than 0.2 M (~1%) salt; some of them can tolerate high concentrations of salts and are called halotolerant; ii) slight halophiles, those which grow best in media with 0.2-0.5 M (~1-3%) salt; iii) moderate halophiles for those that grow best in media containing 0.5-2.5 M (~3-15%) salt; iv) borderline extreme halophiles, if they grow best in media with 1.5-4.0 M (~9-23%) salt, and v) extreme halophiles, if they grow best in media containg 2.5-5.2 M (saturated) salt (~15-32%).

The two most important groups of microorganisms adapted to live in hypersaline habitats are moderately and extremely halophilic bacteria. Halotolerant and slight halophiles seem to play a minor role in these habitats, and at least in habitats with more than 10% salts they constitute a low proportion of the total microbial population [1]. Until the discovery that the archaebacteria (which includes the members of the Halobacteriaceae and some other organisms) were a different phylogenetic branch [8], there were speculations about a possible close relationship between these and moderately halophilic eubacteria. However, now there is clear evidence that they are very different phylogenetic groups with the same ecological niches. On the other hand, moderately halophilic eubacteria are also phylogenetically very diverse, and the few organisms that have been studied have similarities with other organisms from different eubacterial branches [9-12].

In this paper the current status of the taxonomy of moderate to extremely halophilic bacteria will be reviewed. Recent publications on this subject should be consulted for a more detailed description [7,13-16].

MODERATELY HALOPHILIC BACTERIA

From a taxonomic point of view this is a very heterogeneous group that includes a wide variety of microorganisms, in contrast with extremely halophilic bacteria. Historically, they have been less atractive to researchers than halobacteria and the majority of studies were focused on their physiology and almost exclusively used one representative, Vibrio costicola, isolated from cured meats [17]. However, in this decade they have been extensively studied and currently they include a great variety of eubacteria, from heterotrophic to phototrophic, Gram-positive or Gram-negative, and very recently some moderately halophilic archaebacteria have also been isolated and characterized [18-21]. Table 1 lists all the bacterial species currently accepted as valid species and their type strain. References are also included for a detailed description of each species. The differential characteristics can be found in recent reviews [14,15]. Many of these species belong to genera that also include nonhalophilic or marine bacteria; however, some genera (Halomonas, Haloanaerobium, Halobacteroides, Sporohalobacter, Marinococcus, Methanohalophilus and Halomethanococcus) have been created specifically. Only one family, Haloanaerobiaceae, has been proposed for the grouping of moderately halophilic bacteria [11,12]. Flavobacterium halmephilum was included in the Approved Lists of Bacterial Names (46) but after the redefinition of the genus Flavobacterium, it is considered as a species of incertae sedis because it has a high G+C content, possesses ubiquinones and is not related genotypically to other flavobacteria [48]. Sporohalobacter lortetii was described originately as Clostridium lortetii , but based on com-

TABLE 1
Moderately halophilic bacteria accepted as valid species

Species	Type strain	Reference
1. Eubacteria:		
a. Heterotrophic:		
Gram-negative:		
Vibrio costicola	NCMB 701	[17,22]
Flavobacterium halmephilum	ATCC 19717	[23,24]
Deleya halophila	CCM 3662	[25]
Halomonas elongata	ATCC 33173	[26]
Halomonas subglaciescola	UQM 2927	[27]
Halomonas halodurans	ATCC 29686	[28,29]
Haloanaerobium praevalens	ATCC 33744	[30]
Halobacteroides halobius	ATCC 35273	[31]
Sporohalobacter lortetii	ATCC 35059	[12,32]
Sporohalobacter marismortui	ATCC 35420	[12]
Spirochaeta halophila	ATCC 29478	[33]
Paracoccus halodenitrificans	ATCC 13511	[34,35]
Gram-positive:		
Micrococcus halobius	ATCC 21727	[36]
Sporosarcina halophila	DSM 2266	[37]
Marinococcus halobius	ATCC 27964	[38,39]
Marinococcus albus	CCM 3517	[39]
b. Phototrophic:		
Rhodospirillum salexigens	DSM 2132	[40]
Rhodospirillum salinarum	ATCC 35394	[41]
Ectothiorhodospira vacuolata	DSM 2111	[42]
2. Archaebacteria:		
Methanohalophilus mahii	ATCC 35705	[43]
Methanohalophilus zhilinae	DSM 4017	[44]
Halomethanococcus doii	ATCC 43619	[45]

This list was correct on September 1, 1988, and contains all the moderately halophilic bacteria which have been validly published. Due to a typographic error Flavobacterium halmephilum was given incorrectly as F. halmephilium in the Approved Lists [46] and subsequently corrected [47].

parative 16S ribosomal RNA cataloging, its inclusion in a new genus was proposed [12,32].

Besides the moderately halophilic bacteria listed in Table 1, there are other moderately halophilic organisms that have not been validly published, as for example "Pseudomonas halosaccharolytica", "Micrococcus varians subsp. halophilus" or "Methanococcus halophilus" [21,49,50]. Other organisms are not considered members of the genus in which they were originally included and were not listed in the Approved Lists [46], as for example "Chromobacterium marismortui" [23]. Taxonomic studies are required to clarify the exact situation of these and other bacteria tentatively assigned to different genera, as for example Acinetobacter sp. [51], Flavobacterium sp. [51] and Bacillus spp. [52,53].

BORDERLINE EXTREME HALOPHILES

This is an intermediate category that was proposed [7] to include some species that, having an extreme salt requirement, were able to grow at lower salt concentrations than those of the "typical" extreme halophiles (members of the family Halobacteriaceae) described at that time. These bacteria are the actinomycete Actinopolyspora halophila (type strain ATCC 27976) [54], and the photosynthetic species Ectothiorhodospira halophila (type strain DSM 244) [55], Ectothiorhodospira halochloris (type strain DSM 1059) [56] and Ectothiorhodospira abdelmalekii (type strain DSM 2110) [57]. However, some halobacteria defined as extreme halophiles are able to grow in media with lower salt concentrations than classically accepted and thus, for example Haloferax volcanii and Haloferax mediterranei could also be included in this category [7,16]. Besides, some methanogenic bacteria have been isolated with salt responses similar to those of the previously mentioned microorganisms, as well as those of extreme halophiles, although they have not been validly described up to date [21,58]. When more halophilic bacteria are isolated from natural habitats with different characteristics, the inclusion of new isolates in the currently established salt-response categories will be more difficult, reflecting the wide salt spectrum that exists in Nature. For the reasons explained, in the past extreme halophile was synonymous with halobacteria and halococci but now is not, the term "aerobic halophilic archaebacteria" being used to define more precisely the members of the family Halobacteriaceae [13].

EXTREMELY HALOPHILIC BACTERIA

In the current edition of Bergey's Manual, vol. 1 [59], halobacteria are included in the family Halobacteriaceae and two genera are recognized: Halobacterium, for the rod-shaped bacteria and Halococcus for the coccoid forms. The genus Halobacterium includes five species: Halobacterium salinarium, Halobacterium saccharovorum, Halobacterium volcanii, Halobacterium vallismortis and Halobacterium pharaonis, and the genus Halococcus includes only one species: Halococcus morrhuae.

The main characteristics of the family Halobacteriaceae are: i) rods, cocci, a multitude of involution forms (discs, triangles, etc...), ii) requirement for at least 1.5 M (approx. 8%) NaCl for growth, iii) red pigmentation due to the presence of bacterioruberins (C_{50} carotenoids) in the cell, iv) lack of muramic acid-containing peptidoglycan in the cell envelope, v) membrane lipids composed of ether-linked isoprenyl posphoglycerides, vi) insensitivity towards many antibiotics (especially cell wall and protein synthesis inhibitors), vii) occur in habitats with high salt concentration (salt lakes, salterns, etc.) and viii) DNA composed of a major component (mol% G+C= 61-71) and a minor component (mol% G+C= 51-59) [16,59].

Since the publication of volume 1 of the current edition of Bergey's Manual [59] there have been several studies on the taxonomy of the family Halobacteriaceae, which lead to the description of new genera and species. Thus, new haloalkaliphilic strains were isolated by Tindall et al. [60] and they were pla-

ced in two new genera: Natronobacterium (which also included
H. pharaonis) and Natronococcus. The genotypic diversity of the
halobacteria was studied by Ross and Grant [61]. These authors
showed that genotypically they are more heterogeneous than previously reported and on the basis of DNA-16S rRNA hybridization
studies they obtained nine different rRNA groups that could
constitute different genera. Furthermore, Torreblanca et al.
[62] carried out an extensive phenotypic study of 148 non-alkaliphilic halobacteria, including the type strains of named species and isolates from different hypersaline environments. On
the basis of these results, the polar lipid composition of representative strains of each group, and the previous genotypic
studies, these authors proposed the classification of the non-
alkaliphilic halobacteria into three different genera: Halobacterium, Haloarcula and Haloferax. Two of the phenons obtained
in this study were also proposed as new species: Haloarcula
hispanica and Haloferax gibbonsii [63]. Also, two new species
of the genus Halobacterium have been described: H. sodomense,
isolated from the Dead Sea [64] and H. denitrificans, isolated
from a saltern [65].

Summarizing the current situation of the taxonomy of the
family Halobacteriaceae, Table 2 shows all the taxa that have
been validly published and included in the Approved Lists of
Bacterial Names [46] or published validly after January 1, 1980
and included in the additional Lists published in the International Journal of Systematic Bacteriology. The family currently
includes six different genera and probably in the future more
taxonomic groups will be described, in order to accommodate
some strains that do not fit with the existing genera. In a recent review the characteristics of the species included in these six genera are described in more detail [16]. It must be noted that the species Halobacterium halobium and Halobacterium
cutirubrum were included in the Lists [46] but actually all the
authors agree that they should be included in a single species
as Halobacterium salinarium since this name has priority [58,
60,61,74]. Table 3 shows the differential features of the genera.

In the section dedicated to the archaebacteria Volume 3 of

TABLE 2

Validly published taxa of the family Halobacteriaceae

Taxa	Type strain	Reference
Family Halobacteriaceae		[66]
Genus 1. Halobacterium		[67]
Species: Halobacterium salinarium	NRC 34002	[67]
Halobacterium halobium	NCIB 8720	[67]
Halobacterium cutirubrum	NRC 34001	[67]
Halobacterium trapanicum	NRC 34021	[67]
Halobacterium saccharovorum	ATCC 29252	[68]
Halobacterium sodomense	ATCC 33755	[64]
Halobacterium denitrificans	ATCC 35960	[65]
Genus 2. Haloarcula		[62]
Species: Haloarcula vallismortis	ATCC 29715	[62,69]
Haloarcula hispanica	ATCC 33960	[63]
Genus 3. Haloferax		[62]
Species: Haloferax volcanii	NCMB 2012	[62,70]
Haloferax mediterranei	CCM 3361	[62,71]
Haloferax gibbonsii	ATCC 33959	[63]
Genus 4. Halococcus		[72]
Species: Halococcus morrhuae	ATCC 17082	[72]
Genus 5. Natronobacterium		[60]
Species: Natronobacterium gregoryi	NCMB 2189	[60]
Natronobacterium magadii	NCMB 2190	[60]
Natronobacterium pharaonis	DSM 2160	[60,73]
Genus 6. Natronococcus		[60]
Species: Natronococcus occultus	NCMB 2192	[60]

This list was correct on September 1, 1988. It includes all taxonomic designations which have been validly published according to the rules of the International Code of Bacteriological Nomenclature. The first named species of a genus is the type species of the genus. It must be noted that strain NCIB 8720 is not available from this collection and is not maintained by any other culture collection.

TABLE 3

Main differential features between the six genera of the family Halobacteriaceae *

Characteristic	Halobacterium	Haloarcula	Haloferax	Halococcus	Natronobacterium	Natronococcus
Cell morphology	Rods	Pleomorphic rods	Pleomorphic rods	Cocci	Rods	Cocci
pH range for growth	5.0–8.5	5.0–8.5	5.0–8.5	5.0–8.5	8.5–11.0	8.5–11.0
Amino acid requirement	+	–	–	+	n.d.	n.d.
Mg^{2+} requirement (mM)	5	5	10–40	n.d.	<10	n.d.
Minimum salts requirement (%)	20	15	10	15	12	8
Polar lipids:						
PGP and PG	+	+	+	+	+	+
PGS	+	+	–	–	–	–
S-DGD	–	–	+	–	–	–
DGD-1	–	–	+	–	–	–
S-TGD-1 and S-TeGD or GL-2	+	–	–	Tr	–	–
DGD-2 and TGD-2 or TGD-1	+	+	–	+	–	–

PGP, phosphatidylglycerophosphate; PG, phosphatidylglycerol; PGS, phosphatidylglycerosulfate; S-DGD, sulfated diglycosyldiether; DGD-1, diglycosyl diether; S-TGD-1, sulfated triglycosyl diether; S-TeGD, sulfated tetraglycosyl diether; GL-2, unknown glycolipid; DGD-2, unknown diglycosyl diether; TGD-2, triglycosyl diether (glucosyl-mannosyl-glucosyl diether); TGD-1, triglycosyl diether (galactosyl-mannosyl-glucosyl diether).

n.d., not determined; Tr, trace.
* Adapted from reference [16].

Bergey's Manual will include all these changes with respect to volume 1, reflecting the extensive studies that have been carried out recently in this group of microorganisms. Besides, the new order Halobacteriales is created to include them, based on the molecular data available up to date, since there is clear evidence that halobacteria have a taxonomic ranking similar to the orders that comprise the methanogens (W.D. Grant, personal communication). However, more taxonomic studies are still necesary not only to elucidate whether other strains not included as valid species could constitute new taxonomic entities but also to isolate new halobacteria with different properties to those described. Thus, no thermophilic or acidophilic halobacteria have been reported to date.

Recently, 96 extremely halophilic, non-alkaliphilic cocci were isolated from several salterns located in Spain using different culture media and isolation conditions, and they were taxonomically studied together with seven reference strains of the genus Halococcus. A numerical study based on 114 phenotypic features grouped them in four phenons (Figure 1) [75]. The majority of isolates and all the reference strains clustered together in phenon A, identified as Halococcus morrhuae. Phenons B and C showed phenotypic characteristics similar to this species but with higher metabolic versatility and were considered as Halococcus sp. However, the four strains included in phenon D were significantly different from the other phenons. They were able to produce acid from glucose and used a large number of organic compounds as the sole source of carbon and energy. This study showed that there is a wider phenotypic diversity within the genus Halococcus than the previously described [75]. On the basis of their phenotypic characteristics, as well as other chemotaxonomic features recently studied, the strains included in phenon D have been considered a new species of the genus Halococcus, for which the new name Halococcus saccharolyticus is proposed (Montero et al., submitted for publication).

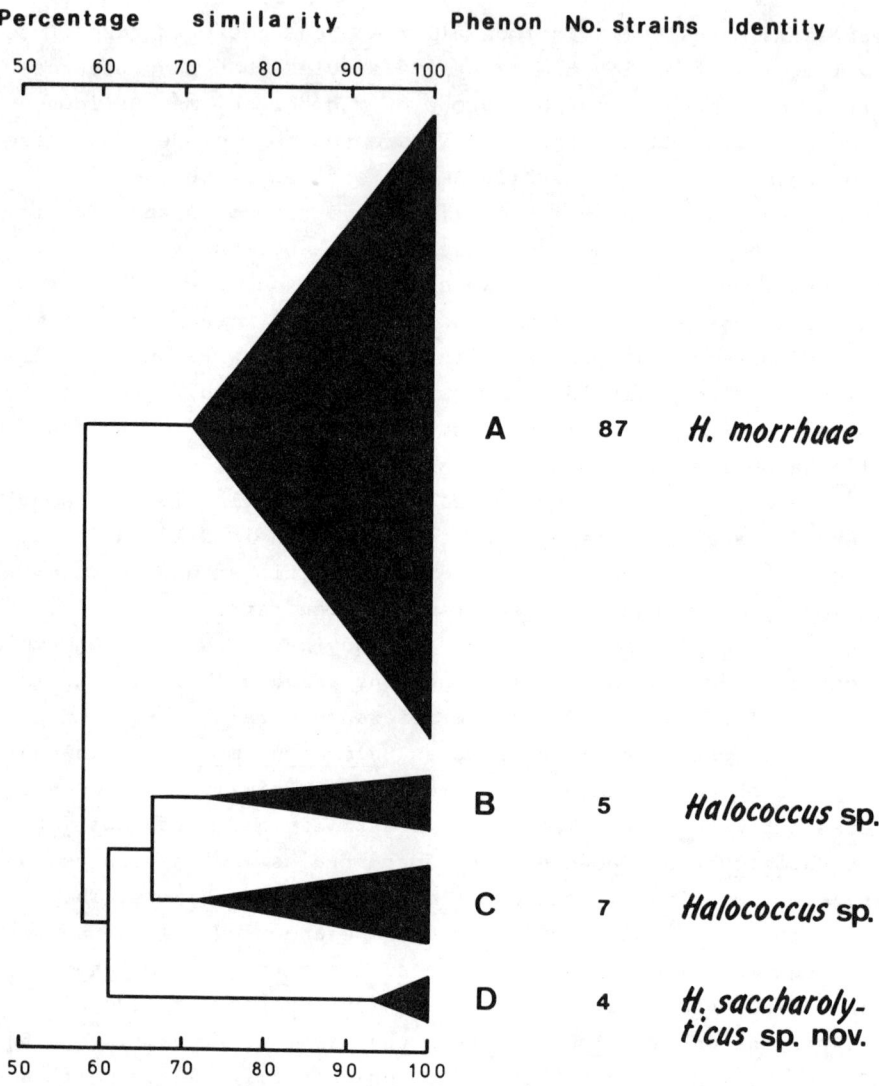

Figure 1. Simplified dendrogram showing the clustering of the strains based on the S_{SM} coefficient and UPGMA clustering, for 96 extremely haliphilic non-alkaliphilic cocci from solar salterns and seven reference strains of Halococcus morrhuae.

ACKNOWLEDGEMENTS

The author thank Dr. F. Rodriguez-Valera for the critical reading of the manuscript, and K. Hernandez for correcting the script.

REFERENCES

1. Rodriguez-Valera, F., Characteristics and microbial ecology of hypersaline environments. In Halophilic Bacteria, vol 1, ed., F. Rodriguez-Valera, CRC Press, Boca Raton, FL, 1988, pp. 3-30.

2. Nissenbaum, A., The microbiology and biogeochemistry of the Dead Sea. Microb. Ecol., 1975, **2**, 139-161.

3. Post, F.J., The microbial ecology of the Great Salt Lake. Microb. Ecol., 1977, **3**, 143-165.

4. Imhoff, J.F., Sahl, H.G., Soliman, G.S.H. and Trüper, H.D., The Wadi Natrun: chemical composition and microbial mass developments in alkaline brines of eutrophic desert lakes. Geomicrobiol. J., 1979, **1**, 219-234.

5. Tindall, B.J., Prokaryotic life in the alkaline, saline, athalassic environment. In Halophilic Bacteria, vol 1, ed., F. Rodriguez-Valera, CRC Press, Boca Raton, FL, 1988, pp. 31-67.

6. Rodriguez-Valera, F., Ventosa, A., Juez, G, and Imhoff, J.F., Variation of environmental features and microbial populations with the salt concentrations in a multi-pond saltern. Microb. Ecol., 1985, **11**, 107-115.

7. Kushner, D.J., The Halobacteriaceae. In The Bacteria, a treatise on structure and function, vol 8, ed., C.R. Woese and R.S. Wolfe, Academic Press, London, 1985, pp. 171-214.

8. Woese, C.R. and Fox, G.E., Phylogenetic structure of the prokaryotic domain: the primary kingdoms. Proc. Natl. Acad. Sci. USA, 1977, **74**, 5088-5090.

9. Paster, B.J., Stackebrandt, E., Hespell, R.B., Hahn, C.M. and Woese, C.R., The phylogeny of the spirochetes. Syst. Appl. Microbiol., 1984, **5**, 337-351.

10. Woese, C.R., Weisburg, W.G., Hahn, C.M., Paster, B.J., Zablen, L.B., Lewis, B.J., Macke, T.J., Ludwig, W. and Stackebrandt, E., The phylogeny of purple bacteria: the gamma subdivision. Syst. Appl. Microbiol., 1985, **6**, 25-33.

11. Oren, A., Paster, B.J. and Woese, C.R., Haloanaerobiaceae: a new family of moderately halophilic, obligatory anaerobic bacteria. Syst. Appl. Microbiol., 1984, 5, 71-80.

12. Oren, A., Pohla, H. and Stackebrandt, E., Transfer of Clostridium lortetii to a new genus, Sporohalobacter gen. nov. as Sporohalobacter lortetii comb. nov., and description of Sporohalobacter marismortui sp. nov. Syst. Appl. Microbiol., 1987, 9, 239-246.

13. Tindall, B.J. and Trüper, H.G., Ecophysiology of the aerobic halophilic archaebacteria. Syst. Appl. Microbiol., 1986, 7, 202-212.

14. Ventosa, A., Taxonomy of moderately halophilic heterotrophic eubacteria. In Halophilic Bacteria, vol. 1, ed., F. Rodriguez-Valera, CRC Press, Boca Raton FL, 1988, pp. 71-84.

15. Imhoff, J.F., Halophilic phototrophic bacteria. In Halophilic Bacteria, vol. 1, ed., F. Rodriguez-Valera, CRC Press, Boca Raton, FL, 1988, pp. 85-108.

16. Juez, G., Taxonomy of extremely halophilic archaebacteria. In Halophilic Bacteria, vol. 2, ed., F. Rodriguez-Valera, CRC Press, Boca Raton, FL, 1988, pp. 3-24.

17. Smith, F.B., An investigation on a taint in rib bones of bacon. The determination of halophilic vibrios (n. spp.). Proc. R. Soc. Queensland, 1938, 49, 29-53.

18. Paterek, J.R. and Smith, P.H., Methanohalophilus mahii gen. nov., sp. nov., a methylotrophic halophilic methanogen. Int. J. Syst. Bacteriol., 1988, 38, 122-123.

19. Mathrani, I.M., Boone, D.R., Mah, R.A., Fox, G.E. and Lau, P.P., Methanohalophilus zhilinae sp. nov., an alkaliphilic, halophilic, methylotrophic methanogen. Int. J. Syst. Bacteriol., 1988, 38, 139-142.

20. Yu, I.K. and Kawamura, F., Halomethanococcus doii gen. nov. sp. nov.: an obligately halophilic methanogenic bacterium from solar salt ponds. J. Gen. Appl. Microbiol., 1987, 33, 303-310.

21. Zhilina, T.N., Methanogenic bacteria from hypersaline environments. Syst. Appl. Microbiol., 1986, 7, 216-222.

22. Garcia, M.T., Ventosa, A., Ruiz-Berraquero, F. and Kocur, M., Taxonomic study and amended description of Vibrio costicola. Int. J. Syst. Bacteriol., 1987, 37, 251-256.

23. Elazari-Volcani, B., Studies on the microflora of the Dead Sea, Ph.D. Thesis, Hebrew University, Jerusalem, 1940.

24. Weeks, O.B., Genus Flavobacterium. In Bergey's Manual of Determinative Bacteriology, 8th ed., ed., R.E. Buchanan and N.E. Gibbons, Williams and Wilkins, Baltimore, 1974, pp. 357-364.

25. Quesada, E., Ventosa, A., Ruiz-Berraquero, F. and Ramos-Cormenzana, A., Deleya halophila, a new species of moderetely halophilic bacteria. Int. J. Syst. Bacteriol., 1984, **34**, 287-292.

26. Vreeland, R.H., Litchfield, C.D., Martin, E.L. and Elliot, E., Halomonas elongata, a new genus and species of extremely salt-tolerant bacteria. Int. J. Syst. Bacteriol., 1980, **30**, 485-495.

27. Franzmann, P.D., Burton, H.R. and McMeekin, T.A., Halomonas subglaciescola, a new species of halotolerant bacteria isolated from Antarctica. Int. J. Syst. Bacteriol., 1987, **37**, 27-34.

28. Rosenberg, A., Pseudomonas halodurans, sp. nov., a halotolerant bacterium. Arch. Microbiol., 1983, **163**, 117-123.

29. Hebert, A.M. and Vreeland, R.H., Phenotypic comparison of halotolerant bacteria: Halomonas halodurans sp. nov., nom. rev., comb. nov. Int. J. Syst. Bacteriol., 1987, **37**, 347-350.

30. Zeikus, J.G., Hegge, P.W., Thompson, T.E., Phelps, T.J. and Langworthy, T.A., Isolation and description of Haloanaerobium praevalens gen. nov. and sp. nov., an obligately anaerobic halophile common to Great Salt Lake sediments. Curr. Microbiol., 1983, **9**, 225-234.

31. Oren, A., Weisburg, W.G., Kessel, M., and Woese, C.R., Halobacteroides halobius, gen. nov., sp. nov., a moderately halophilic anaerobic bacterium from the bottom sediments of the Dead Sea. Syst. Appl. Microbiol., 1984, **5**, 58-70.

32. Oren, A., Clostridium lortetii sp. nov., a halophilic obligatory anaerobic bacterium producing endospores with attached gas vacuoles. Arch. Microbiol., 1983, **136**, 42-48.

33. Greenberg, E.P. and Canale-Parola, E., Spirochaeta halophila sp. n., a facultative anaerobe from a high-salinity pond. Arch. Microbiol., 1976, **110**, 185-194.

34. Robinson, J. and Gibbons, N.E., The effect of salts on the growth of Micrococcus halodenitrificans (n. sp.). Can. J. Bot., 1952, **30**, 147-154.

35. Kocur, M., Genus Paracoccus. In Bergey's Manual of Systematic Bacteriology, vol. 1, ed., N.R. Krieg, Williams and Wilkins, baltimore, 1984, pp. 399-402.

36. Onishi, H. and Kamekura, M., Micrococcus halobius sp. n. Int. J. Syst. Bacteriol., 1972, 22, 233-236.

37. Claus, D., Fahny, F., Rolf, H.J. and Tosunoglu, N., Sporosarcina halophila sp. nov., an obligate slightly halophilic bacterium from salt marsh soils. Syst. Appl. Microbiol., 1983, 4, 496-506.

38. Novitsky, T.J. and Kushner, D.J., Planococcus halophilus sp. nov., a facultatively halophilic coccus. Int. J. Syst. Bacteriol., 1976, 26, 53-57.

39. Hao, M.V., Kocur, M. and Komagata, K., Marinococcus gen. nov., a new genus for motile cocci with meso-diaminopimelic acid in the cell wall; and Marinococcus albus sp. nov. and Marinococcus halophilus (Novitsky and Kushner) comb. nov. J. Gen. Appl. Microbiol., 1984, 30, 449-459.

40. Drews, G., Rhodospirillum salexigens, spec. nov., an obligatory halophilic phototrophic bacterium. Arch. Microbiol., 1981, 130, 325-327.

41. Nissen, H. and Dundas, I.D., Rhodospirillum salinarum sp. nov., a halophilic photosynthetic bacterium isolated from a Portuguese saltern. Arch. Microbiol., 1984, 138, 251-256.

42. Imhoff, J.F., Tindall, B.J., Grant. W.D. and Trüper, H.G., Ectothiorhodospira vacuolata sp. nov., a new phototrophic bacterium from Soda Lake. Arch. Microbiol., 1981, 130, 238-242.

43. Paterek, J.R. and Smith, P.H., Methanohalophilus mahii gen. nov., sp. nov., a methilotrophic halophilic methanogen. Int. J. Syst. Bacteriol., 1988, 38, 122-123.

44. Mathrani, I.M., Boone, D.R., Mah, R.A., Fox, G.E. and Lau, P.P., Methanohalophilus zhilinae sp. nov., an alkaliphilic halophilic, methylotrophic methanogen. Int. J. Syst. Bacteriol., 1988, 38, 139-142.

45. Yu, I.K. and Kawamura, F., Halomethanococcus doii gen. nov. sp. nov.: an obligately halophilic methanogenic bacterium from solar salt ponds. J. Gen. Appl. Microbiol., 1987, 33, 303-310.

46. Skerman, V.B.D., McGowan, V. and Sneath, P.H.A., Approved lists of bacterial names. Int. J. Syst. Bacteriol., 1980, 30, 225- 420.

47. Hill, L.R., Skerman, V.B.D. and Sneath, P.H.A., Corrigenda to the approved lists of bacterial names. Int. J. Syst. Bacteriol., 1984, 34, 508-511.

48. Holmes, B., Owen, R.J. and McMeekin, T.A., Genus Flavobac-

terium. In Bergey's Manual of Systematic Bacteriology, vol. 1, ed., N.R. Krieg, Williams and Wilkins, Baltimore, 1984, pp. 353-361.

49. Hiramatsu, T., Ohno, Y., Hara, H., Yano, I. and Masui, M., Effects of NaCl concentration on the envelope components in a moderately halophilic bacterium, Pseudomonas halosaccharolytica. In Saline Environments, ed., H. Morishita and M. Masui, Business Centre for Academic Societies Japan, Tokyo, 1980, pp. 189-200.

50. Kamekura, M., Hamakawa, T. and Onishi, H., Application of halophilic nuclease H of Micrococcus varians subsp. halophilus to commercial production of flavoring agent 5'-GMP. Appl. Environ. Microbiol., 1982, 44, 994-995.

51. Quesada, E., Valderrama, M.J., Bejar, V., Ventosa, A., Ruiz-Berraquero, F. and Ramos-Cormenzana, A., Numerical taxonomy of moderately halophilic Gram-negative nonmotile eubacteria. Syst. Appl. Microbiol., 1987, 9, 132-137.

52. Onishi, H., Mori, T., Takeuchi, S., Tani, K., Kobayashi, T. and Kamekura, M., Halophilic nuclease of a moderately halophilic Bacillus sp.: production, purification, and characterization. Appl. Environ. Microbiol., 1983, 45, 24-30.

53. Weisser, J. and Trüper, H.G., Osmoregulation in a new haloalkaliphilic Bacillus from the Wadi Natrun (Egypt). Syst. Appl. Microbiol., 1985, 6, 7-11.

54. Gochnauer, M.B., Leppard, G.G., Komaratat, P., Kates, M., Novitsky, T. and Kushner, D.J., Isolation and characterization of Actinopolyspora halophila, gen. et sp. nov., an extremely halophilic actinomycete. Can.J. Microbiol., 1975, 21, 1500-1511.

55. Raymon, J.C. and Sistrom, W.R., Ectothiorhodospira halophila: a new species of the genus Ectothiorhodospira. Arch. Microbiol., 1969, 69, 121-126.

56. Imhoff, J.F. and Trüper, H.G., Ectothiorhodospira halochloris sp. nov., a new extremely halophilic phototrophic bacterium containing bacteriochlorophyll b. Arch. Microbiol., 1977, 114, 115-121.

57. Imhoff, J.F. and Trüper, H.G., Ectothiorhodospira abdelmalekii sp. nov., a new halophilic and alkaliphilic phototrophic bacterium. Zbl. Bakt. I, Abt. Orig. C, 1981, 2, 228-234.

58. Pérez-Fillol, M., Rodriguez-Valera, F. and Ferry, J.G., Isolation of methanogenic bacteria able to grow in high salt concentration. Microbiologia SEM, 1985, 1, 29-33.

59. Larsen, H., Family V. Halobacteriaceae. In Bergey's Manual of Systematic Bacteriology, vol. 1, ed., N.R. Krieg, Williams and Wilkins, Baltimore, 1984, pp. 261-267.

60. Tindall, B.J., Ross, H.N.M. and Grant, W.D., Natronobacterium gen. nov. and Natronococcus gen. nov., two new genera of haloalkaliphilic archaebacteria. Syst. Appl. Microbiol., 1984, **5**, 41-57.

61. Ross, H.N.M. and Grant, W.D., Nucleic acid studies on halophilic archaebacteria. J. Gen. Microbiol., 1985, **131**, 165-173.

62. Torreblanca, M., Rodriguez-Valera, F., Juez, G., Ventosa, A., Kamekura, M. and Kates, M., Classification of non-alkaliphilic halobacteria based on numerical taxonomy and polar lipid composition, and description of Haloarcula gen. nov. and Haloferax gen. nov. Syst. Appl. Microbiol., 1986, **8**, 89-99.

63. Juez. G., Rodriguez-Valera, F., Ventosa, A. and Kushner, D.J., Haloarcula hispanica spec. nov. and Haloferax gibbonsii spec. nov., two new species of extremely halophilic archaebacteria. Syst. Appl. Microbiol., 1986, **8**, 75-79.

64. Oren, A., Halobacterium sodomense sp. nov., a Dead Sea halobacterium with an extremely high magnesium requirement. Int. J. Syst. Bacteriol., 1983, **33**, 381-386.

65. Tomlinson, G.A., Jahnke, L.L. and Hochstein, L.I., Halobacterium denitrificans sp. nov., an extremely halophilic denitrifying bacterium. Int. J. Syst. Bacteriol., 1986, **36**, 66-70.

66. Gibbons, N.E., Halobacteriaceae. In Bergey's Manual of Determinative Bacteriology, 8th ed., ed., R.E. Buchanan and N.E. Gibbons, Williams and Wilkins, Baltimore, 1974, pp. 269-273.

67. Elazari-Volcani, B., Halobacterium. In Bergey's Manual of Determinative Bacteriolgy, 7th ed., ed., R.G. Breed, E.G. D. Murray and N.R. Smith, Williams and Wilkins, Baltimore, pp. 207-212.

68. Tomlinson, G.A. and Hochstein, L.I., Halobacterium saccharovorum sp. nov., a carbohydrate-metabolizing, extremely halophilic bacterium. Can. J. Microbiol., 1976, **22**, 587-591.

69. Gonzalez, C., Gutierrez, C. and Ramirez, C., Halobacterium vallismortis sp. nov., an amylolytic and carbohydrate metabolizing extremely halophilic bacterium. Can. J. Microbiol., 1978, **24**, 710-715.

70. Mullakhambhai, M.F. and Larsen, H., Halobacterium volcanii new species, a Dead Sea Halobacterium with a moderate salt requirement. Arch. Microbiol., 1975, **104**, 207-214.

71. Rodriguez-Valera, F., Juez, G. and Kushner, D.J., Halobacterium mediterranei spec. nov., a new carbohydrate-utilizing extreme halophile. Syst. Appl. Microbiol., 1983, **4**, 369-381.

72. Kocur, M. and Hodkiss, W., Taxonomic status of the genus Halococcus Schoop. Int. J. Syst. Bacteriol., 1973, **23**, 151-156.

73. Soliman, G.S.H. and Trüper, H.G., Halobacterium pharaonis sp. nov., a new extremely haloalkaliphilic archaebacterium with a low magnesium requirement. Zbl. Bakt. Hyg., I Abt. Orig. C, 1982, **3**, 318-329.

74. Ventosa, A. and Grant, W.D., Minutes of the meeting of the Subcommittee on the Taxonomy of Halobacteriaceae. Int. J. Syst. Bacteriol., 1988, **38**, 332.

75. Montero, C.G., Ventosa, A., Rodriguez-Valera, F. and Ruiz-Berraquero, F., Taxonomic study of non-alkaliphilic halococci. J. Gen. Microbiol., 1988, **134**, 725-732.

HALOPHILIC BACTERIA: THEIR LIFE IN AND OUT OF SALT

D.J. KUSHNER
Department of Biology
University of Ottawa
Ottawa, Ontario, K1N 6N5, Canada

ABSTRACT

Some of the practical uses and fundamental properties of halophilic archaebacteria and eubacteria are reviewed, including the adaptation of active transport systems to higher concentrations of NaCl and the correlation of internal ionic concentrations with the salt response of internal enzymes and of in vitro protein synthesis. Intracellular- and exo-enzymes of halophilic archaebacteria and anaerobic halophilic eubacteria function in high salt environments, while this is usually true only for the exo-enzymes of aerobic halophilic eubacteria. The internal enzyme of the latter may, in fact, function best in a low salt environment. This seems largely due to the fact that concentrations of Cl^- ions are much lower inside the aerobic halophilic eubacteria than outside, while for the other halophiles internal and external Cl^- ion concentrations are approximately equal.

Having worked with different kinds of halophilic bacteria over the last twenty-seven years, I have been very gratified to observe how interest in their properties has grown. Once considered mainly biological curiosities, they have made more substantial contributions to basic biological knowledge than we could have first expected. Interest arose in them many years ago through the dramatic ways they announced their presence in the landscape, causing solar salterns to turn bright red. Later, they were studied, especially in Canada, as organisms that cause the spoilage of salted fish and salted hides. Together with such applied studies, physiological and biochemical investigations revealed more of their highly unusual nature. Recent reviews, from which much of the following material is taken, deal with the properties of the halophilic eubacteria and archaebacteria (1-3).

One question, specially relevant for this Symposium, is: of what obvious use are the halophiles? How can we cash in on their "biotechnological potential"? How can we use them to make ourselves rich or, at least, to support our more esoteric research projects?

Admittedly, we are at something of a disadvantage when we compare halophiles with other microorganisms that live in extreme environments. Methanogens produce a clean and valuable fuel. Thermophiles have many obvious uses, some of which are discussed in this Symposium: they provide enzymes that can be included in detergents to break down fats and proteins in dirty clothes; some (especially Sulfolobus) can carry out mine leaching more efficiently than mesophiles such as the thiobacilli; thermophilic DNA polymerases are extremely valuable for carrying out DNA amplification.

What about the halophiles? Some of the uses are very ancient: the bright red color, mainly due to the bacterioruberins of halophilic archaebacteria, of solar salterns was used, perhaps as much as 5000 years ago, to signal that drying sea water was near saturation and could be drawn off into another evaporation ditch for final crystallization. More recently, it has been realized that the bacterial pigments absorb sunlight and promote evaporation, increasing the rate by 30% or more (R. Vreeland, personal communication). Thus, conditions that lead to increased growth of red halophiles may lead to increased evaporation and more effective preparation of solar salts from sea water.

In addition, some very important food products depend on microbial activity in high salt concentrations. In soy sauce manufacture ground wheat and soy grains are suspended in water with about 19% NaCl and incubated for 9 months. A great number of changes occur, due to halophilic and salt-tolerant bacteria and yeasts -- not including, to my knowledge, any halophilic archaebacteria, which is just as well considering how cultures of the latter smell. Soy sauce fermentation takes place in the dark: likely the halophilic archaebacteria can only predominate in bright sunlight, where their pigments give them a selective advantage. I understand (M. Kamekura, personal communication) that one reason the soy sauce fermentation takes so long is because proteolytic activity is rate-limiting. Thus, there would be considerable interest in adding salt-tolerant or salt-requiring proteases to batches of developing soy sauce, or in engineering organisms that can make such proteases while growing in soy sauce.

Others, in this Symposium and elsewhere, discuss special chemicals produced by halophilic organisms. The possibility that the halophilic archaebacteria may be used to screen for anti-tumor drugs has been considered by Forterre's group (4); Kohiyama's group (5) is examining the possibility that some of their products, which seem related to those of mammalian oncogenes, may be used as diagnostic tools for human cancer.

The halophilic archaebacteria also possess the bacteriorhodopsin system, that elegant membrane-associated apparatus that uses light to generate a proton gradient. This system, which is still the delight of biophysicists, has probably provided the best support of the chemi-osmotic hypothesis. Together with related systems, such as halorhodopsin, which serves as a Cl^- pump, and others involved in photosensing and phototaxis, it may have further intellectual treats in store for us. It was once hoped that since small artificial membrane systems containing bacteriorhodopsin could generate small amounts of electric current, much larger membranes might be constructed to provide large amounts of electrical energy from sunlight. Unfortunately, this hope has not been realized.

The most-studied extreme halophiles are charter members of the archaebacteria. It is gratifying to note that despite the exciting properties of the other archaebacteria -- the methanogens and the sulfur-dependent extreme thermophiles -- much current work on archaebacterial biochemistry is carried out with the halophilic ones. This is not because their enzymes are intrinsically easier to study; on the contrary, for many years most of them were difficult or impossible to purify by standard methods, since these depended on exposing enzymes to low salt concentrations which inactivated them, usually irreversibly. (An important exception is the bacteriorhodopsin system which functions very well in the absence of high salt concentrations.)

More recently, methods including hydrophobic chromatography and others have been developed for isolating the enzymes of extreme halophiles. Several have been purified, some studied in great detail, and these studies have revealed especially fascinating properties of these enzymes. Thus, the malate dehydrogenase of such creatures has a large number of carboxyl groups exposed to the surrounding aqueous medium. It has such a high ability to bind water and Na+ or K+ ions that it has been described as an "anionic sponge".

The aerobic halophiles, all of which are heterotrophes, are certainly the easiest archaebateria to grow. The first species studied, isolated on complex media, grew only on such media or on defined media with a number of amino acids. Since then, others have been found that can grow on minimal media with simple carbon sources, such as sugars and organic acids, and NH4+ as the nitrogen source. Such organisms, notably Haloferax (formerly Halobacterium) volcanii (6), are being used for isolation of auxotrophic mutants and studying methods of genetic recombination in the halobacteria (7,8).

My own recent work has mainly dealt with halophilic eubacteria. Though these seem to be less glamorous than the halophilic archaebacteria they have their own curious secrets. They need salts, usually an adequate concentration of NaCl, for growth and often for stability. The moderate halophiles, such as Vibrio costicola and Paracoccus halodenitrificans, normally grow in media containing from 0.5 to 3.5 M NaCl. The salt range of growth, however, and possibly even the salt requirement, may be a function of temperature. An extreme example of this is Planococcus halophilus, which in complex media needs no added NaCl at 20 C, while at 25 C and higher it needs 0.5 M NaCl, which cannot be replaced by other solutes. We have also found that V. costicola needs at least 0.5 M NaCl for growth at 30 C, but can get along with 0.2 M NaCl at 20 C. Though the matter has not been investigated, it seems likely that a number of halophilic eubacteria might prove to be less- or non-halophilic at lower temperatures.

What is the explanation of such a phenomenon? I suspect, without knowing any evidence, that such salt-temperature adaptation involves changes in the cell membrane.

Another curious finding is in the adaptation of transport systems. This may reflect changes at the membrane level or in intracellular solute composition. It was first shown in the very salt-tolerant halophile, Ba1 that growth at higher salt concentrations permitted cells to transport proline at higher salt concentrations as well. This was later shown for α-aminoisobutyrate (AIB) transport in V. costicola and Halomonas elongata.

In V. costicola AIB transport requires Na^+ ions and is dependent on a Na^+/H^+ antiport system, as found in other marine and halophilic bacteria. There is certainly a changed response to NaCl concentrations following growth at different NaCl concentrations. Cells grown in the presence of 1.0 M NaCl are able to transport AIB at much higher concentrations of NaCl (and other salts) than cells grown in the presence of 0.5 M NaCl: in the former, good transport is observed in the presence of 4 M salt, while in the latter it is almost completely inhibited by 3 M salts. Further studies showed that this change could take place in non-growing cells incubated for several hourse in the presence of high NaCl concentrations, if carbon and nitrogen sources were also present. The adaptation was not due to synthesis of new proteins, since it occurred in the presence of chloramphenicol and other antibiotics. Testing effects of a number of substances on development of salt-tolerant transport showed that the greatest effect was produced by glycine betaine (betaine), a substance known to accumulate in halophilic -- and some non-halophilic -- eubacteria at higher external osmolarities. We found there was a good correlation between betaine accumulation in cells growing or incubated at higher NaCl concentrations, and the development of salt-tolerant transport. However, chloramphenicol, which did not block the development of salt-tolerant transport, did block betaine accumulation in V. costicola (M. Klein and D.J. Kushner, unpublished). Thus, betaine cannot be the main mediator of increased tolerance to transport. It is quite possible that another substance, for example ectoine, which accumulates in certain eubacteria (9, and this Symposium), is reponsible for salt-tolerant transport.

In studying transport we are certain of the composition of the external medium. Knowing the exact composition of the internal composition of bacterial cells is more difficult (10). Sometimes, when we consider the effects of salts on intracellular eyzme activity, the situation appears paradoxical. Enzymes of the halophilic archaebacteria usually behave as they "should", from the ionic conditions inside these cells. That is, most depend on high salts (usually NaCl or KCl) for activity or stability or both. In many cases, enzyme activity is the same in NaCl or KCl; in others, KCl is specifically needed for allosteric enzyme regulation, and in vitro protein synthesis requires high KCl concentrations. These observations show that enzymes act best in the kind of ionic environment that direct chemical analysis suggests are present inside the cell.

The earliest assays of most intracellular enzymes of halophilic eubacteria seemed to indicate that these were inhibited by the NaCl or KCl concentrations that analyses of Na^+ and K^+ suggested were present inside the cells. In contrast, exoenzymes and most membrane-associated enzymes either need or can tolerate salt concentrations as high as those in the growth medium. Since the "salts" used in these tests are almost always chlorides of sodium or potassium, these results may well be explained by the fact that in the assays, the intracellular enzymes were being exposed to much higher Cl^- concentrations than they meet inside the cell, while enzymes on the membrane and exoenzymes are normally exposed to the Cl^- concentration found in the growth medium and can function in such concentrations.

This is a major point: many bacteria exclude Cl^- ions, which are able to inhibit a number of intracellular processes. Intracellular cations have been measured more frequently than intracellular anions. In all aerobic halophilic and non-halophilic eubacteria, the intracellular Cl^- ion concentration is 1/5 or less than that outside the cell. In contrast, recent studies of the internal solutes and enzymes of anaerobic halophilic eubacteria suggest that these may resemble those of the halophilic archaebacteria in composition and behaviour. In both Haloanaerobium prevalens and Halobacteroides acetoethylicus, the internal and external Cl^- concentrations are approximately equal over several different external NaCl concentrations; in the latter, a number of internal enzymes required or tolerated up to 3 M NaCl for maximal activity (10).

Our own recent work has been much concerned with in vitro protein synthesis by V. costicola. We had shown that this process is also inhibited by NaCl, KCl, or NH_4Cl concentrations greater than about 0.1 M, but that good activity could be obtained in the presence of 0.6 M or higher Na- or K-glutamate, or other organic anions. We had also shown that these anions partly counteracted the inhibitory effects of Cl^- ions, and that betaine could also counteract this action. We have now been able to determine the site of action of Cl^- ions: it is not on phenylalanyl t-RNA synthetase, nor do these ions cause miscoding of poly-uridylic acid (poly-U), the artificial messenger used in our studies. However these ions do prevent attachment of the artificial messenger to ribosomes. Furthermore, this action, which would, in itself, account for the inhibitory action of Cl^- ions, is partly counteracted by both glutamate and betaine (12).

We are still forced to work with artificial messengers in this, as in other halophilic protein-synthesizing systems: no natural messengers have yet been isolated. We have found a more "natural" system for V. costicola recently: the R_{17} RNA phage of Escherichia coli can serve as an in vitro messenger for V. costicola and can promote the formation of this phage's coat proteins. (C. Choquet and D.J. Kushner, unpublished.) This may prove to be a better system for studying the intimate details of protein synthesis in the halophile, though still not as good as a truly natural one.

Very tentatively, the picture may be emerging that aerobic halophilic eubacteria keep Cl^- ions out while anaerobic halophilic eubacteria (as well as aerobic halophilic archaebacteria) let them in. The latter may do this by special pumping mechanisms or by virtue of a lower membrane potential. If Cl^- ions are freely permeable, a strong outside-positive membrane potential would result in their being much higher on the ouside than in the inside; possibly, anaerobic bacteria have lower membrane potentials and thus do not maintain such a Cl^- ion gradient. (It should be noted that membrane potential is thought to be responsible for the transport of the positively-charged aminoglycoside antibiotics into bacterial cytoplasm; it has been suggested that the potential is high enough to accomplish this transport in aerobic but not in anaerobic bacteria, which are generally insusceptible to the aminoglycosides (13).

We do not know what the counter-anion is for the cell-associated cations in the moderate halophiles, which often reach concentrations of 1 M or higher in terms of cell water. It may be that the cations are associated with the cell envelopes, say on negatively-charged proteins. If this proves to be so, the aerobic eubacteria may exibit halophilic behaviour mainly on the outside of the cytoplasmic membrane, and the anaerobic ones on both sides.

ACKNOWLEDGMENT

The experimental work cited in this review was supported by a grant from the Natural Sciences and Engineering Research Council of Canada.

REFERENCES

1. Kushner, D.J., Molecular adaptation of enzymes, metabolic systems and transport systems in halophilic bacteria. FEMS Microbiol. Rev., 1986, 39, 121-27.

2. Kushner, D.J. The Halobacteriaceae. The Bacteria, Vol. VIII, 1985, pp. 171-214.

3. Kushner, D.J. and Kamekura, M. Physiology of halophilic eubacteria. In Halophilic Bacteria, ed., F. Rodriguez-Valera, CRC Press, Boca Raton, FL, U.S.A., 1988, Vol. I, pp.109-38.

4. Sioud, M., Forterre, P. and de Recondo, A.-M. Effects of the antitumour drug VP16 (etoposide) on the archaebacterial Halobacterium GRB 1.7kb plasmid in vitro. Nucleic Acid Res., 1987, 15, 8217-34.

5. Ben-Mahrez, K., Thierry, D., Sorokine, I., Danna-Muller, A. and Kohiyama, M. Detection of circulating antibodies against c-myc protein in cancer patient sera. Br. J. Cancer, 1988, 57, 529-34.

6. Juez, G. Taxonomy of extremely halophilic archaebacteria. In Halophilic Bacteria, ed., F. Rodriguez-Valera, CRC Press, Boca Raton, FL, U.S.A., 1988, Vol. II, pp. 3-24.

7. Mevarech, M. and Werczberger, R. Genetic transfer in Halobacterium volcanii. J. Bacteriol., 162, 461-2.

8. Charlebois, R.L., Lam, W.L. Cline, S.W., and Doolittle, W.F. Characterization of pHV2 from Halobacterium volcanii and its use in demonstrating transformation of an archaebacterium. Proc. Nat. Acad. Sci. USA, 1987, 84, 8530-4.

9. Galinski, E.A., Pfeiffer, H.P., and Trüper, 1,4,5,6-Tetrahydro-2-methyl-4 -pyrimidine carboxylic acid, a novel cyclic amino acid from halophilic phototrophic bacteria of the genus Ectothiorhodospira. Eur. J. Biochem., 1985, 149, 135-9.

10. Kushner, D.J. What is the "true" internal ionic composition of halophilic and other bacteria? Can. J. Microbiol., 1988, 34, 482-6.

11. Rengipipat, S., Lowe, S.E., and Zeikus, J.G. Effect of extreme salt concentrations on the physiology and biochemistry of Halobacteroides acetoethylicus. J. Bacteriol. 1988, 170, 3065-71.

12. Choquet, C.G., Kamekura, M., and Kushner, D.J. In vitro protein synthesis by the moderate halophile, Vibrio costicola: the site of action of Cl⁻ ions. J. Bacteriol., 1989, in press.

13. Bryan, L.E. Aminoglycoside resistance. In Antimicrobial Drug Resistance, ed., L.E. Bryan, Academic Press, New York, 1984, pp. 241-77.

ECOLOGICAL DISTRIBUTION AND BIOTECHNOLOGICAL POTENTIAL OF HALOPHILIC MICROORGANISMS

ALBERTO RAMOS-CORMENZANA
Department of Microbiology (Faculty of Pharmacy)
University of Granada. Granada (Spain)

ABSTRACT

A summarized review of results from some thalassohaline and athalassohaline habitats (including Great Salt Lake, Dead Sea, Salar of Atacama, and a marine and inland saltern) is made. There is a wide diversity of both halophilic Eubacteria and Archaebacteria in their natural habitats. A great diversity of taxonomic groups are found, and the general characteristics of hypersaline environments affect the types of halophilic bacteria present. The range of salt concentrations for growth are specially analyzed. It seems there are greater differences between soil and water environments, in relation to salt tolerance of the isolates, than when they are compared to the thalassohaline and athalassohaline habitats.

Some considerations to the biotechnological potential of these microorganisms are made. These are their interest in polymeric substances, food and the food industry, enzymes, the pharmaceutical industry, environmental biotechnology and oil recovery.

INTRODUCTION

In relation to this symposium, I am going to talk about Halophilic microorganisms (salt loving organisms), these are heterogenous groups, prokaryotic and eukaryotic organisms, characterized by growth over a wide range of salt concentrations. When microorganisms were isolated first from high salt brines they were called halophiles, however there are many kinds of halophiles, and in my opinion, according to Kushner criterium (1) they could be grouped broadly into four categories: slight halophiles, moderate halophiles, extreme halophiles and halotolerants. Slight halophiles, moderate halophiles and halotolerants (usually including extreme halotolerants) are Eubacteria, but extremely halophilic bacteria belong to the Archaebacteria; when I use the term Halobacteria I will be refering to both eubacterial and archaebacterial halophiles, although some authors (2,3) used the term Halobacteria exclusively for archaebacterial halophiles. In it I will deal two apparently separate aspects: some considerations on the distribution of these organisms and some aspects of their biotechnological potential.

SALINE-HABITATS

Although there exists a broad group of habitats, there are two important types of saline environments to be considered: the hypersaline soils and waters. I am going to refer to these soil and water habitats; for other types such as seasands and seaweeds, salt farm, salted fish, dried fish, others references could be consulted (4).

At first the studied on saline soils were due to the importance on Microbial ecology of Desert Soils. During the past two decades, the economic and agricultural utilization of arid lands has emerged; consequently, biological and environmental research on these soils has increased (5). These are usually characterized by high soil pH and often by high salinity, both of these factors influence the activities of soil microorganisms. The saline soil environment has not been so much studied till now; it has been shown by authors that chemical transformations carried out by soil microorganisms are inhibited by the addition to the soil of small amounts of salts (6,7). Any soil containing significant amounts of soluble marine salts could be considered hypersaline soil (8). The chemical composition of these environments is determined in some cases by the existence of mineral deposits, but mostly by complex precipitation and solubilization processes which happen with water evaporation.

Hypersaline waters are defined as those which have higher concentrations of salts than seawater (9). Classically the hypersaline aquatic environment is classified into two categories: Thalassohaline, if the ions composition is qualitatively similar to those found in sea water;and Athalassohaline if these amounts of salts are markedly different from those of seawater.It is generally believed that in the thalassohaline environment there exists a greater diversity and number of microorganisms than in athalassohaline. It is easy to find examples of thalassohaline waters such as Great Salt Lake,Marine Solar Salterns from Spain,Yugoslavia, Italy and so on; and of the athalassohaline waters such as Dead Sea, Lake Assal, Wadi Natrum, or Texcoco. For numerical data good reviews exist in the literature refering to the composition of athalassohaline waters (8). There are no examples of similar considerations in soils;by extension, we may refer to soils in the same sense, although no attempt appears in the bibliography refering to this concept.

I am going to consider in my talk those which may represent different places among the traditionally best known and those which I have studied more.

These are: Great Salt Lake, Dead Sea, Salar of Atacama, the inland saltern of la Malá and a Mediterranean solar saltern. They represent both thalassohaline (Great Salt Lake, Marine Saltern) and athalassohaline (Inland Saltern of la Malá, Dead Sea) habitats. These environments provide us with good examples of highly specialized and characterized ecological niches. In table 1, are shown the main chemical characteristics of these saline habitats. The halophilic environment of Great Salt Lake and Dead Sea has been studied well (10,11), in a similar way the mediterranean saltern of Santa Pola, has also been studied well (14). Recently studied have been done of la Malá, a peculiar inland saltern that works usually from May to September -depending on the wheater- .The saltern is located in La Malá (20 km from Granada -Spain-); the water emerging from a subterraneous saline well,is led into a system of shallow ponds arranged in series. We have studied the microflora that develops in this habitat (13,15,16).The Salar de Atacama,located in Northen Chile (Antofagasta), is the largest salt deposit in the country due to its location, in the Atacama desert, it has unique environmental conditions(17) and its microbial

TABLE 1
Characteristics of some saline environments

		Dead Sea (10)	Great Salt Lake (11)	Atacama Salar (12)	Marine saltern(*)	La Malá saltern(13)
Sea level (m)		-400	1,280	2,400	10	700
pH		5.9-6.3	7.7-8.4	7.2-8.6	6.8-8.1	6.7-8.0
Temperature (ºC)		21-36	-5-30	12-43	22-42	24-36
Total salts (g/l)		299-332	113-332	137	140-260	170-230
Cations	1st	Mg(40.7)[a]	Na(105)	Na(38)	Na(87)	Na(57)
	2nd	Na(39)	Mg(11)	Ca(0.5)	Mg(8)	Ca(3.6)
	3rd	Ca(17)	K(6.7)	K(0.6)	Ca	K(3.5)
Anions	1st	Cl(212)	Cl(181)	Cl(59)	Cl(169)	Cl(103)
	2nd	Br(5.1)	SO_4(27)	SO_4(2.4)	SO_4(22)	SO_4(4.9)
	3rd	SO_4(0.5)	HCO_3(0.01)	HCO_3(0.06)	HCO_3	–

( )[a] : Concentration g/l
* : unpublished results

communities have never been studied before 1987 (12). When we compare these environments, and if we only analyzed the order of dominance of both cations and anions, apparently there are no big differences between the thalassohaline and athalassohaline habitats, for example the major anion is Cl^-. But if we considered the quantitative values the differences appreciated are considerable. There are also important differences for example the dominant cation that in the Dead Sea is Mg^{2+}, while in the others habitats it is Na^+ (the dominant cation).

DISTRIBUTION OF HALOPHILIC MICROORGANISMS

A great diversity of taxonomic groups have been isolated from these habitats. Table 2 indicates the main taxonomic groups described, and with the increase of scientific work, more and more blank places are to be filled. When we analyzed in detail the types and microorganisms described it was noticed that there was a coincidence between most of the main taxonomic groups described in the Great Salt Lake and Dead Sea and those isolated by us in solar salterns (18), and other habitats (12,13,15,16).

It is generally accepted that in hypersaline environments, there are some characteristic types such as the alga Dunaliella, and the extreme halophilic bacteria, belonging to genus Halobacterium and Halococcus. Although, there are also important differences between the organisms found. The distribution of halophilic eubacteria and archaebacteria in saline environments, based on viable counts, varies considerably with the sample location, depth, isolation medium, in some cases it could be difficult to determine if those differences are due to the variations existing in the diverse habitats or to the methodology and thechniques used by authors in the study of these environments. There is a lack of some determinations, such as the investigation of anaerobic bacteria; although in 1943 an anaerobic halophilic bacterium was isolated from the sediments of Dead Sea, that unfortunately has not been preserved. After the report that certains strains are able of an anaerobic fermentative mode of growth (36), more anaerobic bacteria have been isolated from the Dead Sea (24, 25), and Great Salt Lake (28); recently the ecology and taxonomy of anaerobic halophilic eubacteria has been reviewed (40).

TABLE 2
Types of Halobacteria on saline environments

Categories	Dead Sea	Great Salt Lake	Atacama Salar	Marine saltern	La Malá saltern
Algae					
Dunalliella	(10,11)	(11,27)		(*)	(*)
Chlorophyta	(10)				
Chrysophyta	(10)				
Cyanobacteria					
Aphanocapsa	(10)				
Amoebobacter	(20)				
Ectothiorhodospira	(20)				
Archaebacteria					
Halobacterium	(10,21,22)	(11,20)	(12)	(30)	(15)
Halococcus	(23)	(11)	(18)	(15)	
Sarcina-like	(10)				
Haloarcula				(31)	
Haloferax				(31)	
Eubacteria aerobes or facultatives					
Flavobacterium	(19)		(12)		(16)
Pseudomonas	(19)				(16)
Chromobacterium	(19)				
Alteromonas					(15)
Marinomonas			(12)		
Acinetobacter			(12)	(32)	(16)
Deleya				(33)	(16)
Vibrio			(12)	(32)	(16)
Marinococcus			(12)	(33)	
Bacillus			(12)		
Eubacteria anaerobes					
Clostridium	(24)				
Haloanaerobium		(28)			
Halobacteroides	(25)				
Sporohalobacter	(26)				
methanogens		(29)			
sulfate reducers	(10)	(23)			

*: unpublished results

Larsen (28), considers that the bottom floor of the shallow ponds of marine salterns, beneath the mat of cyanobacteria if present, often there is black mud regardless of concentration of the brine in the pond. This indicated anaerobiosis in the mud beneath the brine and activity of sulfate reducing bacteria.So I consider that it is really a pity that other authors, including our group, had not tried to isolate anaerobic halobacteria from marine solar salterns. So it could be said that the data obtained is limited to a number of specific groups of prokaryotes, wich reflects the scientific bias of the groups working on these habitats.

TABLE 3
Salt Response of some Extreme Halobacteria

Microorganisms	Range of growth	Optimal growth
Halobacterium halobium	10-20	*
H. salinarium	5-30	15
H. saccharovorum	5-30	10-15
Halococcus morrhuae (Delf)	5-25	10-25
H. morrhuae (CCM 537)	1-30	10-15
Halococcus sp. (NCMB 757)	10-30	10-15
Haloferax mediterranei	15-30	*
Ectothiorhodospira halochloris (37)	10-35	14-27
E. abdelmalekii (38)	5-30	15
Natronobacterium (39)	12-30	20-25
Natronococcus (39)	8-30	20-22

*: Difficult to apreciate

General examples from salt and solute-tolerant microorganisms isolated in diverse location are given on tables 3, 4, and 5. Brock in an excellent review on the Ecology of Saline Lakes analyzed the factors influencing the chemical composition and origin of saline lakes(20).

It was clear how temperature and other factor should influence the presence of microorganisms. So, it seems obvious that the microbiota and ecological relations found in saline soils and waters will depend primarily on their salinity and a question easily emerges: is salt concentration the really predominant factor on the distribution of these organisms?. When I tried to compare the isolated by different authors in relation to salt requirement, I was a little disappointed because in some cases most of the data were lacking, the same thing happened in relation to other factors such as limits for temperature of growth,and nutrients requirements. Kushner and Kamekura (46) in a recent review establish range of salt concentration for growth described in the bibliography, I include personal data obtained in our Department with the main groups of halophiles; these results, with the inclusion of some data taken from the references are given in tables 4,5 and 6. If we observe these results and look carefully at the categories of the groups described by Kushner (1) it looks evident that a lot of organisms could not be easily adscribe to these groups. Onishi et al. (4) classified Halophilic Bacteria into nine types from halotolerant to extremely halophilic, based on the salt-response pattern of growth. Although we have not tested in the same conditions than Onishi et al. (4) it seems clear that the organisms tested could be grouped in none of these nine groups.

TABLE 4
Salt Response of Some Moderate Halobacteria

Microorganisms	Range of growth	Optimal growth
Acinetobacter sp.	1 - 30	3 - 10
Chromobacterium marismortis	1 - 30	3 - 10
Deleya halophila	0.5 - 30	5 - 15
Flavobacterium sp.	1 - 25	3 - 7.5
F. halmephilum	3 - 30	7.5 - 12
Haloanaerobium praevalens (28)	2 - 30	13
Halobacteroides halobius (25)	8.4 - 16.8	
Micrococcus varians	0 - 30	5 - 7.5
M. halobius	3 - 30	5 - 10
Paracoccus halodenitrificans	0 - 30	7.5
Planococcus halophilus	0 - 30	7.5
Rhodospirillum salexigens (41)	5 - 20.5	6 - 8.5
Rhodotermus marinus (42)	0.5 - 6	2
Sporohalobacter lortetii (26)	5.5 - 20	5.5 - 18
Sporosarcina sp.	0 - 25	0.5 - 5
Spirochaeta halophila (43)	0.3 - 7.5	4.5
Vibrio costicola	1 - 25	7.5 - 10

Perhaps our unpublished results in relation with the data taken from the literature, shown a more broad range of salt concentration for growth inclusive for the extreme halophilic archaebacteria, as we will consider later on. We have also observed how in some cases the salt range for growth changes at different temperatures, for example this happen in the extreme halophile Halococcus morrhuae, Halobacterium halobium and Halobacterium salinarium; in the moderate halophiles Vibrio costicola, Planococcus halophilus and Pseudomonas halosaccharolytica; and in the slight halophiles Pseudomonas nautica, Pseudomonas doudorofii and Deleya marina among others, usually the wider range of tolerance occur among higher temperatures (unpublished results). There are also fewer differences in relation to nutrients.

The important thing is that both halophilic eubacteria and archaebacteria coexist in these habitats. The general characteristics of hypersaline environments seem to affect the types of halophilic bacteria present (47).

Although it has been considered that each type of halophile has been isolated from specific saline materials: extreme halophiles from solar salterns and salt lakes, moderate halophiles from curing brines, hypersaline soils, and slight halophiles from marine samples. We have observed in some cases (35) that no correlation was found between the isolation habitat or the salt range in which growth occured and the taxonomic position of the strains found. A wide diversity of both halophilic Eubacteria and Archaebacteria there exist in their natural habitats (12,13). To clarify some aspects of the competition between the groups of moderate and extreme halophiles in their natural habitats, a study on behaviour of mixed populations of Halobacteria in continuous culture was done (48). We used continuous cultures to provide a changing spectrum of conditions of salt concentration, temperature and nutrient concentration. It was found how other factors should take place to explain how distribution could be realized. Temperatures seems to be the decisive factor within the range of salt concentration 20-30%(wt/vol).Moderately halobacteria were favoured by low temperatures.

TABLE 5
Salt Response of Some Halotolerant and Slight Halobacteria

Microorganisms	Range of growth	Optimal growth
Agmenellum quadriphcatum (45)	0 - 10.8	3
Alcaligenes aquamarinus	0 - 25	0.5 - 10
Bacillus pantothenicus	0 - 20	0 - 5
B. pumilus (ATCC 7061)	0 - 30	0 - 3
B. pumilus (ATCC 14884)	0 - 15	0 - 7.5
Beneckea natriengens	0.5 - 10	1 - 10
Be. neptuna	0.5 - 10	3 - 7.5
Deleya aesta	0 - 15	3 - 7.5
D. cupida	0.5 - 25	1 - 7.5
D. marina	0 - 25	0.5 - 10
D. pacifica	0 - 25	1 - 7.5
D. venusta	0 - 25	0.5 - 7.5
Flavobacterium oceanosedimentum	0 - 10	0 - 5
F. okeanokoides	0 - 15	0.5 - 7
F. marinolyticum	0 - 10	0.5 - 3
Pseudomonas doudorofii	0 - 15	0 - 7.5
Ps. nautica	0 - 20	0.5 - 7.5
Vibrio alginolyticus	0 - 7.5	1 - 5

Another important paper on nutrients and salts requirements demostrated how moderate halophiles and strains of the extreme halophiles Haloarcula and Halococcus grew on most of the substrated tested in the experiment (49); other tests from the same authors demonstrated that none of the halobacteria grew well in brines which harbour the densest populations of these bacteria in solar salterns; the high concentrations of Na^+ and Mg^{2+} found in saltern crystallizer brines limited bacterial growth, but the concentration of K^+ found in these brines had little effect. $MgSO_4$ was relatively more inhibitory to the extreme halophiles than was $MgCl_2$, but the reverse was true for the moderate halophiles (49). If we refer again to tables 3, 4 and 5 we can explain to growth ability of a considerable numbers of microorganisms to develop to different salt concentrations. Most of the halotolerant or the actually considered slight halophiles are able to grow in up to 20 or 25% salt concentration, above this concentration the moderate halophiles are able to grow in up to 30%, and the extreme halophiles are able to grow a relatively low concentration, such as 1% in Halococcus morrhuae. This also could explain the distribution of these bacteria. The evolution of halophilic microbiota has been studied by different authors, Grant and Ross (2) in a recent review analyzed the microbial successions in saline lakes of halophilic archaebacteria. The selection for moderately halophilic bacteria by gradual salinity increased has been done by Ventosa et al. (50). Recently Del Moral et al. (51) interpreted what might happen to the composition of a population of bacteria from saline environments when the salt concentration increased or decreased. The fact that Extreme halophiles has been isolated from Seawater (52) suggest that these organisms could also have been isolated from fresh water rivers, at least in those with a relatively high salts content.In the province of Granada there are a lot of rivers with different physico-chemical characteristics; we tried to isolate halophilic bacteria from these rivers, results obtained up to date are shown on table 6, the rivers samples were plated in specific medium for halophilic bacteria; the halophilic eubacteria isolated belong to a variety of taxonomic groups. Until now we did not succeed in the isolation of halophilic archaebacteria. In relation to the halophilic eubacteria isolated from Granada rivers, apparently it could be easy to interpret the results obtained, if the rivers had a high salt concentration, but it is more difficult to explain this situation when there is a relatively low salt concentration in these habitats. We have analyzed about 20 rivers and when we tried to correlate the results it was exciting that the isolates did not correspond precisely to rivers with high salt concentrations, in some cases it was exactly with a very low salt concentrations 0.5 g/l. It could be observed that sometimes exists a relation with temperature, because these rivers were originated from termal springs, according to that temperature seems to affect the presence of these halobacteria.

TABLE 6
Salt Requirements of Moderate Halobacteria Isolated from Granada Rivers

Number of strains	Range of growth	Optimal growth
9	3 - 15	7.5
3	3 - 15	10
2	3 - 10	7.5
1	3 - 20	5

What are the origins of these organisms? How long may they persist in low salt concentration?. Perhaps looking at the taxonomic diversity some of the halophilic eubacteria could be adapted to the environmental conditions, Quinn (53) suggest model of haloadaptation in bacterial membranes; Russell and Kogut (54), and Russell et al. (55) show that phospholipid changes are essential for phenotypic adaptation. We proved the effect of growth temperature and salt concentration on the fatty acid composition of moderate halophilic bacteria (56,57,58).Recently Monteoliva-Sanchez et al. (59), observed that the moderate halophile Deleya halophila exhibited phenotypic adaptation to changes in salt concentration. This adaptation process involved changes in the fatty acid composition specially in the relative amounts of cyclopropane fatty acids. These adaptations could also affect the morphological appearance of these halobacteria. And effectively there are marked changes in the morphology working on different salts concentration (5, 10 and 15%). From these unpublished observations we deduce two facts: the first one was the clear influence of salt concentration on the morphology of halophilic bacteria (at least in those isolated from La Malá); and secondly was the difficulty to observe these microorganisms directly from culture media; it seems easier to detect them when the samples are observed directly from environmental samples.Kushner (60) in a recent review presented the special properties and behaviour of enzymes, ribosomes, metabolic systems, protein turnover and active transport systems that are associated with the ability of halophilic bacteria to grow in different salt concentrations, other adaptations could also take place but all these things should be proved in environmental habitats.

However, there is another hypothesis about how these organisms could be dispersed. It has been proved that these organisms may survive in NaCl salt crystals for up to six months after entrapment (3); as these bacteria are able to form other crystals such as carbonate (calcite) and phosphate (struvite) (61,62), the entrapment could also be possible in these kind of crystals, and the survival of Halobacteria within inclusions because of its solubility it could retain viable for longer time, although especulative, this is another very interesting suggestion.

It is worthwhile to indicate the suggestions made in the study of Henis and Eren (63) who consider that microorganisms are unable to multiply in the saline soil; and the microbiota of soil saline habitats are as passive inhabitants possibly brought there by the wind. Our results are contradictory: on the one hand the results obtained in the Laguna of Tevenquiche (Salar of Atacama) in relation to different samples (water, soil, rhizosphere and sediments) from the same location; shown great differences (12). These results are exposed on table 7. It could be clearly

TABLE 7
Distribution of number of Moderately and Extremely Halophilic Strains isolated to different concentrations media and sampling sites

Sampling sites	5 M	5 E	10 M	10 E	15 M	15 E	20 M	20 E	25 M	25 E
ALT	5	-	62	-	26	-	12	-	6	18
SA	-	-	-	-	1	-	-	-	-	-
SB	-	-	1	-	1	-	-	-	-	-
SC	-	-	2	-	-	-	-	-	-	-
TR	2	-	10	-	6	-	27	-	10	-
TS	15	-	38	-	44	-	65	-	28	57

M: Number of moderately halophilic strains isolated
E: Number of extremely halophilic strains isolated
ALT: Water of Laguna Tevenquiche
SA: Soil of Rio Salado
SB: Soil of Laguna Tevenquiche
SC: Soil of Tilopozo
TR: Rhizosphere of Tevenquiche
TS: Sediment of Tevenquiche

observed that more microorganisms has been isolated from water and sediments than from soil or soil rhizosphere. The only conclusion to be taken, in these studies, is that fewer microorganisms were isolated from soil than from water or sediments. But, on the other hand, other results corresponding to the microbiota content from soil and saltern ponds samples, proved that the highest and lowest bacterial contents correspond to soil and shallow ponds, respectively (64).

Finally, we clearly observed (unpublished results) how the majority of these soil isolated grew quiet well at a wide range of salt concentration between 1-20% (w/v), which differs from water halophile isolates, which scarcely grew at 1%, these results agree with those obtained by Rodriguez-Valera (65), who suggested this could be due to the heterogeneity of the soil habitat where the salinity can obviously change markedly with distance and time.

The difficulty with discussing the distribution of Halobacteria in saline environments is that there is no single saline environment. Generalizations are difficult to make and they are always open to criticism.

BIOTECHNOLOGICAL POTENTIAL

The biotechnological possibilities of these organisms are summarized as follows:
1.- Their use in production of polymeric substances.
2.- Their use in food and the food industry.
3.- To be used in enzyme production.
4.- In the Pharmaceutical Industry.
5.- As Biological Response Modifiers.
6.- In oil recovery.
7.- In environmental biotechnology.

I only make some reflexions related to the main topics existing in the literature, but I do not exclude other possibilities of their use such as the recovery of metals by microbial accumulations and others.

Polymeric Substances

The applications of biotechnology to the production and recovery of polysaccharides is actually increasing. They are used in coatings, pharmaceuticals, feed stocks, substrates, foods, adhesives, and biotechnological separations. The extreme halophile Haloferax mediterranei has been described (66) to produce of extracellular polymeric substance (EPS), similar to other bacterial exopolysaccharides, with the properties of an acidic character due to the presence of uronic acids, and a high percentage of sulfate. The reologic study shows a high viscosity and pseudoplastic properties. Its viscosity is resistant to pH, temperature, and of course, salinity, which suggests a broad possibilities of use. The marine bacteria were described to produce xanthans, the PAVE (polysaccharide adhesive viscous exopolymer) and so on (67). We have observed (unpublished results) that the moderate halophilic bacteria also produce extracellular polymeric substances.

Other production of biological polymers from extreme halophilic bacteria, has been described recently by Garcia-Lillo and Rodriguez-Valera (68), the polyester poly-β-hidroxybutirate (PHB), could be easily obtained as plastic material. PHB possesses the important properties of thermoplasticity, biodegradability and biocompatibility, which have attracted considerable commercial attention.

Food and the Food Industry
Food Proteins

There is evidence to show that the natives in Mexico (Lake Texcoco), or in other places, as lake Chad area, used cyanobacteria species as food (69). There are some advantages for the use Spirulina spp.as unconventional source of protein: it is easy to cultivate, and growths on high pH and salt concentration. This alkaline and saline environment prevents the invasion and growth of other contaminant organisms. A pond devoted totally to the growth of Spirulina sp. can produce 125 times as much protein as the same amount of area devoted to corn, 70 times as much protein as to fish farming, and 600 times as much as to cattle ranching (70). The use of seawater as a suitable growth medium, has been investigated and this is a promising appeal because it is abundant and utilizes the photosynthetic machinery for the production of food. It has shown that magnesium are toxic for Spirulina spp. production (71).

The other genera of interest is Dunaliella. Strains of the genus Dunaliella, D. bardawil and D. salina, have been clearly shown to produce three valuable commercial products: glycerol, β-carotene and a high protein containing feed material (72).

The interesting thing is the accumulation and synthesis of β-carotene is improved under conditions of limited growth, and growth can be limited by high salt concentrations. It can be said that B-carotene is used commercially both as provitamin A and as natural food colouring agent.

Food aditives

There are numerous halophilic bacteria, isolated from food, Micrococcus varians, and two spore forming rods, were isolated from sauce mash, Vibrio costicola, from bacon, and so on, most of the studies have been done trying to search the contaminants to avoid the destruction of tons of feed. My suggestion is realize a more extensive study of these microorganisms to be used to contribute in flavoring these aliments. Antecedents exist in the work of Kamekura et al. (73), with the application of an extracellular nuclease from Micrococcus varians ssp. halophilus to commercial production of flavoring agents.

Food control

It is known the importance of the Halobacteria in food poisoning, and I do not want to talk too much about this, but certainly studies have been increasing with the recent reports on histamine-forming halobacteria (74,75).

Enzymes

Enzymes from halophilic bacteria have been interesting scientists. The problem was that authors have been investigating the effects of salts on activity and function (76) instead of its biotechnological application. Relatively few papers have appeared on the extracellular production of the halophiles.

Recently, Kamekura (77) reviews the production and function of enzymes of eubacterial halophiles;on table 8 are shown some of the enzymes produced by halophilic eubacteria and archaebacteria.

The activity of these enzymes could be realized at a very high salt concentration. This is an important role for the enzymes of Halophilic Bacteria, only the intracellular enzymes of halophilic eubacteria may be inhibited by salt concentrations. In some cases, for example the extracellular collagenase produced by the slight halophile Vibrio alginolyticus, does not appear to require salt for activity (78). On the list of described enzymes that is given in table 8, we would like to emphasize to those that behave as if they are more active in ionic conditions approximating those of the external medium; and perhaps the most promising, are those exoenzymes such nucleases and amylases. Kamekura and Onishi (79) found that Micrococcus varians ssp. halophilus, produces and extracellular nuclease. Good and Hartman (80) reported first time the production of extracellular amylase by the archaebacterium Halobacterium halobium, Onishi (81) described the production of a dextrinogenic amylase by the halophilic eubacterium Micrococcus halobius, and the response of amylase produced in media of different salt concentrations (82). Kamekura and Onishi (83) found that Micrococcus varians spp. halophilus, produces an extracellular nuclease, and after the application to commercial production of flavoring agent (73).

Recently Honda et al. (84) described the production of extracellular xylanase of alkalophilic Bacillus sp. by Escherichia coli, it is advisable to transfer biotechnological useful properties found in such organisms into the biotechnological "work horses". These are easy to grow and also the viceversa because of its difficulties in being contaminated, due to the salts requirements of these organisms.

TABLE 8
Enzymes of Halobacteria

Enzyme	Organisms	Salt conc.	Ref.
Amylase	Acinetobacter	1.2-3.6%	85
Amylase	Micrococcus halobius	3.0-12%	81,82
Amylase	Halobacterium halobium	6%	80
Amylase	Halobacterium sodomense		86
Dehydrogenase	Listeria denitrificans	9%	87
β-Lactamase	Actinopolyspora halophila	20%	88
Nitrate reductase	Paracoccus halodenitrificans	0%	89
Nuclease	Micrococcus varians	30%	83
Nuclease	Bacillus sp.	12-18%	90
Nuclease	Halococcus acetoinfaciens		91
Carboxylase	Ectothiorhodospira halophila	0%	92
Carboxylase	Aphanothece halophytica	1.5%	93
Phosphorylase	Vibrio costicola	3.6%	94
Protease	Halobacterium halobium		95
Protease	Halobacterium salinarium	12-24%	96
Protease	Bacillus sp.		97
Protease	Pseudomonas sp.		98
Protease	Caulobacter, Flavobacterium		99
Superoxide dismutase	Halobacterium cutirrubrum	12%	100
Xylanase	Bacillus sp.		84

Finally, a patent refering to a process for producing restriction enzyme has been made (91). This patent covers a process to produce a restriction endonuclease capable of recognizing the nucleotide sequence 5'↓GATC↑ and specifically cleaving at the arrow-marked positions, was obtained from the microorganism Halococcus acetoinfaciens FERM BP-942. We may conclude that a field is wide open for the near future.

Pharmaceutical Industry

There are a lot of possible uses of the microorganisms in the pharmaceutical industry, I am going to show some examples:

1. Pharmaceutical surfactants: some compounds derived from halophilic bacteria have surfactant properties, so they could be used as pharmaceutical surfactants. The lipid composition suggest its use in liposomes preparation.

2. Antimicrobial compounds: Many marine bacteria are known to produce antibacterial substances (101), and although the salt tolerance of these bacteria has not been studied in detail, many of them turn out to growth in a wide range of salt concentration. Forsyth et al. (102) found that

the great majority of their isolates grew in the range of 5 to 20% NaCl. Isolation and characterization of antibiotic components have been carried out by various researchers, and although other halophilic bacteria have not been studied as much, recent studied (103) on marine bacteria could indicate the possible use of some of these halobacteria to produce antibiotics.

3. Hormone-like material: Recently, (104) it has been reported the presence of hormones-like material in Halobacterium salinarium, they have detected insulin-related material from Halobacterium, that was present in amounts similar to that found for Escherichia coli. Although the molecule has not yet been isolated and purified, the biological and immunochemical evidence for its existence is strong.

4. A revision on the potential use of algae, including the halotolerants, for drugs, based on the chemical identification, extraction, isolation and biological valoration, has been made by Accorinti (105).

5. Other possibilities do exist. I referred to the use of Dunalliela to the commercial use as pro-vitamin A. Medical studies pointed to an inverse correlation between human cancer risks and dietary intake of β-carotene (106). We have recently assayed the production of vitamins and aminoacids by moderately halophilic bacteria isolated from Atacama Salar (unpublished results), it is too early to conclude their applicability to produce these compounds.

Another interesting product is eicosapentaenoic acid, a polyunsaturated fatty acid that seems to reduce cholesterol levels in blood and could reduce the risk of heart disease and artherosclerosis (106).

6. Finally, we got some promising results which suggest their possible use as Biological Response Modifiers.

Biological Response Modifiers
Researchers all over the world describe immune modifiers derived from, or analogous to, microbially produced metabolites or structural peptidoglycans. In relation with the Biological Response Modifier, in a communication that we presented in this symposium (107) we have tested extreme halophilic bacteria belonging to genera Halobacterium (H. salinarium, H. saccharovorum, H. mediterranei), and Halococcus (Hc.morrhuae) two of these bacteria affected significatively the modulation of humoral response in mice by two different ways: stimulating and suppressing. In the figures of the mentioned reference we can observe these results. It is clear how H. saccharovorum, estimulated, in all the cases tested, and Hc. morrhuae, suppresses the biological response, also in all the cases tested. This effect was inverted when weii studied the Delayed Hypersensitivity response; with both organisms the results were statistically significant. This opens a door on the use of these microorganisms as biological response modifiers.

Environmental Biotechnology
Other possibilities exist such as the use of halophilic bacteria as biological phosphate removal in saline or alkaline industrial waters. The removal of phosphates is too expensive by chemical methods.

We have proved the formation of struvite by moderately halophilic bacteria so there is a suggestion for the use of these microorganisms as biological precipitation of phosphate.

The halobacteria could also be used in the recovery of saline soils. Time ago experiments have been done in Argentina utilizing halotolerant fixing nitrogen organisms (Molina, personal communication); but surely more studies should be done in relation with this aspect. Surely other possibilities exist, for example the use as contaminant indicators in hypersaline environments.

Oil Recovery
This has been suggested recently by Post and Al-Harjan (108) on studying the surface activity of Halobacteria and the potential use in microbial enhanced oil recovery.

The properties of the other substances produced by halophilic bacteria indicates that they also could be used in the oil recovery.

CONCLUDING REMARKS

Ecology and Distribution
On completing this presentation, we are again impressed with the wide diversity of both halophilic eubacteria and archaebacteria. The halophilic eubacteria are much more diverse than the halophilic archaebacteria.

It also seems that there are greater differences between soil and water environments, in relation to salt tolerance of the isolates, than when we compare thalassohaline and athalassohaline habitats.

I want to complain about the fact that there is a lack of comparative results about the distribution of bacteria in saline environments, primarily due to the difficulty in comparing the results obtained, because of the different methodology used by authors. The aims of this symposium could be to encourage scientists to obtained comparable results.

Biotechnological Potential
Up to date the use of halophilic bacteria has not been concrete, however the halophilic bacteria offers a number of different possibilities, that I may show mainly in the following aspects: food industry, as enzyme producers, in oil recovery, in environmental biotechnology, and in the pharmaceutical industry. It is may hope that in presenting this talk, others who have not yet studied these aspects, may be inspired to do so.

ACKNOWLEDGMENTS

I would like to thank Drs. Quesada, Ferrer, Monteoliva-Sánchez, Del Moral and Gomez, from the group of our department working on halophiles for encouragement and dayly help.

REFERENCES

1. Kushner, D.J., Life in High Salt and Solute Concentration: Halophilic Bacteria. In Microbial Life in Extreme Enviroments, ed., D.J. Kushner, Academic Press, London, 1978, 8, pp. 317-68.
2. Grant, W.D. and Ross, H.N.M., The ecology and taxonomy of halobacteria. FEMS Microbiol. Rev., 1986, 39, 9-15.
3. Norton, C.F. and Grant, W.D., Survival of halobacteria within fluid inclusions in salt crystals. J. Gen. Microbiol., 1988, 134, 1365-73.
4. Onishi, H., Fuchi, H., Konomi, K., Hidaka, O. and Kamekura, M. Isolation and distribution of a variety of halophilic bacteria and their classification by salt- response. Agric. Biol. Chem., 1980, 44, 1253-58.
5. Skujins, J., Microbial Ecology of Desert Soil. In Advances in Microbial Ecology, ed., K.C. Marshall, Plenum Press, New York, 1984, 7, pp.49-91.
6. Greaves, J.E., The influence of salts on the bacterial activities of the soil. Soil Sci., 1916, 2, 448-80.
7. Brown, P.E. and Johnson, D. R., The effect of certain alkali salts on ammonification. Iowa State Res. Bull., 1918, 44, 3-24.
8. Rodriguez-Valera, F., Characteristics and Microbial Ecology of Hypersaline environments. In Halophilic Bacteria, ed., F. Rodriguez-Valera, CRC Press, Boca Raton (Florida), 1988, 1, pp. 3-30.
9. Edgerton, M.E. and Brimblecombe, P., Thermodynamics of halobacteria environments. Can. J. Microbiol., 1981, 27, 899-909.
10. Nissenbaum, A., The microbiology and biogeochemistry of the Dead Sea. Microb. Ecol., 1975, 2, 139-61.
11. Post, F.J.,Microbiology of the Great Salt Lake north arm. Hydrobiol., 1981, 81, 59-69.
12. Prado, B., Estudio taxonómico de bacterias halófilas moderadas aisladas del Salar de Atacama, Chile. Tesis Doctoral, Granada, 1987.
13. Del Moral, A., Estudio de la flora bacteriana halófila de una salina interna. Tesis Doctoral, Granada, 1986.
14. Rodriguez-Valera, F., Estudio de las relaciones entre los distintos tipos de halofilismo bacteriano en medios salinos. Tesis Doctoral, Granada, 1978.
15. Del Moral, A., Quesada, E., and Ramos-Cormenzana, A., Distribution and types of bacteria isolated from an inland saltern. Ann. Inst. Pasteur/ Microbiol., 1987, 138, 59-66.
16. Del Moral, A., Prado, B., Quesada, E., Garcia, T., Ferrer, R., and Ramos-Cormenzana, A., Numerical taxonomy of moderately halophilic Gram-negative rods from and inland saltern. J. Gen. Microbiol., 1988, 134, 733-41.
17. Moraga, A.B., Chong, G.D., Fortt, M.A. and Henriquez, A.H., Estudio geológico del Salar de Atacama, provincia de Antofagasta, Boletin nº 29 del Instituto de Investigaciones Geológicas. Departamento de Geociencias Universidad del Norte. Chile, 1974.

18. Rodriguez-Valera, F., Ruiz-Berraquero, F. and Ramos-Cormenzana, A., Characteristics of the heterotrophic bacterial populations in hypersaline environments of different salt concentrations. Microb. Ecol. 1981, 7, 235-43.

19. Elazari-Volkani, B., Bacteria in the bottom sediments of the Dead Sea. Nature, 1943, 152, 274-5.

20. Brock, T.D., Ecology of Saline Lakes. In Strategies of Microbial Life in Extreme Environments, ed., M. Shilo, Verlag Chemie, Berlin, 1979, pp. 29-47.

21. Mullahhanbai, M.F. and Larsen, H., Halobacterium volcanii spec. nov., a Dead Sea Halobacterium with moderate salt requirement. Arch. Microbiol., 1975, 104, 207-14.

22. Oren, A., Halobacterium sodomense sp. nov., a Dead Sea halobacterium with an extremely high magnesium requirement. Int. J. Syst. Bact., 1983, 33, 381-6.

23. Larsen, H., Ecology of Hypersaline Environments. In Developments in Sedimentology, ed., H.Nissenbaum, Elsevier, Amsterdam, 1980, 28 pp. 23-9.

24. Oren, A., Clostridium lortetii sp. nov., a halophilic obligatory anaerobic bacterium producing endospores with attached gas vacuoles. Arch. Microbiol., 1983, 136, 42-8.

25. Oren, A., Weisburg, A.W.G., Kessel, M. and Woese, C.R., Halobacteroides halobius gen. nov., sp. nov., a moderately halophilic anaerobic bacterium from the bottom sediments of the Dead Sea. Syst. Appl. Microbiol., 1984, 5, 58-70.

26. Oren, A., Pohla, H. and Stackebrandt, E., Transfer of Clostridium lortetii comb. nov., and description of Sporohalobacter marismortui sp. nov. Syst. Appl. Microbiol., 1987, 9, 239-46.

27. Post, F.J., The microbial ecology of the Great Salt Lake. Microb. Ecol., 1977, 3, 143-65.

28. Zeikus, J.G., Hegge, P.W., Thomson, T.E. and Langworthy, T.A., Isolation and description of Haloanaerobium praevalens gen. nov. and sp. nov. and obligately anaerobic halophile common to Great Salt Lake sediments. Curr. Microbiol., 1983, 9, 225-34.

29. Paterak, J.R. and Smith, P.H., Isolation and characterization of a halophilic methanogen from Great Salt Lake. Appl. Environ. Microbiol., 1985, 50, 877-81.

30. Rodriguez-Valera, F., Juez, G. and Kushner, D.J., Halobacterium mediterranei sp. nov., a new carbohydrate utilizing extreme halophile. Syst. Appl. Microbiol., 1983, 4, 369-81.

31. Torreblanca, M., Rodriguez-Valera, F., Juez, G., Ventosa, A., Kamekura, M. and Kates, M., Classification of non-alkaliphilic halobacteria, based on numerical taxonomy and polar lipids composition, and description of Haloarcula gen. nov., and Haloferax gen. nov. Syst.Appl. Microbiol., 1986, 8, 89-99.

32. Quesada,E.,Bejar,M.V.,Valderrama,M.J.,Ventosa,A.,Ruiz-Berraquero,F. and Ramos-Cormenzana, A., Numerical taxonomy of moderately halophilic Gram-negative nonmotile eubacteria. Syst. Appl.Microbiol.,1987, 9,132-7.

33. Quesada, E., Ventosa, A., Rodriguez-Valera, F., Megias, L. and Ramos-Cormenzana, A., Numerical taxonomy of moderately halophilic Gram-negative bacteria from hypersaline soils. J. Gen. Microbiol., 1983, 129, 2649-57.

34. Garcia, M.T., Ventosa, A., Ruiz-Berraquero, F. and Kocur, M., Taxonomic study on amended description of Vibrio costicola. Int. J. Syst. Bact., 1987, 37, 251-6.

35. Ventosa, A., Ramos-Cormenzana, A. and Kocur, M., Moderately halophilic Gram-positive cocci from hypersaline environments. Syst. Appl.Microbiol. 1983, 4, 564-70.

36. Hartman, R., Sickinger, H.D. and Oesterhelt, D., Anaerobic growth of halobacteria. Proc. Natl. Acad. Sci.USA, 1980, 77, 3821-5.

37. Imhoff, J.F. and Truper, H.G., Ectothiorhodospira halochloris sp. nov., a new extremely halophilic phototrophic bacterium containing bacteriochlorophyll b. Arch. Microbiol., 1980, 114, 115-21.

38. Imhoff, J.F. and Truper, H.G., Ectothiorhodospira abdelmalekii sp. nov., a new extremely halophilic phototrophic bacterium. Zentralbl.Bakteriol. Microbiol. Hyg. Abt. 1: Orig. C., 1981, 2, 228-34.

39. Tindall, B.J., Ross, H.N.M. and Grant, W.D., Natronobacterium gen. nov. and Natronococcus gen. nov., two new genera of haloalkaliphilic Archaebacteria. Syst. Appl. Microbiol., 1984, 5, 41-57.

40. Oren, A., The ecology and taxonomy of anaerobic halophilic eubacteria. FEMS Microbiol. Rev., 1986, 39, 23-9.

41. Drews, G., Rhodospirillum salexigens spec. nov., an obligatory halophilic phototrophic bacterium. Arch. Microbiol., 1981, 130, 325-7.

42. Alfredsson, G.A., Kristjansson, J.K., Hjorleifsdottir, S. and Stetter, K.O., Rhodothermus marinus gen. nov. sp. nov., a thermophilic, halophilic bacterium from submarine hot springs in Iceland. J. Gen. Microbiol., 1988, 134, 299-306.

43. Greenberg, E.P. and Canale-Parola, E., Spirochaeta halophila sp. nov., a facultative anaerobe from a high salinity pond. Arch. Microbiol., 1976, 110, 185-94.

44. Brock, T.D., Microbial growth under extreme conditions. Symposia of the Society for General Microbiology, 1969, 9, 15-41.

45. Golubic, S., Halophily and halotolerance in cyanophytes. Origins Life, 1980, 10, 169-83.

46. Kushner, D.J. and Kamekura, M., Physiology and Halophilic Eubacteria. In Halophilic Bacteria, ed., F. Rodriguez-Valera, CRC Press, Boca Raton, 1988, 5, pp. 109-38.

47. Rodriguez-Valera, F., Ventosa, A., Juez, G. and Imhoff, J.F.,Variation of environmental features and microbial populations with salt concentration in a multi-pond saltern. Microb. Ecol., 1985, 11,107-15.

48. Rodriguez-Valera, F., Ruiz-Berraquero, F.and Ramos-Cormenzana, A., Behaviour of mixed populations of halophilic bacteria in continuous cultures. Can. J. Microbiol., 1980, 26, 1259-63.

49. Javor, B.J.,Growth potential of halophilic bacteria isolated from solar salt environments carbon sources and salt requirements. Appl. Environ. Microbiol., 1984, 48, 352-60.

50. Ventosa, A., Rodriguez-Valera, F., Poindexter, J.S. and Reznikoff, W.S. Selection for moderately halophilic bacteria by gradual salinity increases. Can. J. Microbiol., 1984, 30, 1279-82.

51. Del Moral, A., Quesada, E., Bejar, V. and Ramos-Cormenzana,A. Evolution of bacterial flora from a subterranean saline well by graduated salinity changes in enrichment media. J. Appl. Bacteriol., 1987, 62, 465-71.

52. Rodriguez-Valera, F., Ruiz-Berraquero, F. and Ramos-Cormenzana, A., Isolation of extreme halophiles from sea water. Appl. Environ. Microbiol., 1979, 38, 164-5.

53. Quinn, P.J., Models of haloadaptation in bacterial membranes. FEMS Microbiol. Rev., 1986, 39, 87-94.

54. Russell, N.J. and Kogut, M., Haloadaptation: salt sensing and cell-envelope changes. Microbiol. Sci., 1985, 2, 345-50.

55. Russell, N.J., Adams, R., Bygraves, J. and Kogut, M., Cell envelope phospholipid changes in a moderate halophile during phenotypic adaptation to altered salinity and osmotic stress. FEMS Microbiol.Rev., 1986, 39, 103-7.

56. Monteoliva-Sanchez, M.and Ramos-Cormenzana, A., Effect of growth temperature and salt concentration on the fatty acid composition of Flavobacterium halmephilum CCM 2831. FEMS Microbiol.Lett.,1986,33,51-4.

57. Monteoliva-Sanchez, M. and Ramos-Cormenzana, A., Cellular fatty acid composition of Planococcus halophilus NRCC 14033 as affected by growth temperature and salt concentration. Curr. Microbiol., 1987, 15, 133-6.

58. Monteoliva-Sanchez, M. and Ramos-Cormenzana, A., Cellular fatty acid composition in moderately halophilic Gram-negative rods. J. Appl. Bacteriol., 1987, 62, 361-6.

59. Monteoliva-Sanchez, M., Ferrer, M.R., Ramos-Cormenzana, A., Quesada, E. and Monteoliva, M., Cellular fatty acid composition of Deleya halophila: effect of temperature and salt concentration. J. Gen. Microbiol., 1988, 134, 199-203.

60. Kushner, D.J., Molecular adaptation of enzymes,metabolic systems and transport systems in halophilic bacteria. FEMS Microbiol. Rev., 1986, 39, 121-7.

61. Del Moral, A., Roldan, E., Navarro, J., Monteoliva-Sanchez, M. and Ramos-Cormenzana, A., Formation of calcium carbonate crystals by moderately halophilic bacteria. Geomicrobiol. J., 1987, 5, 79-87.

62. Ferron-Vilches, M.C., Rivadeneyra, M.A., Perez-Garcia, I. and Ramos-Cormenzana, A., Precipitación de calcita y estruvita por bacterias aisladas de las aguas del pantano de Cubillas. Ars Pharm.,1984,25,341-8.

63. Henis, Y. and Eren, J., Preliminary studies on the microflora of a highly saline soil. Can. J. Microbiol., 1963, 9, 902-4.

64. Quesada, E., Bejar, V., Valderrama, M.J., Ventosa, A. and Ramos-Cormenzana, A., Isolation and characterization of moderately halophilic nonmotile rods from different saline habitats.Microbiologia,1985,1,89-96

65. Rodriguez-Valera,F.,The ecology and taxonomy of aerobic chemoorganotrophic halophilic eubacteria. FEMS Microbiol.Rev., 1986, 39, 17-22.

66. Anton, J., Meseguer, I. and Rodriguez-Valera, F., Composicion,reología y producción de un exopolisacarido producido por Haloferax mediterranei

Biotec 88/Abstracts (2nd Spanish Conference on Biotechnology).1988, 274
67. Weiner, R.M., Colwell, R.R., Jarman, R. N., Stein, D.C., Somerville,Ch.C. and Bonar,D.B.,Applications of Biotechnology to the production recovery and use of marine polysaccharides. Biotechnol., 1985, 3, 899-902.
68. Garcia-Lillo, J.A. and Rodriguez-Valera, F., Plásticos biodegradables a partir de halobacterias. Biotec 88/Abstracts (2nd Spanish Conference on Biotechnology), 1988, 276.
69. Ciferri, O., Spirulina, the edible microorganism.Microbiol. Rev., 1983, 47, 551-78.
70. Contreras, A., Herbert, D.C., Grubbs, B.G. and Cameron, I.L., Blue-green alga, Spirulina, as the sole dietary source of protein in sexually maturing rats. Nutr. Rev. Intern., 1979, 19, 749-63.
71. Tindall, B.J., Prokaryotic Life in the Alkaline, Saline, Athalassic Enviroment. In Halophilic Bacteria, ed., F. Rodriguez-Valera, CRC Press Boca Raton (Florida), 1988, 1, pp. 31-67.
72. Ben-Amotz, A. and Avron,M.,Accumulation of metabolites by halotolerant algae and its industrial potential. Ann. Rev. Microbiol.,1983,37,95-119.
73. Kamekura, M., Hamakawa, T. and Onishi, H., Application of halophilic nuclease H of Micrococcus varians subsp. halophilus to commercial productions of flavoring agent 5'-GMP. Appl. Environ. Microbiol., 1982, 44, 994-5.
74. Okuzumi, M., Okuda, S. and Awano, M., Occurrence of psychrophilic and halophilic histamine-forming bacteria (N-group bacteria) on/in red meat fish. Bull. Jap. Soc. Sci. Fish., 1982, 48, 799-804.
75. Okuzumi, M., Okuda, S. and Awano, M., Isolation of psychrophilic and halophilic histamine-forming bacteria from Scomber japonicus. Bull. Jap. Soc. Sci. Fish., 1981, 47, 1591-8.
76. Kushner, D.J., The Halobacteriaceae. In The Bacteria, eds., C.R. Woese and R.S. Wolfe. Academic Press, Orlando Fl., 1985, VIII, pp.171-214.
77. Kamekura, M., Production and function of enzymes of eubacterial halophiles. FEMS Microbiol. Rev., 1986, 39, 145-50.
78. Reid, G., Woods, D.R. and Robb, F.T., Peptone induction and rifampin-insensitive collagenase production by Vibrio alginolyticus. J.Bacteriol. 1980, 142, 447-54.
79. Kamekura, M. and Onishi, H., Floculation and adsorption of enzymes during growth of a moderate halophile Micrococcus varians var. halophilus. Can. J. Microbiol., 1978, 24, 703-9.
80. Good, V.A. and Hartman, P.A., Properties of the amylase from Halobacterium halobium. J. Bacteriol., 1970, 104, 601-3.
81. Onishi, H., Halophilic amylase from a moderately halophilic Micrococcus J. Bacteriol., 1970, 109, 570-4.
82. Onishi, H., Salt response of amylase produced in media of different NaCl or KCl concentration by a moderately halophilic Micrococcus. Can. J. Microbiol., 1970b, 18, 1617-20.
83. Kamekura, M. and Onishi, H., Halophilic nuclease from a moderately halophilic Micrococcus varians. J. Bacteriol., 1974, 119, 339-44.

84. Honda, H., Kudo, T. and Horikoshi, K., Production of extracellular xylanase of alkalophilic Bacillus sp. strain C-125 by Escherichia coli carrying pCX311. AMS meeting, 1988, 424.

85. Onishi, H. and Hidaka, O., Purification and properties of amylase produced by a moderately halophilic Acinetobacter sp. Can. J. Microbiol. 1978, 24, 1017-23.

86. Oren, A., A thermophilic amyloglucosidase from Halobacterium sodomense, a halophilic bacterium from the Dead Sea. Curr. Microbiol., 1983, 8, 225-30.

87. Hong, Y.K. and Seu, J.H., Isolation and physiological properties of a moderately halophilic bacterium Listeria denitrificans HB-38 (korean). Bull. Korean Fish Soc., 1983, 16, 68-74.

88. Jonhson, K.G. and Lanthier, P.H., β-lactamase from Actinopolyspora halophila, an extremely halophilic actinomycete. Arch. Microbiol., 1986, 143, 379-87.

89. Rosso, J.P., Forget, P. and Pichinoty, F.,Les nitrates-reductases bacteriennes.Solubilisation, purification et properties de l'enzyme A de Micrococcus halodenitrificans. Biochim. Biophys. Acta, 1973, 321, 443-51.

90. Onishi, H., Mori, T., Takeuchi, S., Tani, K., Kobayashi, T. and Kamekura, M., Halophilic nuclease of a moderately halophilic Bacillus sp.: production, purification, and characterization. Appl. Environ. Microbiol., 1983, 45, 24-30.

91. Obayashi, A., Hiraoka, N., Kita, K., Nakajima, H. and Shuzo, T., Process for producing restriction enzymes. U. S. Patent, 1988, 4, 724,209, US Cl. 435/199.

92. Tabita, F.R. and MacFadden, B.A., Molecular and catalytic properties of ribulose 1,5-bisphosphate carboxylase from the photosynthetic extreme halophile Ectothiorhodospira halophila. J. Bacteriol., 1976, 126, 1271-7.

93. Codd, G.A., Cook, C.M. and Stewart, W.D.P., Purification and subunit structure of D-ribulose 1,5-biphosphate carboxylase from the cyanobacterium Aphanothece halophytica. FEMS Microbiol. Lett., 1979, 6, 81-6.

94. Harry, K., Sharma, N. and Fitt, P.S., Preparation and properties of highly-purified Vibrio costicola polynucleotide phosphorylase. Biochim. Biophys. Acta, 1985, 828, 29-38.

95. Izotova, L.S., Strongin, A.Y., Chekulaeva, L.N., Sterkin, V.E., Lyunblinskaya, V.I., Timokhina, E.A. and Stepanov, V.M., Purification and properties of serine protease from Halobacterium halobium. J. Bacteriol., 1983, 155, 826-30.

96. Norberg, P. and Hofsten, B.V., Proteolytic enzymes from extremely halophilic bacteria. J. Gen. Microbiol., 1969, 55, 251-6.

97. Kamekura, M. and Onishi, H., Protease formation by a moderately halophilic Bacillus strain. Appl. Microbiol., 1974, 27, 809-10.

98. Makino, K., Koshikawa, T., Nishira, T.H., Ichikawa, T. and Kondo, M. Studies on protease from marine bacteria. 1 Isolation of marine Pseudomonas sp. 145-2 and purification of protease. Microbios, 1981, 31, 103-12.

99. Van Qua, D., Simidu, U. and Taga, N., Occurrence and genetic composition of protease producing moderately halophilic bacteria in neritic seawater around Japan. Bull. Jap. Soc. Fish., 1981, 47, 359-64.
100. May, B.P. and Dennis, P.P., Superoxide dismutase from the extremely halophilic archaebacterium Halobacterium cutirrubrum. J. Bacteriol., 1987, 169, 1417-22.
101. Baam, B.R., Gandhi, N.M. and Freitas, Y.M., Antibiotic activity of marine microorganisms. Helgol. Wiss. Meeresunters, 1966, 13, 181-5.
102. Forsyth, M.P., and Kushner, D.J., Nutrition and distribution of salt response in populations of moderately halophilic bacteria. Can. J. Microbiol., 1970, 16, 253-61.
103. Nair, S. and Simidu, U., Distribution and significance of heterotrophic marine bacteria with antibacterial activity. Appl. Environ. Microbiol., 1987, 53, 2957-62.
104. Rubinovitz, Ch. and Shiloach, J., Insulin-related material in prokaryotes. FEMS Microbiol. Lett., 1985, 29, 53-8.
105. Accorinti, J., Recursos marinos: ALGAS fuente potencial de nuevos fármacos. Dirección Nacional del Antártico. Inst. Antártico Argentino. Buenos Aires, 1987.
106. Klausner, A., Algaculture: Food for thoght. Biotechnol., 1986, 4, 947-53.
107. Ruiz, C., Monteoliva-Sanchez, M. and Ramos-Cormenzana, A.,Modification of immune response by extreme halophilic bacteria. Symposium on the Microbiology of Extreme Environments and its Biotechnological Potential, 1988.
108. Post, F.J. and Al-harjan, F.A., Surface activity of halobacteria and potential use in microbially enhanced oil recovery. 8th International Symposium on Environmental Biogeochemistry. Abstracts, 8. Nancy. (France),1987.

Polyol Accumulation in Yeasts in Response to Water Stress.

Milton S. da Costa and M. Fernanda Nobre
Departamento de Zoologia, Universidade de Coimbra
3049 Coimbra Codex, Portugal.

Introduction

Microorganisms can be found growing in aqueous environments ranging from fresh water with extremely low concentrations of ions to saturated brines or extremely concentrated sugar solutions; nevertheless most microorganisms grow in environments which are dilute in terms of salts or sugars. As with all extreme environments, microbial diversity decreases as the salt or sugar concentration of an aqueous environment increases and only a few adapted species grow under conditions of extreme water stress (Trüper and Galinski, 1986). Yet many other microorganisms appear to be able to adjust physiologically, within limits of intrinsic tolerance, to periodic alterations in the concentration of osmotically active solutes which occur in some aqueous environments and influence growth or survival.

An increase in the concentration of low molecular weight solutes of an aqueous environment will cause a decrease in the water available to the microorganism, with concomitant alterations in the intracellular water, resulting in a decrease of the cell volume and/or the turgor pressure. In order to adjust to the new environment of high solute concentration, microorganisms must accumulate an intracellular osmoregulatory substance (osmolyte) to retain as much cellular water as possible, and increase the turgor pressure and/or the cell volume.

Under extreme water stress, the osmoregulatory substance can attain very large intracellular concentrations and may, therefore, affect enzyme activity. The accumulation of an osmolyte must, as pointed out by Brown (1976), serve to maintain the cell volume and/or turgor pressure and counterbalance the external osmotic pressure or water activity (a_w) (for discussion of water activity, see Brown,1978) and at the same time be compatible with the enzymatic activities of the cell causing as little inhibition as possible. These low molecular weight osmolytes were, therefore, termed compatible solutes. The incapability of adjusting osmotically to increasing solute concentrations of the aqueous environment (decreasing a_w) will result in cessation of growth or death.

It should be noted that frequently the a_w of an aqueous environment may increase due to dilution, resulting in an increase in the turgor pressure and/or the cell volume, and the intracellular compatible solute must be lost or metabolized until adjustment with the extracellular a_w is achieved.

It is not our purpose to discuss the terminology used to describe the adjustment of microorganisms to environments of low a_w, imposed by salts or sugars, and we will continue to use the term osmoregulation defined by Brown *et al.* (1986) as "the maintenance of turgor pressure and/or cell volume within the limits necessary for growth and multiplication of the organism". This working definition seems adequate for our discussion, although as noted by Reed (1984; 1986) there is need to standardize terminology, and it should be remembered that other terms, equally descriptive, such as osmotic adjustment, appear in the literature.

Xerotolerant yeasts

We usually think of eubacteria, and especially archaebacteria, as the organisms most capable of exploiting extreme environments. While this is true in most cases, some eukaryotic microorganisms are also capable of growing in environments of extreme salinity, and some yeasts and molds are unsurpassed in their ability to grow at extremely low a_w imposed by sugars (Hocking, 1988).

Barnett *et al.* (1983) gave a large list of yeast species which grow in laboratory media containing 50% glucose, a few of which, also grow in media with 60% glucose. Other authors, such as Onishi (1963) and Spencer and Spencer (1978), have also described yeasts capable of growing in media of lowered a_w. Nevertheless, detailed studies on yeast osmoregulation has involved very few species, namely *Saccharomyces cerevisiae, Zygosaccharomyces rouxii, Debaryomyces hansenii* and more recently, *Hansenula anomala* (Brown, 1978; Adler and Gustafsson, 1980; Nobre and da Costa, 1985b; André *et al.*, 1988; Bellinger and Larher, 1988; van Eck, 1988).

Most yeast species studied show no special requirements for elevated concentrations of salts or sugars. They grow in laboratory media containing no added salt, and with low concentrations of sugars as carbon sources. Their growth rate, and in some cases the final biomass, decreases as the concentration of solute (salts or sugars) in the medium is increased. Some yeasts can, nevertheless, grow in media of extremely low a_w, the range of tolerance depending, among other environmental factors, on the strain and the solute used to lower the a_w. Initially, yeasts capable of growing at very low a_w were termed osmophilic; this term, although useful in distinguishing certain yeasts capable of growth at low a_w, lead to some confusion, since the majority of these organisms are not osmophilic in the sense that they require high salt or sugar concentrations for growth or demonstrate enhanced growth at lower a_w. Brown (1976), therefore proposed the term xerotolerant to describe yeasts capable of growth in media of low a_w imposed both by salts or sugars, but which do not require a low a_w for growth. A few strains of yeasts do appear to be xerophilic, and some have been described which require salt or sugars in fairly large concentrations for growth. These strains include a mutant of *Z. rouxii* (Koh, 1975), *Candida (Torulopsis) halonitratophila* described by Onishi (1960) as halophilic at 30°C, but not at 20°C, and *Z. (nectarophilus) rouxii*, which also has a requirement for low a_w at 30°C, but not at 16-23°C (Anand and Brown, 1968; Brown, 1978).

Debaryomyces hansenii is a typical xerotolerant yeast which is capable of growing in media of very low a_w imposed by salts or sugars. Norkrans (1966) was able to grow strain 26, isolated from sea water, in medium containing 24 % (about 4.1 M) NaCl. The strain we have used (IGC 2968) is less tolerant; increasing the NaCl concentration of the medium caused a progressive decrease in the growth rate until growth ceased or was very slow in 3 M NaCl medium, and as expected, the lag phase also became longer as the NaCl was increased (Fig.1). The same progressive decrease in the growth rate of this strain occurred as KCl was added in increasing concentrations to the growth medium, with irregular growth occurring in 3.5 M KCl (Nobre and da Costa, unpublished results). Growth in media containing large concentrations of non-ionic solutes, such as glucose, fructose, sucrose and polyethylene glycol M.W. 200 (PEG 200) also lead to an increase in the lag period and a decrease in the growth rate (Nobre and da Costa, 1985b). These results are in accordance with other authors and clearly indicate the lack of salt or sugar dependent growth in this yeast. In addition to a decrease in the growth rate and an increase in the lag phase, Hobot and Jennings (1981) also demonstrated a clear decrease in the final cell number of *D. hansenii* concomitant with decreasing a_w.

Other yeast species have the same behaviour, for example, *Candida cacaoi, Candida magnoliae, Hansenula anomala, Pichia sorbitophila* and *Zygosaccharomyces bisporus* have lower growth rates as the a_w of the medium is decreased with glucose or NaCl from 0.998 to 0.920 (van Eck, 1988). The yeast-like fungus, *Geotrichum candidum,* although not xerotolerant, grows in media made up with 1 M NaCl or 25% glucose. Inoculation of 2% glucose-derived arthrospores in medium containing 25% glucose caused a delay of several hours in germination, as compared to the germination in medium containing 2% glucose, but did not affect the viability of the spores, which was close to 100% in both cases. The delay in germination was reflected by an increase in the lag phase in the 25% glucose medium (da Costa and Niederpruem, 1980).

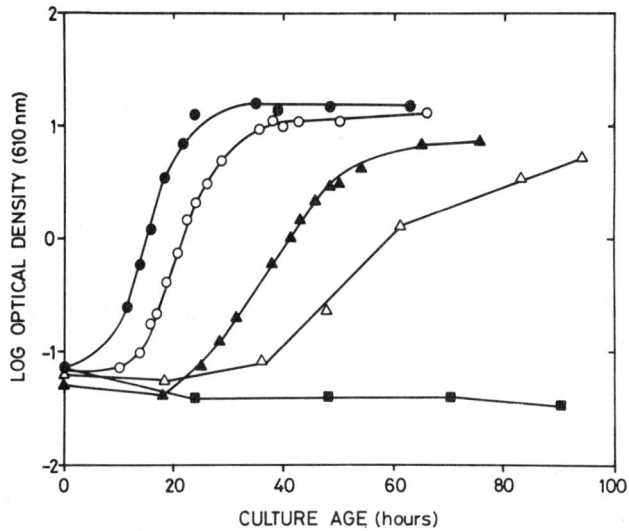

Fig.1. Growth of *D. hansenii* (IGC 2968) in 1% glucose-YEP medium containing various concentrations of NaCl. No NaCl added (●); 1 M NaCl (○); 2 M NaCl (▲); 2.5 M NaCl (△) and 3 M NaCl (■). Stationary phase cells from 1% glucose-YEP medium were used as inoculum. Growth conditions described by Nobre and da Costa (1985a).

Xerophilic filamentous fungi

Some xerotolerant and a few xerophilic fungi have been studied in detail (Pitt and Hocking, 1977; Luard, 1982a, b, c; Hocking, 1986, and Andrews and Pitt, 1987). Some of the species studied are true xerophiles which require low a_w for growth or demonstrate enhanced growth in media with high concentrations of salts or sugars. Andrews and Pitt (1987) presented results on the germination and growth of eight xerophilic fungi on NaCl, glucose/fructose or glycerol. This study illustrates the difference between xerotolerance common in yeasts and the xerophilic nature of some filamentous fungi, and demonstrates the effect of specific solutes on the water relations of these organisms. Two strains of *Basipetospora halophila*, demonstrated superior growth in media with NaCl, and were considered halophilic as compared to growth in media with glucose/fructose or glycerol, although there was no obligate requirement for this salt. The same was true for *Exophiala werneckii* and *Polypaecilum pisce*, the former of which grew in media saturated with NaCl. In contrast, *Aspergillus penicilloides* was considered by these authors to be xerophilic since it grew equally well in media containing large concentrations of NaCl and non-ionic solutes. *Eurotium* strain FRR 2471 was considered an obligate xerophile growing to low water activities in the non-ionic solutes tested; in addition, another xerophile, *Geomyces* FRR 2375 was very sensitive to NaCl and very tolerant to non-ionic solutes. Other extremely xerophilic fungi have been studied, namely *Xeromyces bisporus*, which may well be the most xerophilic fungus known (Hocking,1988).

Compatible solutes

The intracellular accumulation of an organic compatible solutes is a predominant physiological adaptation to water stress in the eubacteria and eukaryotic microorganisms studied, although other cellular alterations are probably also necessary for growth at low a_w.

A vast variety of organisms accumulate organic compatible solutes in response to water stress, yet the number of different types of compatible solutes is very small (Trüper and Galinsky, 1986). Discussing osmoregulation from an evolutionary view, Yancey *et al.* (1982) showed the number of different organic osmolytes (compatible solutes) to be limited to polyols, polyol derivatives, some sugars, a few free amino acids and amino acid derivatives, urea and methylamines and proposed that the limited types of compatible solutes encountered in these organisms, reflect a "ubiquitous set of physicochemical interactions between solutes, water and macromolecules that establish which types of solutes are compatible with macromolecular structure and function".

Most authors also agree that potassium chloride, which accumulates to high levels in halophilic archaebacteria, functions as a compatible solute equivalent to the organic compatible solutes of eubacteria and eukaryotic microorganisms (Brown, 1976; Yancey *et al.* 1982; Trüper and Galinsky, 1986; Hocking, 1988), even though the proteins of these organisms are also highly modified in amino acid content and require high salt concentrations for enzyme activity (Kushner, 1978;1988). On the other hand, most eubacterial and eukaryotic enzymes have maximal activity at low salt concentrations. Some enzymes of halotolerant and halophilic eubacteria have been described which tolerate or require salt for activity, but as Kushner (1988) points out, a distinction must be made between the salt relations of intracellular and extracellular or membrane-bound enzymes. Furthermore, this author suggests that most of the cell-associated cations of these organisms are, in fact, bound to the envelope, while the cytoplasm is relatively salt-free.

Yeasts and fungi respond to water stress by the accumulation of glycerol, and more rarely other polyols, although in at least one oomycetous fungus, *Phytophothora cinnamomi*, proline accumulates in response to water stress (Luard, 1982).

The accumulation of glycerol in response to water stress, as well as its role as a compatible solute has been discussed in detail by Brown (1978), Brown *et al.* (1986) and Hocking (1988), and it has become clear that glycerol accumulates, in various yeasts and fungi, when the a_w is decreased with salts, metabolizable or non-metabolizable sugars and polyethylene glycol M.W.200 (PEG 200). These observations suggest that a general mechanism, not normally dependent on the type of controlling solute, exists for the accumulation of glycerol in response to decreased a_w.

Glycerol accumulation in *Debaryomyces hansenii*

Debaryomyces hansenii accumulates glycerol after water stress with various small molecular weight solutes such as NaCl, KCl, glucose, fructose and PEG 200 (Adler and Gustafsson, 1980; Nobre and da Costa, 1985b; Larson and Gustafsson, 1987; André *et al.*, 1988).

In conventional media of high a_w, glycerol is not detected intracellularly or is found in very low concentrations, nevertheless increasing the solute concentration of a medium slightly form 1% glucose to 8% glucose (Nobre and da Costa, 1985a), or adding 0.25% NaCl to 1% glucose medium (André *et al.*, 1988) causes glycerol accumulation. This indicates that very small decreases in the a_w elicits an osmotic response with intracellular glycerol accumulation in *D. hansenii*. Decreasing the a_w further with NaCl, KCl, glucose, fructose and PEG 200, causes a concomitant increase in the glycerol accumulation during exponential growth, as shown in 1% glucose medium containing 1 M and 2 M NaCl or KCl (Fig.2). As the culture progresses to stationary phase, the intracellular glycerol decreases to very low levels under all conditions tested (Adler and Gustafsson,1980; Nobre and da Costa, 1985a, b; André *et al.*, 1988). Extracellular glycerol also decreases at this time,as demonstrated by Adler and Gustafsson (1980), which leads to the conclusion that this polyol is consumed after depletion of the carbon source from the medium.

These results were confirmed by Reed *et al.* (1987) who demonstrated that glycerol is the major compatible solute during exponential growth of *S. cerevisiae*, *Z. rouxii* and *D. hansenii* in response to water stress imposed by 5% NaCl, but conclude that in basal medium with no added salt, or medium containing 1% NaCl, glycerol did not appear to be the major solute counterbalancing the external osmotic pressure. Since no other organic solutes were detected by natural-abundance ^{13}C nuclear magnetic resonance (NMR), these authors proposed that intracellular inorganic ions counterbalanced the remaining external osmotic pressure, especially in media with low NaCl concentrations. Other authors have shown that monovalent cations (K^+ and Na^+) may have a role in the osmoregulation of *D. hansenii*, but conclude that glycerol is the major compatible solute (Norkrans and Kylin, 1969; Hobot and Jennings, 1981).

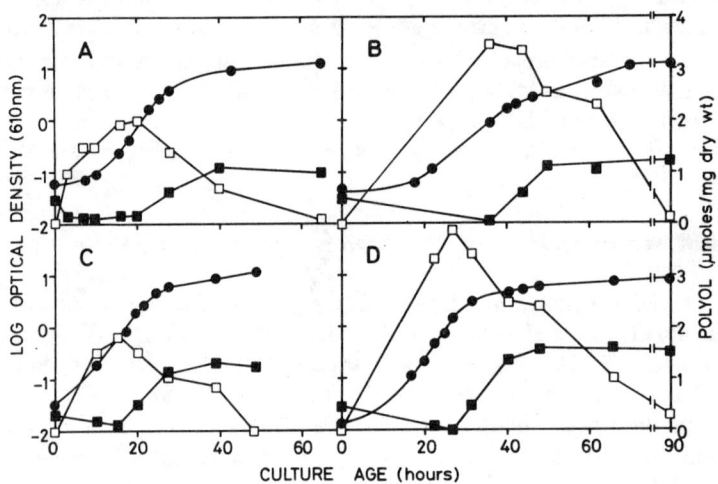

Fig.2. Growth of *D. hansenii* (IGC 2968) in 1% glucose-YEP medium with (A) 1 M NaCl; (B) 2 M NaCl; (C) 1 M KCl; (D) 2 M KCl. Log O.D. (●). The intracellular glycerol (□) and arabitol (■) were determined during growth as described by Nobre and da Costa (1985a).

Under the conditions described above, arabitol accumulates to very low levels during exponential growth (Adler and Gustafsson, 1980; Nobre and da Costa, 1985a,b; André *et al.*, 1988), but increases in concentration during the stationary phase. We have generally used stationary phase cells, containing arabitol, as inocula for experimental cultures, and found that there is an initial decrease in arabitol, irrespective of the carbon source or solute concentration, to low levels during exponential growth. Arabitol increases again during stationary phase in all media tested, except when mannitol is used as a carbon source. In this medium, mannitol accumulates during the stationary phase with little or no arabitol detected by paper chromatography (Nobre and da Costa, 1985a;b).

The accumulation of arabitol during the stationary phase of growth is, in contrast with glycerol, not related to the solute concentration of the medium in this and other yeasts studied (Adler and Gustafsson, 1980; Edgley and Brown,1978; 1983; Nobre and da Costa, 1985b), although slightly higher levels of arabitol may be found, during stationary phase, in some yeasts when the a_w is decreased. Other factors, such as type of carbon source (Nobre and da Costa, 1985a), depletion of a nitrogen source or, a decrease in the growth rate may influence the accumulation of arabitol in *D. hansenii* (Veríssimo-Pires and da Costa, unpublished observations).

Glycerol accumulation in *Candida famata* and *Pichia farinosa*

The intracellular accumulation of glycerol during the exponential phase of growth in media with low a_w, its subsequent decrease, as well as the concomitant accumulation of another polyol or trehalose during the stationary phase has been observed in other yeasts (see Brown *et al.*, 1986). For example, *Candida famata*, the asexual state of *D. hansenii*, and *Pichia farinosa* have a similar behaviour as the a_w of the medium is decreased (Fig.3;Fig.4). In conventional media of high a_w, there was, in

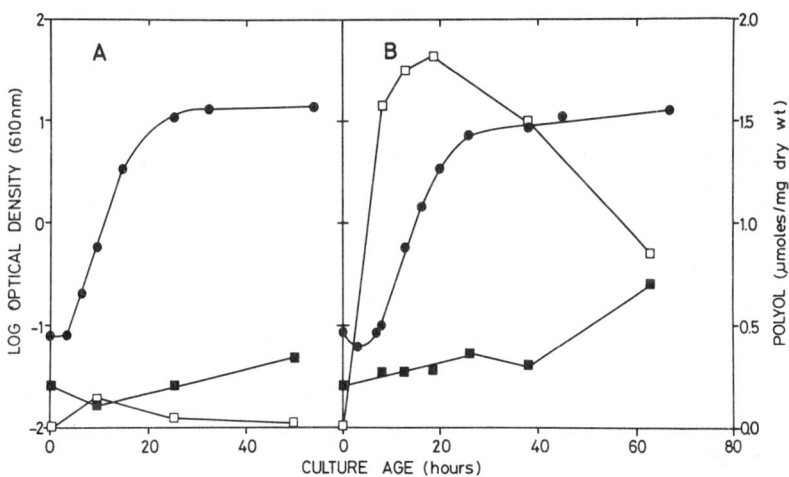

Fig.3. Growth of *C. famata* (IGC 3524) in 1% glucose-YEP medium (A) without added NaCl and (B) with 1 M NaCl. Log O.D. (●). The intracellular glycerol (□) and arabitol (■) were determined as described by Nobre and da Costa (1985a). Stationary phase cells from 1% glucose-YEP medium were used as inoculum.

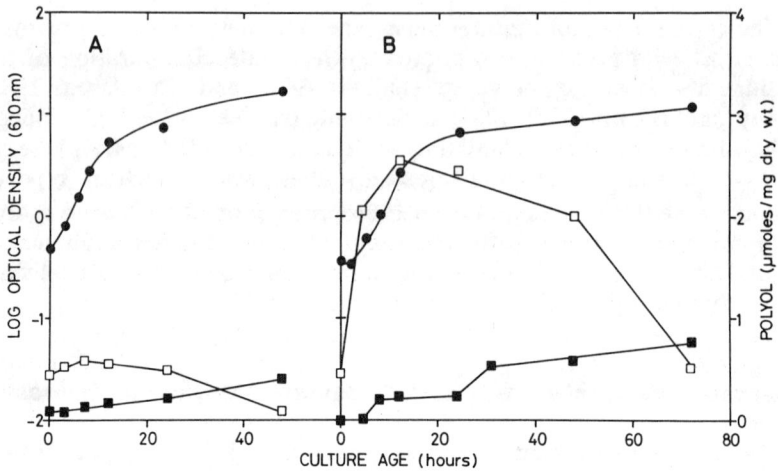

Fig. 4. Growth of *P. farinosa* (IGC 3021) in 2% glucose-YNB-CA medium (A) without added NaCl and (B) with 1 M NaCl. Log O.D. (●). The intracellular glycerol (□) and arabitol (■) were determined as described previously. Traces of erythritol were detected by paper chromatography during the stationary phase of growth. Exponential phase cells from 2% glucose-YNB-CA medium were used as inoculum. YNB-CA: 1.7 g Yeast Nitrogen Base w/o amino acids and ammonium sulphate (Difco,0335-15), 2.0 g casamino acids (Difco), 0.46 g KH_2PO_4 and 1.0 g K_2HPO_4 in 1 L deionized water, pH 6.7.

contrast to *D. hansenii*, a slight accumulation of glycerol during the exponential growth phase of both yeasts, which was especially significant in *P. farinosa*. In this yeast, glycerol reached 0.53 µmol/mg of the dry weight. Arabitol increased during the stationary phase of growth concomitant with a decrease in intracellular glycerol in these strains.

Growth of these yeasts in glucose medium containing 1 M NaCl lead, as expected, to a very large intracellular accumulation of glycerol during exponential growth reaching 1.82 µmol/mg of the dry weight in *C. famata* and 2.57 µmol/mg of the dry weight in *P. farinosa*. The intracellular glycerol of both yeasts decreased during the stationary phase of growth with the concomitant accumulation of arabitol; trace levels of erythritol were also detected by paper chromatography, during the stationary phase, in *P. farinosa*.

Since glycerol appears to act as a compatible solute and represents a major physiological adaptation to water stress, the cells must, as suggested by André et al. (1988), release and/or catabolize the compatible solute upon transfer of the culture to a medium of higher a_w, thereby decreasing the turgor pressure. These authors demonstrated that glycerol accumulated in high salinity medium by *D. hansenii* was lost very rapidly upon transfer to a low salinity medium, although some glycerol was retained during the largest shockdowns indicating, perhaps, that some of the glycerol pool was not free to move out of the cells. Comparable results were obtained with *G. candidum* arthrospores derived from 25% glucose medium, and containing elevated levels of arabitol and mannitol, upon resuspension in water. These polyols were rapidly lost to the levels found in 2% glucose-derived arthrospores, and this fraction could only be released by boiling (da Costa and Niederpruem, 1982). Arthrospores from 2% glucose medium released no detectable mannitol for at least 60 minutes after resuspension in water. In neither of these organisms was the viability affected by the shockdowns.

In the yeasts studied, arabitol does not appear to be a primary compatible solute under extreme water stress,but it can nevertheless, due to its intracellular concentration, function as a secondary osmoregulator or compatible solute, as can be inferred from the results obtained by Brown and coworkers (Brown *et al.*, 1986; Mackenzie *et al.*, 1986; 1988) on the water stress plating hypersensitivity of some yeasts. This phenomenon is characterized by a massive loss in viability when a sensitive yeast, for example *Saccharomyces cerevisiae* or *Candida krusei*, is transferred from a exponentially growing culture in a medium of high a_w to a plating medium of low a_w. Non-sensitive yeasts, including yeasts which are not normally considered xerotolerant, do not exhibit this loss of viability. Furthermore, the plating hypersensitivity is eliminated during the stationary phase or when the sensitive yeast is grown to exponential phase in 2% NaCl medium.

As shown by these authors, the resistant species accumulated one or more polyols during exponential growth in media with high a_w, while the sensitive strains did not. Water stress plating hypersensitivity was eliminated during stationary phase cultures of *S. cerevisiae* due, presumably, to the accumulation of trehalose, as well as the accumulation of glycerol during exponential growth in 2% NaCl medium. They propose that polyols, trehalose and probably other intracellular compounds accumulated to fairly high levels, even if not in response to osmotic adjustment (e.g. trehalose), will protect the cells against this type of stressing procedure.

We consider these experiments to be of extreme importance to our understanding of osmoregulation in microorganisms and particularly in yeasts, since it illustrates the role of intracellular solutes,such as arabitol, which do not respond to changes in the solute concentration of the medium, but which can contribute to osmotic adjustment.

Sugars, such as glucose, fructose or sucrose, do not appear to serve as compatible solutes in yeasts, since their intracellular concentration is always vestigial even when the a_w of the medium is decreased with large concentrations of these sugars (Nobre and da Costa, 1985b; van Eck, 1988).

Alternate compatible solutes in yeasts

In addition to glycerol, other polyols have a role as compatible solutes in fungi and yeasts. For example, Gadd et al. (1984) reported the accumulation of glycerol as the primary compatible solute during growth of *Penicillium ochro-chloron* in high concentrations of NaCl and $CuSO_4$, however, erythritol, arabitol, hexitol (mannitol and sorbitol) and the disaccharide trehalose were also present in large intracellular concentrations and would necessarily contribute to the total osmolyte content of the cells. In fact, erythritol and the hexitols were always found in higher concentrations than glycerol.

In *D. hansenii* and *P. farinosa* erythritol, used as a carbon source, appears to replace glycerol as the primary compatible solute.

Growth of *D. hansenii* in complex medium with 1% *meso*-erythritol as a carbon source, lead to the accumulation of erythritol to 0.77 µmol/mg of the dry weight, while growth in medium with erythritol and 1 M NaCl lead to the accumulation of erythritol to 1.92 µmol/mg of the dry weight during the exponential phase. On the other hand, glycerol accumulated to 0.69 µmol/mg of the dry weight, as compared to 1.97 and 1.82 µmol/mg of the dry weight in 1% glucose medium containing 1 M NaCl or KCl, respectively (Nobre and da Costa, 1985b).

The same phenomenon was observed with *P. farinosa* grown in erythritol-based media (Fig.5). Growth of the yeast in 1% *meso*-erythritol resulted in the accumulation of this polyol to 1.16 µmol/mg of the dry weight during exponential growth followed by decreasing intracellular concentrations towards the stationary phase. Glycerol accumulation, observed in this yeast during growth in 1% glucose medium, was detected in vestigial concentrations.

Growth in erythritol medium containing 1 M NaCl lead to the intracellular accumulation of erythritol to 2.15 µmol/mg of the dry weight; glycerol, which accumulated initially to a lower concentration than in the corresponding glucose medium, decreased very rapidly as the erythritol accumulation continued. Arabitol accumulation was also largely inhibited and could only be detected, in trace amounts, by paper chromatography at the last experimental points in both cases. Contamination of the ethanolic extracts by exogenous erythritol would not have accounted for the values obtained, since the supernatant of the last washing step was always checked for polyol and could not contribute more than 0.07 μmol/mg of the dry weight to the total cellular polyol.

In these experiments, not only was the intracellular concentration of erythritol very high during growth in erythritol-1 M NaCl medium, but intracellular glycerol was also very low. The decrease in glycerol accumulation reinforces the notion that erythritol has an osmoregulatory role in these yeasts, and that some inhibitory effect is exerted on the synthesis of glycerol, the ability to retain it, or both.

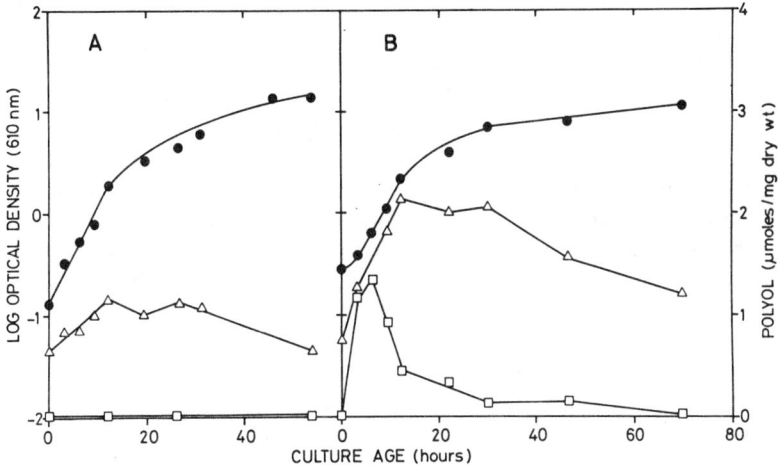

Fig. 5. Growth of *P. farinosa* (IGC 3021) in 1% erythritol- YNB-CA medium (A) without added NaCl and (B) with 1 M NaCl. Log O.D. (●). The intracellular glycerol (□) and erythritol (△) were determined as described by Nobre and da Costa (1985a). Trace levels of arabitol were detected at the last experimental time points of both experiments. Exponential phase cells from 1% erythritol- YNB-CA medium were used as inoculum.

In *D. hansenii*, exogenous mannitol has an effect similar to erythritol, although less pronounced. The moderate accumulation of mannitol, during the exponential phase in 1 M NaCl medium, caused a decrease in the accumulation of glycerol, as compared with the corresponding medium with 1% glucose as carbon source, and suggests that mannitol can also contribute to the osmoregulation of this yeast (Nobre and da Costa, 1985b).

These results strongly indicate that erythritol can replace glycerol as a compatible solute when the former polyol is used as a carbon source. However, Adler and Gustafsson (1980), using a mutant of *D. hansenii*, less xerotolerant because it retained less intracellular glycerol than the wildtype (strain 26), found exogenously supplied mannitol, arabitol, and glycerol to stimulate growth in media with 3.1 M NaCl, while erythritol had no effect. These authors also reported that exogenous glycerol had no effect on the growth of the wildtype in high salt media. Using another mutant of the same wildtype strain, André et al., (1988) showed that mannitol, arabitol, proline, trimethylaminoxide and betaine, in contrast to glycerol, had no effect in promoting growth in high salt medium. Furthermore, van Eck (1988) did not obtain accumulation of erythritol in the yeast *Hansenula anomala* in response to decreased a_w in minimal medium containing NaCl and erythritol as the carbon source.

Arabitol accumulation in *Geotrichum candidum*

Arabitol does not appear to be a primary compatible solute during the exponential phase in the yeasts, or most of the extremely xerotolerant and xerophilic fungi studied. Yet this polyol was shown to respond to increases in solute concentration of the growth medium imposed by NaCl, KCl, glucose, sorbose, and sucrose in the yeast-like fungus *G. candidum*, while mannitol was found primarily in the arthrospores and did not appear to respond to decreased a_w (da Costa and Niederpruem, 1982).

We have recently studied another strain of *G. candidum* (DZ-Y2) which also accumulates arabitol in response to increased NaCl in the growth medium (Fig.6). Mannitol, arabitol and trehalose were found at very low intracellular levels during the exponential phase in 2% glucose medium. Mannitol and trehalose increased to 0.45 μmol/mg and 0.10 μmol/mg of the dry weight, respectively, as the culture entered stationary phase and formed arthrospores, while intracellular glycerol was vestigial throughout the growth cycle. In 2% glucose medium plus 1 M NaCl there was a rapid accumulation of arabitol during exponential growth reaching 1.25 μmol/mg of the dry weight, while mannitol accumulated towards the end of growth and reached 0.20 μmol/mg of the dry weight. In this medium, there was a slight accumulation of glycerol during the exponential phase, reaching 0.12 μmol/mg of the dry weight.

In an attempt to induce glycerol accumulation we grew this strain in 1% glycerol and 1% glycerol plus 0.5 M NaCl (Fig.7). Growth in 1% glycerol led to the accumulation of mannitol and trehalose during the stationary phase to 0.51 μmol/mg and 0.13 μmol/mg of the dry weight, respectively. Arabitol was low throughout the growth cycle and glycerol accumulated to low levels primarily during exponential growth. In 1% glycerol medium plus 0.5M NaCl there was rapid increase in the intracellular arabitol concentration to 0.64 μmol/mg of the dry weight, while glycerol did not accumulate beyond 0.10 μmol/mg of the dry weight. Mannitol and trehalose increased during the stationary phase.

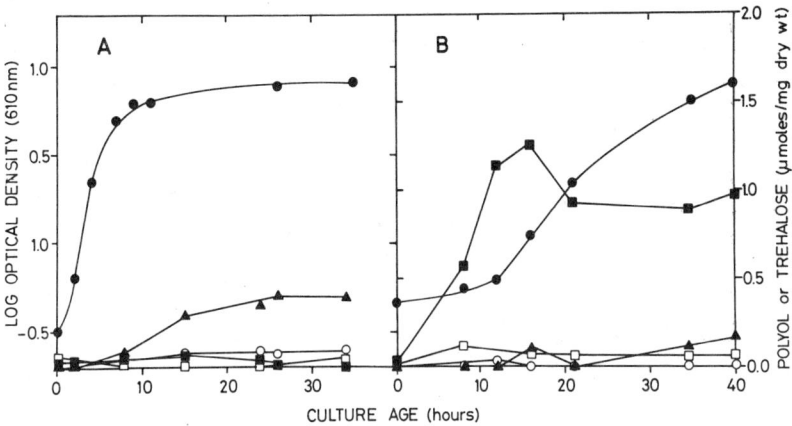

Fig.6. Growth of *G. candidum* (DZ-Y2) in 2% glucose-YNB-CA medium (A) without added NaCl and (B) with 1 M NaCl. Log O.D. (●). Glycerol (□) was determined by the glycerol kinase method; arabitol (■), mannitol (▲) and trehalose (○) were determined by high pressure liquid chromatography (HPLC) using a Varian CHO-620 column maintained at 90°C using ultrapure water as eluent with a flow rate of 0.4 mL min.$^{-1}$. Exponential phase cells from 2% glucose- YNB-CA medium without salt were used as inoculum. Culture conditions were as described by Nobre and da Costa (1985a).

The results show, that under these experimental conditions, glycerol does not substitute for arabitol as a compatible solute in this strain of *G. candidum*, perhaps because glycerol cannot be retained intracellularly.

The role of the cell membrane on the osmoregulation of yeasts is still obscure, primarily because there have been very few studies on the alterations in the lipid composition in relation to changes in the a_W. In the few published studies no differences in lipid composition conclusively related to the effect of the a_W, and which could influence the ability to retain glycerol or other polyols, have been found (Tunblad-Johansson and Adler,1987; Hocking,1988;). In halotolerant and moderately-halophilic eubacteria, on the other hand, several studies have shown that physiological adaptation to increased salt concentrations has a significant effect on the polar lipid, and fatty acid composition of several species (Russell et al., 1986; Monteoliva-Sanchez and Ramos-Cormenzana,1987; Monteoliva-Sanchez et al.,1988).

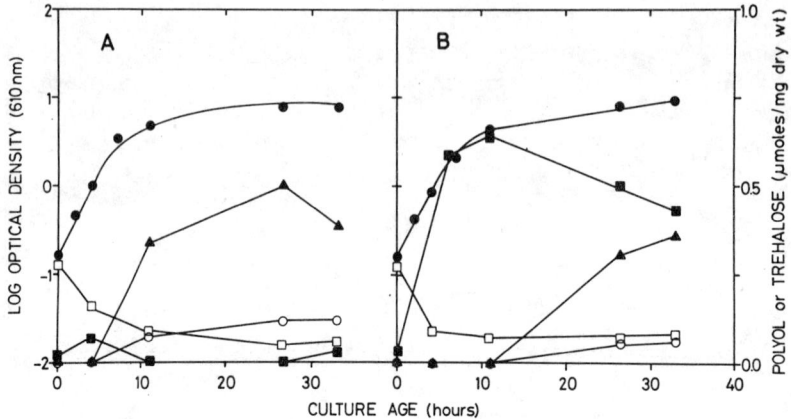

Fig.7. Growth of *G. candidum*(DZ-Y2) in 1% glycerol-YNB-CA medium (A) without added NaCl and (B) with 0.5 M NaCl. Log O.D. (●). Glycerol (□) was determined by the glycerol kinase method; arabitol (■), mannitol (▲) and trehalose (O) were determined by high pressure liquid chromatography (HPLC). Exponential phase cells from 1% glycerol-YNB-CA medium without salt were used as inoculum.

Other membrane-related mechanisms, presumably, involved in the accumulation of compatible solutes are active transport systems. Their importance are suggested by Adler *et al.* (1985), who demonstrated that glycerol accumulated against a very large concentration gradient in *D. hansenii*. Furthermore, the results which show that other xerotolerant yeasts, such as *Z. rouxii* and *H. anomala* retain glycerol more efficiently than the non-xerotolerant yeast, *S. cerevisiae*, also imply the existence of active transport systems in these yeasts (Edgley and Brown,1983; Larsson and Gustafsson,1987; Bellinger and Larher, 1988).

Research in the area of active transport systems will, no doubt, reveal the importance of these systems in maintaining high intracellular concentrations of compatible solutes in yeasts which tolerate extreme water stress.

Acknowledgements

Research support of our studies on the osmoregulation in yeasts has been provided by the Instituto Nacional de Investigação Científica (INIC) and the Junta Nacional de Investigação Científica e Tecnológica (JNICT).

The yeast strains used in these studies were kindly donated by Prof. N. van Uden, Instituto Gulbenkian de Ciência (IGC), 2781, Oeiras, Codex, Portugal. We would also like to thank the technical personnel of the Quality Control Department of UNICER (União Cervejeira), Leça do Bailio, Portugal for performing the HPLC on the *G. candidum* extracts. We would also like to thank Eduardo Seleiro for editing the text.

References

Adler, L., and L. Gustafsson. 1980. Polyhydric alcohol production and intracellular amino acid pool in relation to halotolerance of the yeast *Debaryomyces hansenii*. Arch. Microbiol. **124**: 123-130.

Adler, L., A. Blomberg, and A. Nilsson. 1985. Glycerol metabolism and osmoregulation in the salt-tolerant yeast *Debaryomyces hansenii*. J. Bacteriol. **162**: 300-306.

Anand, J. C., and A. D. Brown. 1968. Growth rate patterns of the so-called osmophilic and non-osmophilic yeasts in solutions of polyethylene glycol. J. Gen. Microbiol. **52**:205-212.

Andrews, S., and J. I. Pitt. 1987. Further studies on the water relations of xerophilic fungi, including some halophiles. J. Gen. Microbiol. **133**:233-238.

André, L., A. Nilsson, and L. Adler. 1988. The role of glycerol in osmotolerance of the yeast *Debaryomyces hansenii*. J. Gen. Microbiol. **134**: 669-677.

Barnett, J. A., R. W. Payne, and D. Yarrow. 1983. "Yeasts: Characteristics and Identification". New York and Cambridge. Cambridge University Press.

Bellinger, Y., and F. Larher. 1988. A ^{13}C comparative magnetic resonance study of organic solute production and excretion by the yeasts *Hansenula anomala* and *Saccharomyces cerevisiae* in saline media. Can. J. Microbiol. **34**: 605-612.

Brown, A. D. 1976. Microbial water stress. Bacteriol. Rev. **40**: 803-846.

Brown, A. D. 1978. Compatible solutes and extreme water stress in eukaryotic microorganisms. Adv. Microbial Physiol. **17**: 181-242.

Brown, A. D., K. F. Mackenzie, and K. K. Singh. 1986. Selected aspects of microbial osmoregulation. FEMS Microbiol. Rev. **39**: 31-36.

da Costa, M. S., and D. J. Niederpruem. 1980. Temporal accumulation of mannitol and arabitol in *Geotrichum candidum*. Arch. Microbiol. **126**:57-64.

da Costa, M. S., and D. J. Niederpruem. 1982. Arabitol accumulation in *Geotrichum candidum*. Arch. Microbiol. **131**: 283-286.

Edgley, M., and A. D. Brown. 1978. Response of xerotolerant and non-tolerant yeasts to water stress. J. Gen. Microbiol. **104**: 343-345.

Edgley, M., and A. D. Brown. 1983. Yeast water relations: physiological changes induced by solute stress in *Saccharomyces cerevisiae* and *Saccharomyces rouxii*. J. Gen. Microbiol. **129**: 3453-3463.

Gadd, G. M., J. A. Chudek, R. Foster, and R. H. Reed. 1984. The osmotic responses of *Penicillium ochro-chloron* : changes in internal solute levels to copper and salt stress. J. Gen. Microbiol. **130**: 1969-1975.

Hobot, J. A., and D. H. Jennings. 1981. Growth of *Debaryomyces hansenii* and *Saccharomyces cerevisiae* in relation to pH and salinity. Exp. Mycol. **5**: 217-228.

Hocking, A. D. 1986. Effects of water activity and culture age on the glycerol accumulation patterns of five fungi. J. Gen. Microbiol. **132**: 269-275.

Hocking, A. D. 1988. Strategies for microbial growth at reduced water activities. Microbiol. Sciences. **5**: 280-284.

Koh, T. Y. 1975. Sudies on the "osmophilic" yeast *Saccharomyces rouxii* and obligate osmophilic mutant. J. Gen. Microbiol. **88**: 101-114.

Kushner, D. J. 1978. Life in high salt and solute concentrations: halophilic bacteria. "Microbial Life in Extreme Environments". (Ed. D. J. Kushner), pp. 317-368. London . Academic Press.

Kushner, D. J. 1988. What is the "true" internal environment of halophilic and other bacteria? Can. J. Microbiol. **34**: 482-486.

Larson, C., and L. Gustafsson. 1987. Glycerol production in relation to the ATP pool and heat production rate of the yeasts *Debaryomyces hansenii* and *Saccharomyces cerevisiae* during salt stress. Arch. Microbiol. **147**: 358-363.

Luard, E. J. 1982a. Accumulation of intracellular solutes by two filamentous fungi in response to growth at low steady state osmotic potential. J Gen. Microbiol. **128**: 2563-2574.

Luard, E. J. 1982b. Effect of osmotic shock on some intracellular solutes in two filamentous fungi. J. Gen. Microbiol. **128**: 2575-2581.

Luard, E. J. 1982c. Growth and accumulation of solutes by *Phytophothora cinnamomi* and other lower fungi in response to changes in external osmotic potential. J. Gen. Microbiol. **128**: 2583-2590.

Mackenzie, K. F., A. Blomberg, and A. D. Brown. 1986. Water stress plating hypersensitivity of yeasts. J. Gen. Microbiol. **132**: 2053-2056.

Mackenzie, K. F., K.K. Singh, and A. D. Brown. 1988. Water stress plating hypersensitivity of yeasts: protective role of trehalose in *Saccharomyces cerevisiae*. J. Gen. Microbiol. **134**: 1661-1666.

Monteoliva-Sanchez, M., and A. Ramos-Cormenzana. 1987. Cellular fatty acid composition of *Planococcus halophilus* NRCC 14033 as affected by growth temperature and salt concentration. Curr. Microbiol. **15**: 133-136.

Monteoliva-Sanchez, M.,M. R. Ferrer, A. Ramos-Cormenzana, E. Quesada, and M. Monteoliva. 1988. Cellular fatty acid composition of *Deleya halophila* : effect of growth temperature and salt concentration. J. Gen. Microbiol. **134**: 199-203.

Nobre, M. F., and M. S. da Costa. 1985a. Factors favouring the accumulation of arabinitol in the yeast *Debaryomyces hansenii*. Can. J. Microbiol. 31: 467-471.

Nobre, M. F., and M. S. da Costa. 1985b. The accumulation of polyols by the yeast *Debaryomyces hansenii* in response to water stress. Can. J. Microbiol. 31: 1061-1064.

Norkrans, B. 1966. Studies on marine occurring yeasts: growth related to pH, NaCl concentration and temperature. Archiv. für Mikrobiol. 54: 374-392.

Norkrans, B., and A. Kylin. 1969. Regulation of the potassium to sodium ratio and of the osmotic potential in relation to salt tolerance in yeasts. J. Bacteriol. 100: 836-845.

Onishi, H. 1960. Studies on osmophilic yeasts. Part IX. Isolation of a new obligate halophilic yeast and some considerations on halophilism. Bull. Agric. Chem. Soc. Japan. 24: 126-130.

Onishi, H. 1963. Osmophilic yeasts. Adv. Food Res. 12: 53-94.

Pitt, J. I., and A. D. Hocking. 1977. Influence of solute and hydrogen ion concentration on the water relations of some xerophilic fungi. J. Gen. Microbiol. 101: 35-40.

Reed, R. H. 1984. Use and abuse of osmo-terminology. Plant, Cell and Environ. 7: 165-170.

Reed, R. H. 1986. Halotolerant and halophilic microbes. " Microbes in Extreme Environments". (Eds. R. A. Herbert and G. A. Codd), pp. 55-81. Society for General Microbiology. London. Academic Press.

Reed, H. R., J. A. Chudek, R. Foster, and G. M. Gadd. 1987. Osmotic significance of glycerol accumulation in exponentially growing yeasts. App. Environ. Microbiol. 53: 2119-2123.

Russell, N. J., R. Adams, J. Bygraves, and M. Kogut. 1986. Cell envelope phospholipid changes in a moderate halophile during phenotypic adaptation to altered salinity and osmotic stress. FEMS Microbiol. Rev. 39: 103-107.

Spencer, J. F. T., and D. M. Spencer. 1978. Production of polyhydroxy alcohols by osmotolerant yeasts. "Economic Microbiology". (Ed. A. H. Rose), Vol. 2, pp. 393-425. London. Academic Press.

Trüper, H. G., and E. A. Galinsky. 1986. Concentrated brines as habitats for microorganisms. Experientia. 42: 1182-1187.

Tunblad-Johansson, I., and L. Adler. 1987. Effects of sodium chloride concentration on phospholipid fatty acid composition of yeasts differing in osmotolerance. FEMS Microbiol. Letters. 43: 275-278.

van Eck, J. H. 1988. Water relations of polyhydroxy alcohol production by yeasts. Master of Science Dissertation. University of the Orange Free State, Bloemfontein. Republic of South Africa.

Yancey, P. H., M. E. Clark, S. C. Hand, R. D. Bowlus, and G. M. Somero. 1982. Living with water stress: evolution of osmolyte systems. Science. 217: 1214-1222.

DENITRIFICATION OF CONCENTRATED SODIUM NITRATE SOLUTIONS BY THE MODERATE HALOPHILIC DENITRIFIER BACILLUS HALODENITRIFICANS

GERARD DENARIAZ [1], WILLIAM J. PAYNE [2] and Jean LEGALL [1]
Departments of Biochemistry [1] and Microbiology [2]
The University of Georgia, Athens, GA 30602, USA

ABSTACT

Denitrification by Bacillus halodenitrificans in batch cultures in saline semi-defined and complex media was studied . Anaerobic growth on nitrate resulted in an accumulation of nitrite, to a medium-dependent concentration, and nitrous oxide. Nitrous oxide evolution was not affected by high concentrations of nitrate. Nitrate was reduced all the way to elemental nitrogen by transfering the nitrous oxide produced into a culture of Pseudomonas stutzeri. Several substrates were used as electron donors for the reduction of nitrate by cell suspensions. The effects of pH, ionic strength and temperature on nitrate and nitrite reductase activities in washed cells were determined. A packed bed reactor filled with immobilized B. halodenitrificans cells displayed a high activity at the start up when denitrifying solutions containing 10 to 100 g/l sodium nitrate. Prolonged operation was not possible, however, with solutions supplemented only with a simple electron donor. Regeneration of the cells in the bed could be obtained by incubation in a complete growth medium.

INTRODUCTION

Effluents generated by nuclear industries during uranium processing contain concentrations of nitrate often greater than 50g NO_3^--N/l [1]. Many studies have been concerned with the removal of nitrate from waste waters using mixed populations of bacteria able to perform denitrification. This process calls upon the property of some bacteria to carry out, during anaerobic growth, the reduction of nitrate, via nitrite, to gaseous products, namely elemental nitrogen (N_2), and in some cases nitrous oxide (N_2O) as well. Nitrate is thus utilized in place of oxygen as the terminal electron acceptor in the oxidation of organic matter [2]. While most applications have been successful in sewage plants in the treatment of domestic wastes, i.e. with nitrate concentrations less than 50 mg NO_3^--N/l, only a few have been directed toward the denitrification of high-nitrate wastes [3,4,5]. In most systems, depending on the added carbon source, the denitrifier population comprises one or more of the following bacterial species: Hyphomicrobium vulgare, Paracoccus denitrificans, Alcaligenes faecalis, Pseudomonas aeruginosa and Pseudomonas mendocina [6,7,8]. None of those is recognized as a halotolerant organism. Consequently, in order to avoid inhibition of denitrifying bacteria, excessive dilutions are required, especially when sodium is the complementary cation. As Francis and Hancher [1] have pointed out, "probably the most important factor in the biological denitrification of high nitrate wastes is the establishment of a microbial culture that is tolerant to high ionic strength". Their culture seeded from an estuary marsh sediment indeed performed as predicted, as they observed denitrification rates 2 to 6 times higher than those obtained with non-selected mixed populations. Unfortunately, no information was given on which bacteria were present in the chosen sludge. Even with so specialized a population, the authors observed a toxic influence of sodium ions on the denitrification process. Studies on the microbiology of denitrification in high sodium nitrate solutions could certainly bring about improvements in the efficiency of the system.

To begin our work, we decided to look for an organism that would withstand high concentrations of both nitrate and sodium ions and carry out denitrification at acceptable rates under high as well as low salt concentrations, thus permitting greater flexibility in the system. Enrichment cultures from a sample from a solar saltern yielded a moderate halophilic denitrifier [designated Bacillus (B.) halodenitrificans, G.Denariaz, W.J.Payne and J.LeGall, submitted for publication] that tolerates concentrations of 1 M $NaNO_3$ and actively denitrifies. In this paper, we focus on factors involved in denitrification by B. halodenitrificans and present preliminary results concerning its utilization in the treatment of high sodium nitrate-containing solutions.

MATERIAL AND METHODS

Culture Media. The semi-defined medium (YA) contained, per liter of distilled water: Na_2HPO_4, 3.8g; NaH_2PO_4, 1.43g or KH_2PO_4, 1.3g; $(NH_4)_2SO_4$, 1g; yeast extract (Difco), 1g; Na acetate.$3H_2O$ 20g; $Mg(NO_3)_2.6H_2O$, 1g; pH 7.2 with KOH. The complex medium (NY) contained, per liter of distilled water: nutrient broth (Difco), 8g; yeast extract, 5g; $MgSO_4.7H_2O$, 1g; pH 7.4 with KOH. NaCl and $NaNO_3$ were added at the concentrations specified in the text. In some cases, a metal salt solution (D salts) was added to the media. Its composition was (per liter): $CaCl_2.2H_2O$, 2g; $FeSO_4.7H_2O$, 1g; $MnCl_2.4H_2O$, 0.5g; $Na_2MoO_4.2H_2O$, 0.1g; $CuSO_4.5H_2O$, 0.1g; concentrated HCl, 8.5ml. Media supplemented with 10ml of this solution are referred to as DYA or DNY media in the text. The medium used for culturing Pseudomonas stutzeri consisted of (per liter): tryptone, 10g; yeast extract, 5g; NaCl, 5g; glucose, 1g; 1M Tris-HCl (pH 7.6), 10ml and $MgSO_4.7H_2O$, 0.12g (TYG medium). Anaerobiosis was obtained either by cooling of the media under argon or by continuous nitrogen bubbling.

Analyses. Nitrate was measured colorimetrically by formation of nitrosalicylic acid as described by Cataldo et al. [9]. Nitrite was determined according to Parsons et al [10]. Gas samples were analyzed with a Varian Vista series 4600 equipped with a ^{63}Ni electron capture detector. Samples were injected onto a column (1/8 in. by 12 ft) of Porapak Q at 45°C and eluted with

a mixture of argon/methane (95/5) at a flow rate of 15 ml/mn
[11]. In order to take N_2O solubility into account, N_2O
calibration was done by injecting pure N_2O into anaerobic vials
identical to, and containing the same liquid volumes as those
used in each experiment.

Cultural conditions. Anaerobic cultures to be assayed for
gas production and/or nitrate and nitrite reduction were
incubated in 150ml rubber-capped and stoppered vials in which
the 100ml headspace was filled with argon. When growth was to
be measured as well, Erlenmeyer flasks modified by the addition
of a side arm were utilized. Ten liter carboys fitted for gas
sparging and liquid sampling were used for cell production.
Incubation temperature was 37°C unless otherwise specified.

Assay of denitrification with washed cells. All manipulation
and incubation were carried out with cells under argon. Cells
grown anaerobically to stationary phase were harvested by
centrifugation at 4°C for 10 mn at 5,000xg, washed twice in
10 mM potassium phosphate buffer, pH 7.0, containing 5% NaCl
and kept on ice until used (no more than 24 hours). Reaction
mixtures consisted of potassium phosphate buffer, pH 7.0,
0.1 M; NaCl, 0.9 M; Na acetate.$3H_2O$, 50 mM; $NaNO_3$, 20 mM (nitrate
reduction and N_2O production); $NaNO_2$, 10 mM (nitrite reduction)
and washed cells, 2 mg dry weight/ml (nitrate and nitrite
reduction) or 0.2 mg dry weight/ml (N_2O production), and were
incubated at 30°C in 9 ml vials sealed with a rubber septum.
Cells to be assayed for nitrate reductase were preincubated with
10 mM sodium diethyldithiocarbamate (DDC) for 40 mn at 4°C. At
different times, samples were withdrawn by needle and syringe
from the liquid or gas phase and assayed for nitrite or N_2O
concentrations. Dry weights were determined by reference to a
standard curve relating dry weights of cells in suspension and
their optical densities at 660 nm.

Immobilized-cell preparations. All manipulations were
effected under argon. Cells grown to stationary phase in DNY
medium, containing 1% $NaNO_3$ and 3% NaCl, were harvested by

centrifugation at 4°C for 10 mn at 6,000xg and washed twice in 50 mM Tris-HCl buffer, pH 7.6, NaCl, 3% (TS). Immobilization in agar gel was performed as follows. The cell suspensions (10 g wet weight in 30 ml TS) were warmed to 48°C for 10 min and mixed with 170 ml of TS containing 8 g of agar (Difco) itself maintained at 48°C. The resulting pasty preparation was immediately transferred onto a glass plate to give a coating about 2 mm thick. The gel was hardened at 4°C for 30 mn, cut into small cubes (2mm sides) and washed twice in cold TS to remove small particles and free cells. Immobilization in calcium alginate was performed at room temperature. The cell suspension was mixed with 170 ml of a 2% Na alginate solution in TS. The mixture was forced through a hypodermic needle (21.5 gauge) into a 0.1 M $CaCl_2$ solution stirred by a magnetic bar. The resulting beads (2mm ∅) were allowed to harden in the calcium-rich solution for two hours.

Both immobilized cells preparations were washed twice with the nitrate solution chosen for the continuous operation before being packed into a jacketed column. This solution was fed at the bottom of the column with a peristaltic pump. The calcium alginate beads softened somewhat when placed in the presence of high concentrations of sodium ions but proved stable enough for our experiment.

RESULTS - DISCUSSION

As B. halodenitrificans stops the reduction of nitrate at N_2O, gas production always refers to N_2O production.

Denitrification by growing cultures. B. halodenitrificans was relatively insensitive to elevated nitrate concentrations. The amounts of N_2O produced during growth on DNY or DYA medium supplemented with 0.5% $NaNO_3$ + 5% NaCl or with 5% $NaNO_3$ were not vastly different (Table 1).

Table 1
N_2O evolution in cultures containing different nitrate concentrations

medium	N_2O evolved (mM)		
	10h	24h	100h
complex (DNY)			
+ 0.5% $NaNO_3$, 5% NaCl	5.1	14.4	ND
+ 5% $NaNO_3$	4.5	10.6	ND
semi-defined (DYA)			
+ 0.5% $NaNO_3$, 5% NaCl	ND	ND	20
+ 5% $NaNO_3$	ND	ND	16.4

ND = not determined

Growth and N_2O production were still observed on DYA medium at $NaNO_3$ concentration of 1.64 M (14%), but both at diminished rates (data not shown). B. halodenitrificans could also denitrify in NY medium containing concentrations of nitrite up to 0.3 M (0.6 M in aerobiosis; G.Denariaz, W.J.Payne and J. LeGall, submitted for publication). B. halodenitrificans thus tolerates high concentrations of sodium, nitrate and nitrite ions and does not seem greatly affected by variations in the amount of nitrate or nitrite present in its culture medium. Functioning of most denitrifiers is sensitive to the presence of at least one of the above ions. High concentrations of sodium ions generally inhibit non halotolerant organisms and nitrite is a well known bacteriostatic agent. Nitrate has been repeatedly reported to inhibit the reduction of N_2O and/or the reduction of nitrite [12]. The precise mechanism of such inhibitory effects is not well understood as both control of synthesis as well as regulation of enzyme function can be involved. Some denitrifiers, e.g., Ps. perfectomarina, reduce nitrite only when most nitrate has been utilized and are therefore rapidly inhibited by toxic levels of nitrite [2].

The semi-defined medium, whose composition could be controlled, was chosen for the optimization of denitrification by cultures of B. halodenitrificans. Denitrification proceeded best at pH ranging from 7 to 8 and at temperatures between 30 and 35°C (data not shown). An addition of copper sulfate (4 μM) to the semi-defined medium resulted in a doubling of N_2O evolution and a decrease of almost one order of magnitude in nitrite concentration in the medium, while pH increased one unit (Table 2). The effect of copper is attributable to the presence

Table 2
Effect of metal ions on denitrification in culture [a]

additions	(mg/l)	N_2O (mM)	NO_2^- (mM)	pH	A_{660}
control		6.4	21	7.3	0.23
$CuSO_4 \cdot 5H_2O$	0.1	10.8	7.4	8.0	0.24
	1	14.5	3	8.3	0.26
	10	0.12	ND	7.3	0
$FeSO_4 \cdot 2H_2O$	0.01	5.2	17	7.4	0.21
	0.1	6.4	20.5	7.4	0.25
	1	7.2	ND	7.4	0.21
	10	5.5	ND	7.3	0.2
$Na_2MoO_4 \cdot 2H_2O$	0.01	5.2	16.5	7.4	0.2
	1	6.8	ND	7.4	0.19
Cu:Fe:Mo (1:0.1:0.1)		15	1.55	8.4	0.24
D salts		14.8	5.8	8.2	0.29

[a] All results are for 100h cultures in YA medium, $NaNO_3$ 9%.
ND = not determined

of a copper nitrite reductase (G.Denariaz, W.J.Payne and J.LeGall, manuscript in preparation). Adding other transition metals reported to be present in denitrification enzymes (iron and molybdenum)[13] was without effect. It is possible that an extra amount of copper was needed in the medium to overcome complexion with components of the yeast extract [14], although an increase from 0.1 to 0.3% yeast extract in the medium did not affect N_2O evolution in the absence of added copper. From then on, all media were supplemented with 4 µM copper sulfate.

Several carbon sources were tested as alternatives to acetate for both cell yield and denitrification rates. Glucose, sucrose, glycerol, mannitol, gluconate and lactate provided no significant improvements. In every case, cell densities were lower by a factor 2 to 3 than those obtained with the complex medium. The latter was therefore preferred for cell production.

Both reduction of nitrite and evolution of N_2O appeared to be linked to the end of the log phase (Fig.1a), suggesting that the cells primarily used the reduction of nitrate to nitrite for early growth. The amount of nitrite accumulated by the end of the log phase depended on the growth medium rather than on the amount of nitrate present, as concentrations of 3 to 5 mM were found in DYA medium compared to 15 to 20 mM in DNY medium, irrespective of the initial concentration of nitrate (data not shown). The appearance of nitrite reductase (NIR) activity was delayed about two hours in time relative to nitrate reductase (NAR) activity (Fig.1b). B. halodenitrificans may need an accumulation of nitrite to activate synthesis of its NIR. More research will be needed to understand the mechanism involved.

NIR activity decreased rapidly after having reached its maximum, while NAR activity was essentially stable during the stationary phase. Both the changes in pH and nitrite concentration were telling on the activity of the cells. The pH increase corresponded to the onset of NIR and seemed to account for this activity. The increase of one pH unit obtained by the addition of copper to the YA medium (Table 2) is consistent with this assumption. It follows that the denitrifying activity of the cells in culture could be assessed solely by the evolution of pH. The accumulation of nitrite in stationary phase corresponded to a decrease in NIR activity relative to NAR activity. NIR activity never dropped down to zero, as N_2O evolution proceeded for many hours (Fig.1a). Although the

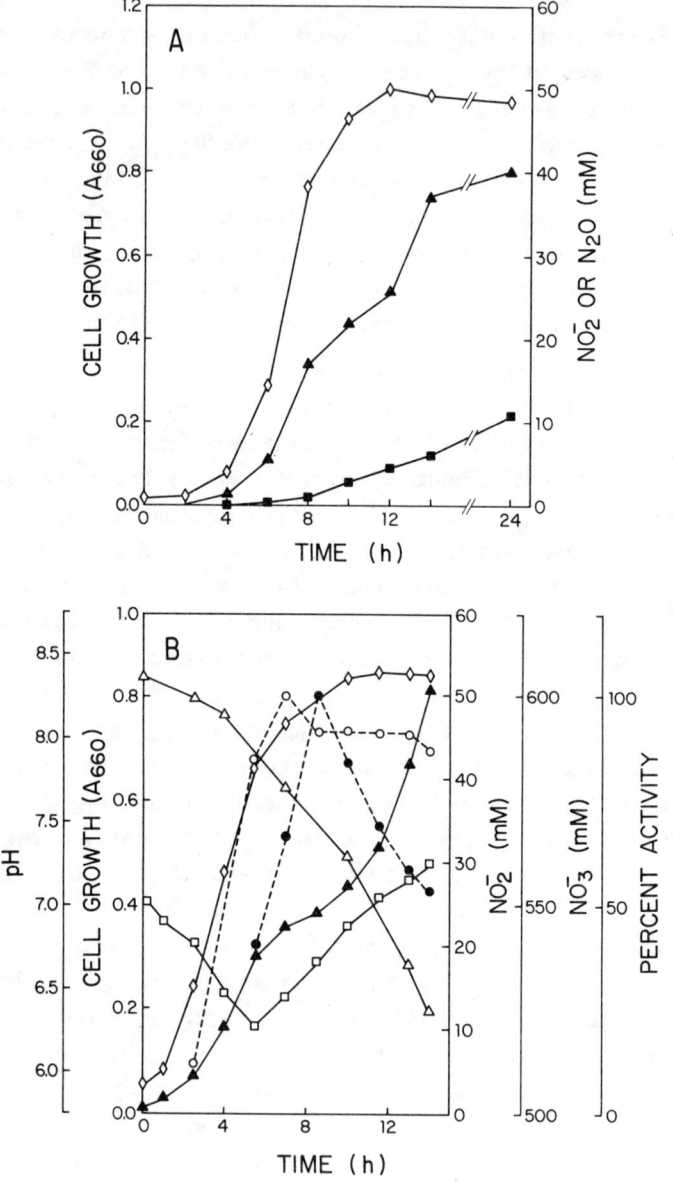

Figure 1. Anaerobic growth of B. halodenitrificans in complex medium containing 5% $NaNO_3$. (A) N_2O and nitrite evolution. (B) nitrate and nitrite reductase activities. ◇ A_{660}; △ NO_3^-; ▲ NO_2^-; ■ N_2O; ○ nitrate reductase; ● nitrite reductase; □ pH.

decrease in NIR activity in stationary phase cultures was not always rapid, cells were usually harvested when denitrifying activity was highest, i.e., 2 to 3 hours after the increase in pH.

Denitrification by washed cells. These studies were directed toward the determination of experimental conditions to be used for continuous systems. In order to single out NAR activity in whole cells, an inhibitor specific for NIR was needed. Based on the effect of copper on N_2O production with denitrifying cultures, we decided to test the action of sodium diethyldithiocarbamate (DDC), a known inhibitor of copper nitrite reductase [15]. This compound was effective (Fig.2). Its use enabled us to measure NAR and NIR activities in washed cells

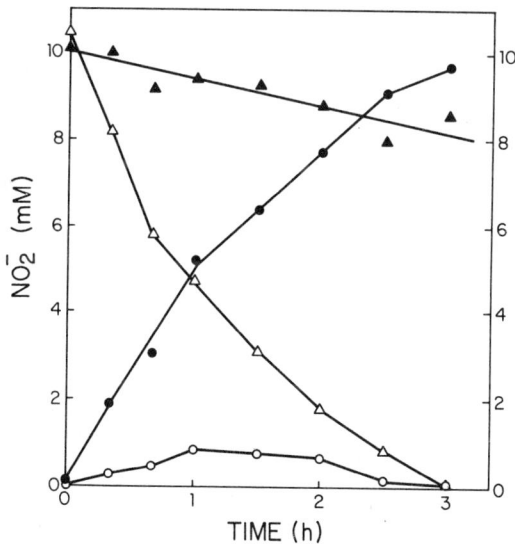

Figure 2. Effect of DDC on nitrate and nitrite reduction by washed cells. O NO_3^-; ● NO_3^- + DDC; △ NO_2^-; ▲ NO_2^- + DDC.

simply by following nitrite appearance or consumption. Nitrate reduction in the absence of DDC resulted in only a transient accumulation of nitrite, thence nitrite reduction was essentially faster than nitrate reduction.

In these cells, NIR was less sensitive to acidic conditions than NAR (Fig.3a). Approximate pH optima were 7.4 and 7.7 for NIR and NAR, respectively. Both enzymes were still active at pH 9.4. A pH increase resulting from the reduction of nitrite to N_2O should therefore not be deleterious to the cells, at least for pH values <9.5. NAR was also more sensitive to high ionic strengths (>1.5 M NaCl) than NIR, but both behaved similarly with salt concentrations <1.4 M (Fig.3b). Both activities were maximal at 0.1 M NaCl and were above 50% of their maximum at 1 M NaCl. NAR and NIR had different apparent temperature optima: 40 and 30°C, respectively (Fig.3c). NAR activity slowed to 25% at 30°C whereas NIR retained 85% of its maximum activity at 40°C. Thus one can expect to change the NAR/NIR ratio simply by adjusting the temperature.

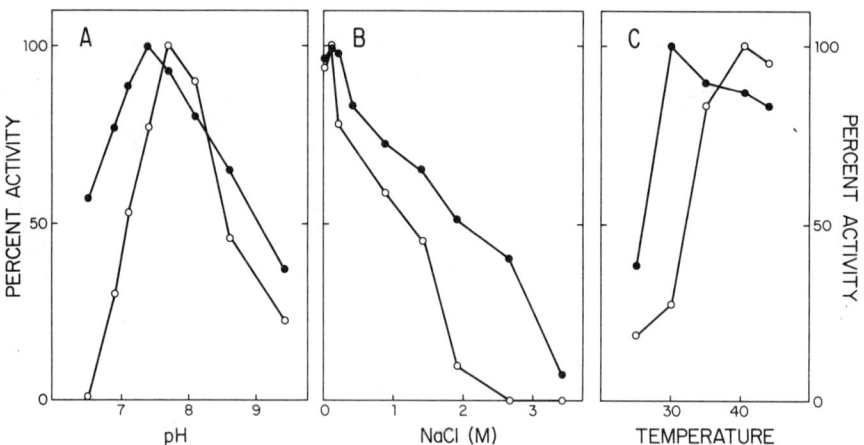

Figure 3. Effect of pH (A), ionic strength (B) and temperature (C) on nitrate and nitrite reductase activities in washed cells. O nitrate reductase; ● nitrite reductase.

Several substrates could serve as electron donors for reduction of nitrate to N_2O (Table 3). Among them, glycerol, glucose and acetate gave the highest activities. N_2O production with acetate depended upon the cell batch used, but no typical pattern has been found so far to explain these variations.

Table 3
N_2O production by washed cells at the expense of different electron donors [a]

electron donor	nmol N_2O/mn/mg dry weight
endogenous	4.3
formate	4.6
succinate	5.8
acetate	7-20 *
lactate	6.4-6.8 *
aspartate	5.4
glutamate	5.5
ethanol	6
glycerol	12.15
glucose	13.3-15.5 *

[a] Cells grown on DNY medium. * Values obtained with two different cell preparations.

Biological reduction of nitrous oxide to dinitrogen. Nitrous oxide evolved by nitrate reducing cells of B. halodenitrificans has to be eliminated as its release in the atmosphere is not desirable [12]. We therefore decided to study the possibility of using a second denitrifier to reduce this gaseous compound to elemental nitrogen. As N_2O is released in the gas phase, a halotolerant bacterium is not required to accomplish this reaction. Pseudomonas stutzeri, a bacterium reported to reduce N_2O rapidly [16], was chosen for this experiment. N_2O accumulated in one flask by a culture of B. halodenitrificans was transferred, via gas impermeable tubing to a second flask containing an oxygen-free medium (TYG) inoculated with Ps. stutzeri. The connection between the two flasks was closed again and growth of Ps. stutzeri as well as N_2O concentration were monitored. N_2O disappeared rapidly from the gas phase, paralleling growth of Ps. stutzeri. Because the amount of N_2O transferred to the Ps. stutzeri culture was set up to be growth limiting, a growth yield could be derived from this experiment and was found to be 10.6 g cells (dry weight)/mol N_2O. A control culture grown with pure N_2O (analytical grade) gave a cell yield of 10.8. These values are higher than the value of 5.1 reported by Bryan et al. [17], who used a different strain and a minimal medium. The absence of oxygen in our system was assured by the observation that Ps. stutzeri did not grow before the transfer of N_2O from B. halodenitrificans culture.

Denitrification by immobilized cells. Immobilized cells of Ps. denitrificans have been reported to denitrify for 2 months when fed with a solution containing 100 mg NO_3^-/l and supplemented with ethanol as electron donor [18]. Use of a non-growth medium was also found to be satisfactory for denitrification by immobilized cells of Paracoccus denitrificans [19]. We therefore set out to check how B. halodenitrificans cells would behave when denitrifying high nitrate solutions in a non-growing state.

B. halodenitrificans cells immobilized in agar were utilized for the denitrification of a solution containing $NaNO_3$ 10.7 g/l and Na acetate·$3H_2O$, 10 g/l. This solution was buffered with Tris-HCl, 50 mM, pH 7.55, and heat sterilized. The nitrate solution was run through the bed at a flow rate of 20 ml/h. The temperature of the column was regulated at 37°C. Gas production (N_2O + CO_2) was very active during the first 2 days, as bubbles could be seen throughout the bed. After 17 hours of operation, the nitrate concentration in the effluent was down to 7.3 mM while the pH was up to 7.77 (Table 4). However, about half of the nitrate had been reduced only to nitrite. The activity of the bed subsequently slowed down to almost zero after 62 hours. The nitrate reducing activity of the bed after the first day of operation could be estimated at 29 nmol NO_3^-/mn/mg cell dry weight.

Table 4
Continuous denitrification by B. halodenitrificans cells immobilized in agar

time (h)	effluent composition		
	NO_3^- (mM)	NO_2^- (mM)	pH
0	126	0	7.55
17	7.3	55.4	7.77
36	47	50	7.71
46	64.7	40.7	7.68
62	103	18.7	7.58

Cells immobilized in alginate displayed the same activity when permitted to denitrify a solution containing $NaNO_3$, 104 g/l (1.23 M), glucose 100 g/l, $CaCl_2$, 0.5 g/l, in Tris-HCl, 50 mM, pH 7.22, at a temperature of 30°C. After 35 hours of operation (flow rate = 8.7 ml/h), the concentration of nitrate and nitrite in the effluent were 940 mM and 61 mM, respectively, and the pH was 7.72. Thus 80% of the nitrate utilized had been reduced entirely to N_2O. By the fourth day, the activity of the bed had stopped. Agar-immobilized cell regenerated by incubation in NY medium, $NaNO_3$ 100 g/l for 30 hours once again reduced nitrate actively only for the first two days, when fed with the solution containing 10% $NaNO_3$. Therefore, changing the immobilization method, the concentration of nitrate, the nature of the electron donor and the temperature did not result in dramatic changes in the denitrifying activity of immobilized cells. Stability was always low, although the nitrate reducing activity of the bed at the start up was very good compared to other processes (Table 5). It appears that denitrification of high nitrate solutions requires the utilization of growing cells rather than resting cells due to the stringent demand imposed on their metabolism.

Table 5
Denitrification by various immobilized bacterial species

organism	immobilization procedure	initial NO_3^--N concentration mg/l	activity[a]	ref.
Paracoccus denitrificans	crosslinking + entrappment	18	10.5	[19]
mixed population	crosslinking + hardening	22.6	0.5-1.5	[20]
Pseudomonas denitrificans	entrapment in alginate	22.6	12	[18]
Bacillus halo-denitrificans	entrapment in agar	1764	24.3	[b]
	entrapment in alginate	17220	23.5	[b]

[a] activity expressed in mg NO_3^--N/h/mg cell (dry weight).
[b] this work.

Reactivation of the bed by incubation in a growth medium revealed that immobilized cells were still alive after more than 60 hours of operation and that the drop in activity was probably due to the depletion of one or several metabolite(s), cofactor(s), or enzyme(s). It should be mentioned that the typical pink-orange color of the cells, attributable to cytochromes of the b- and c-type, disappeared progressively during the operation.

NIR activity was not as high as expected, as nitrite accumulated to 50-60 mM. This concentration, presumably not toxic to B. halodenitrificans cells that tolerate up to 300 mM NO_2^- in culture, seemed to remain constant as long as the cells in the bed displayed good activity, whatever the concentration of nitrate present in the feed. It is possible that onset of nitrite reduction by immobilized cells awaits accumulation of nitrite to a critical concentration, a phenomenon already observed with anaerobic cultures. Hancher and Perona [21] reported an increase in nitrite concentration in the effluent from a fluidized bed as nitrate concentration in the feed went up to 70 mM. Strangely enough, a further increase in nitrate resulted in a decrease in nitrite concentration. This phenomenon could be related to the selective enrichment of different organisms within the bacterial population as salt concentration rose.

CONCLUSIONS

The present study demonstrates that with a selected bacterium, growth and denitrification can be as effective with high as with low sodium nitrate concentrations. Although the moderate halophile, B. halodenitrificans, stops the reduction of nitrate at nitrous oxide, a behavior to be expected anyhow with high $NaNO_3$ concentrations [12], we have shown that this gas can be easily dealt with by use of another denitrifier, Ps. stutzeri. Several substrates could be used as electron donors for the reduction of nitrate but did not permit prolonged denitrification of high sodium nitrate solutions in a continuous process with immobilized cells. As the activity of the cells was high at the beginning, it was concluded that denitrification of high sodium nitrate solutions with B. halodenitrificans should be carried out with a medium permitting cell growth in order to maintain the metabolism, and especially to keep the respiratory chain intact. Use of a culture medium should circumvent such problems as cofactor depletion, enzyme degradation, and respiratory control. A medium giving limited cell growth, such as the semi-defined medium, would be best suited for this purpose as cell release in the effluent must be kept at a minimum. It is actually possible that the required nutrients, except the electron donor, will be present in a real feed, as is the case for most nitrate wastes treated in denitrification processes.

High concentrations of nitrite were detected in denitrifying cultures and with immobilized cells. High concentrations of sodium nitrate have been reported to give abnormally high nitrite concentrations in effluents from denitrification processes [1,7]. We have observed that the reduction of nitrite by B. halodenitrificans was more related to the kind of medium utilized than to the concentrations of sodium or nitrate present. Consequently, if the factor(s) responsible for this behavior can be found, one can expect to control the concentration of nitrite in the effluent without the problems of inhibition by sodium or nitrate inherent in other processes. This study is now under way.

Acknowledgement: This work was supported by a grant from the Commissariat à l'Energie Atomique, France N°C/VD 4036400 and NSF grant DMB-8718646.

REFERENCES

1. Francis, C.W. and Hancher, C.W., Biological denitrification of high nitrate wastes generated in the nuclear industry. In Biological fluidized bed treatment of water and waste water, ed., P.F. Cooper and B. Atkinson, Ellis Horwood ltd, Chichester, 1981, pp. 234-50.
2. Payne, W.J. Denitrification. John Wiley and sons, New York, NY, 1981.
3. Francis, C.W. and Callahan, M.W., Biological denitrification and its application in the treatment of high-nitrate waste water. J. Environ. Qual., 1975, **4**, 153-63.
4. Ibrahim, A.B., Denitrification of high nitrate waste water by anaerobic filter. J. Rubb. Res. Inst. Malaysia, 1978, **26**(2), 85-8.
5. Blaszczyk, M., Przytocka-Jusiak, M., Kruszewska, U. and Mycielski, R., Denitrification of high concentrations of nitrites and nitrates in synthetic medium with different sources of organic carbon. I: acetic acid. Acta Microbiol. Polon., 1981, **30**(1), 49-58.
6. Claus, G. and Kutzner, H.J., Denitrification of nitrate and nitric acid with methanol as carbon source. Appl. Microbiol. Biotechnol., 1985, **22**, 378-81.
7. Blaszczyk, M., Galka, E., Sakowicz, E. and Mycielski, R., Denitrification of high concentrations of nitrites and nitrates in synthetic medium with different carbon sources.

III: methanol. Acta Microbiol. Polon., 1985, **34**, 195-206.
8. Mycielski, R., Jaworowska-Deptuch, H. and Blaszczyk, M., Quantitative selection of denitrifying bacteria in continuous cultures and requirement for organic carbon. I: starch. Acta. Microbiol. Polon., 1985, **34**(1), 67-79.
9. Cataldo, D.A., Haroon, M. Schrader, L.E. and Young, V.L., Rapid colorimetric determination of nitrate in plant tissue by nitration of salicylic acid. Commun. Soil Sci. and Plant. Analysis, 1975, **6**(1), 71-80.
10. Parsons, T.R., Maita, Y. and Lalli, C. M., A manual of chemical and biological methods for seawater analysis, Pergamon Press, Oxford, New York, 1984.
11. Kaspar, H.F. and Tiedje, J.M., Response of electron capture detector to hydrogen, oxygen, nitrogen, carbon dioxide, nitric oxide and nitrous oxide. J. Chromato., 1980, **193**, 142-7.
12. Knowles, R., Denitrification. Microbiol. Rev., 1982, **46**(1), 43-70.
13. Payne, W.J. Diversity of denitrifiers and their enzymes. In Denitrification in the nitrogen cycle, ed., H.L. Golterman, Plenum Press, New York, NY, 1985, pp. 47-66.
14. Chan, Y.-K. and Marshall, P.R., Strain-dependent inhibition of nitrous oxide in denitrifiers by yeast extract. Can. J. Microbiol., 1987, **33**, 1032-7.
15. Shapleigh, J.P. and Payne, W.J., Differentiation of cd_1 cytochrome and copper nitrite reductase production in denitrifiers. FEMS Microbiol. lett., 1985, **26**, 275-279.
16. Carlson, C.A. and Ingraham, J.L., Comparison of denitrification by Pseudomonas stutzeri, Pseudomonas aeruginosa, and Paracoccus denitrificans. Appl. Environ. Microbiol., 1983, **45**(4), 1247-53.
17. Bryan, B.A., Jeter, R.M. and Carlson, C.A., Inability of Pseudomonas stutzeri denitrification mutants with the phenotype of Pseudomonas aeruginosa to grow on nitrous oxide. Appl. Environ. Microbiol., 1985, **50**(5), 1301-3.
18. Nilsson, I. and Ohlson, S., Columnar denitrification of water by immobilized Pseudomonas denitrificans cells. Eur. J. Microbiol. Biotechnol., 1982, **14**, 86-90.
19. Kokufuta, E., Shimohashi, M. and Nakamura, I., Immobilization of Paracoccus denitrificans cells with polyelectrolyte complex and denitrifying activity of the immobilized cells. J. Ferment. Technol., 1986, **64**(6), 533-8.
20. Cizinska, S., Vojtisek, V. Maixner, J., Barta, J. and Krumphanzl, V., Cell aggregates exhibiting denitrifying activity. Biotechnol. Lett., 1985, **7**(10), 737-42.
21. Hancher, C.W. and Perona, J.J., Kinetic model for a fluidized-bed bioreactor for denitrification of waste waters. In Biotechnology and Bioengineering Symposium, vol. 12, ed., C.D. Scott, John Wiley and sons, New York, NY, 1982, pp. 317-26.

ALKALIPHILES

WILLIAM D. GRANT, Department of Microbiology,
University of Leicester, Leicester LE1 9HN and
KOKI HORIKOSHI, The Superbugs Project,
The Riken Institute, Wako, Saitama 351 01, Japan.

INTRODUCTION

Organisms with pH optima for growth in excess of pH 8, usually between 9 and 10 are properly defined as alkaliphiles (or sometimes alkalophiles). Obligate alkaliphiles are incapable of growth at pH values around neutrality, and the term alkalitolerant is reserved for those organisms that are capable of growth under alkaline conditions, but have pH optima for growth outwith the alkaline side of the pH scale.

Most of the microorganisms described as growing under very alkaline conditions to date are prokaryotes, but, often, whether or not an organism is truly alkaliphilic is not always rigorously established. Obligate alkaliphiles are often capable of growth at pH values in excess of 11 although growth much above pH 11.5 is doubtful. Almost all of the biochemical work on alkaliphiles has been carried out on alkaliphilic Bacillus spp. Horikoshi and Akiba (1) list a range of other bacterial types that are alkaliphilic or alkalitolerant, including Pseudomonas, Micrococcus and Corynebacterium spp. Exiguobacterium aurantiacum (2) isolated from alkaline potato processing waste, is one of the few isolates that has been subjected to a detailed taxonomic analysis - it is probable that many alkaliphiles currently ascribed to extant genera for convenience will turn out to represent new taxa.

THE ISOLATION OF ALKALIPHILES

In order to maximise the isolation of alkaliphilic rather than alkalitolerant isolates, it is preferable to culture in/on media at pH 10 or above. Obligate alkaliphiles are unlikely to be isolated at pH values lower than 10. High pH is usually achieved by the addition of Na_2CO_3 at between 1% and 5% (w/v). An inevitable consequence of such pH is that Mg^{2+} and Ca^{2+} are rendered essentially insoluble by precipitation as carbonates, so it is pointless to add significant amounts of these cations to such media. The equilibrium between NH_4^+ and NH_3 at high pH results in the generation of NH_3 from high pH media containing NH_4^+ salts, so nitrogen sources other than NH_4^+ should be used. Commonly used media are listed by Horikoshi and Akiba (1) and by Grant and Tindall (3).

A further problem arises with pH control in such media. Grant and Tindall (3) list a number of buffers that might be used other than the $Na_2CO_3/Na(HCO_3)_2$ system which arises on addition of Na_2CO_3. However, in practice, for prolonged incubation periods, it is extremely difficult to prevent the pH from dropping quite rapidly due to the absorption of atmospheric CO_2. Thus, even Na_2CO_3 containing media, although typically initially having a pH of 10.5 at 1% (w/v), equilibrate to pH 9.5 over a few days as a consequence of the $Na_2CO_3/Na(HCO_3)_2/CO_2$ equilibrium. The problem is particularly pronounced for agar media with a large surface area. Reliable, long-term maintenance of high pH (>9.5) requires liquid media with regular (preferably automated) addition of alkali (normally NaOH).

Most of the media devised to date have a relatively nutritious formulation, usually containing yeast extract and peptone or casein products as carbon and nitrogen sources. The medium originally formulated by Horikoshi (1) has been extensively used and yields a diverse population of bacteria (and occasionally fungi and actinomycetes) from a variety of source material. In practice, it should be possible to formulate a selective medium for alkaliphilic representatives of any physiological group. The discovery of the alkaliphilic halobacteria (4) illustrates the point. It is also worth noting that the alkaliphiles isolated to date have in the main been isolated under aerobic conditions, and it is certain that an equally diverse group of anaerobic alkaliphiles remains to be explored. The recent reports of alkaliphilic methanogens (5) and a range of alkaliphilic anaerobic heterotrophs (6) point the way forward.

WHY STUDY ALKALIPHILES?

1. The bioenergetic problem

Alkaliphilic and alkalitolerant microorganisms present an interesting problem for the physiologist since they, superficially at least, appear to challenge the tenets of the chemiosmotic theory of energy generation.

The theory supposes that prokaryotes generate an electrochemical potential ($\Delta\mu_H^+$) across the cell membrane (or envelope) by the active extrusion of H^+ through respiration. Both the H^+ gradient itself (ΔpH-outside acid) and the transmembrane electric potential ($\Delta\psi$) generated by charge partition (inside negative) contribute to the electrochemical potential in the following relationship:

$\Delta\mu_H^+ = \Delta\psi - Z\Delta pH$ where the constant Z serves to convert $\Delta\mu_H^+$ into electrical units.

$\Delta\mu_H^+$ powers ATP synthesis, solute transport and other energy-dependent processes such as flagellum movement. If one assumes that cytoplasmic pH is regulated within fairly narrow limits close to neutrality, alkaliphiles must grow under conditions where there is a large 'reversed' pH gradient across the membrane (inside acid) and so must compensate in some way for the depression of $\Delta\mu_H^+$ so caused (assuming that the theory is correct).

There seems little reason to doubt that alkaliphiles regulate their internal pH within the range 7-9 (7,8) although almost all measurements have been made on a limited number of Bacillus isolates. Intracellular pH is measured by assaying the distribution of weak bases across the cell membrane. pH homeostasis clearly requires that H^+ should be returned across the membrane (to compensate for H^+ extruded by respiration amongst other cytoplasmic alkalinization effects). An Na^+/H^+ antiporter is strongly implicated in pH homeostasis in alkaliphiles and membrane vesicle preparations show Na^+-dependent movement of H^+ in those alkaliphiles where this has been examined (7,9). Accordingly, alkaliphiles have a significant requirement for Na^+ in the growth medium.

Given an internal pH several units below that of the growth medium, the challenge to the chemiosmotic theory is clear - a simple calculation indicates that a 'reversed' pH of 2.5 units (ie. an alkaliphile growing at around pH 11) generates a 'reverse' force of around 150 mV (7,9) and yet these microorganisms continue to grow normally. A simple solution might suppose that $\Delta\psi$ could be elevated in some way to counteract the large opposing ΔpH. 'Neutrophilic' bacteria have a total $\Delta\mu_H^+$ in the range 150-200 mV (7) composed of a additive combination of ΔpH and

$\Delta\psi$. An alkaliphile might then be expected to have to compensate for the debit in ΔpH by elevating $\Delta\psi$ up to 300 mV or so in order to achieve $\Delta\mu_H^+$ values commensurate with normal energetic processes.

Accurate $\Delta\psi$ measurements are clearly central to any analysis of alkaliphile bioenergetics. It is possible to measure $\Delta\psi$ in a similar way to ΔpH since freely permeable ions will electrophorese across a membrane powered by and at equilibrium, the concentration ratios of the ion between the outside and inside is proportional to $\Delta\psi$.

Measurements of $\Delta\psi$ in alkaliphiles, largely by Krulwich and colleagues, indicate no significant differences between alkaliphile and 'neutrophile' (7,9) (normal $\Delta\psi$ values of 100-200 mV). Assays of $\Delta\psi$ using 'membrane' vesicle preparations (9,10) yield similar results and 'bulk' measurements of this kind (which cannot detect localized H^+ movements) lead to the inexorable conclusion that $\Delta\mu_H^+$ values for alkaliphile are unacceptably low (assuming a straightforward adherence to the chemiosmotic theory).

A number of recent publications (11-15) have suggested that different alkaliphilic or alkalitolerant bacteria may have adopted different strategies to cope with the problem of 'reversed' Δ pH.

Alkalitolerant (or slightly alkaliphilic) marine Vibrio spp. appear to harness an Na^+ motive force at high pH. This is functionally analogous to a H^+ motive force where $\Delta\mu_{Na}^+$ is made up of $\Delta\psi$ (inside negative) and ΔpNa ($[Na^+]_{out} > [Na^+]_{in}$). Skulachev and colleagues (11,15) have shown that some of these organisms at least have a primary Na^+ pump coupled to respiration and probably an Na^+-linked ATPase. The evidence for the Na^+/ATPase is indirect and based on inhibition studies. Cells grown in the presence of inhibitors such as carboxyl cyanide M-chlorophenyl hydrazone (CCCP) which inhibits $\Delta\mu_H^+$ still make ATP provided $[Na^+]_{out} > [Na^+]_{in}$. Furthermore, ATP synthesis is abolished by an artificially produced reverse ΔpNa ($[Na^+]_{in} > [Na^+]_{out}$) under conditions where $\Delta\Psi$ generation is unchanged. A small amount of residual CCCP-sensitive ATP synthesis was found under certain conditions, suggesting both $\Delta\mu_H^+$ and $\Delta\mu_{Na}^+$ powered cell functions in organisms of this type.

The idea of an Na^+-motive force in alkaliphiles is attractive since apart from the key role of Na^+ in pH homeostasis, Na^+ solute symporters have been extensively documented in alkaliphilic Bacillus spp. (7). Non-alkaliphilic mutants devoid of the Na^+/H^+ antiporter lose the capacity to translocate solutes in the presence of Na^+, leading to speculation that Na/H^+ antiporter and Na^+/solute symporters have a common subunit (7) Na^+ is also required for motility in certain bacteria (13). Accordingly, various authors have considered the Na^+ motive force a plausible general mechanism central to alkaliphily (16,17).

However, Krulwich and coworkers have repeatedly failed to find any evidence for an Na^+-motive force in a range of alkaliphilic Bacillus spp. Their attempts, most notably reported in a recent publication (18) have included a detailed comparison of ATP synthesis comparing the effects of imposed H^+ and Na^+ gradients in membrane preparations in the presence of inhibitors that abolish one or other of the motive forces. Imposed H^+ gradients always energized ATP synthesis, whereas imposed chemical and electrochemical gradients of Na^+ did not result in ATP synthesis in B. firmus RAB. Furthermore, ATP synthesis was abolished in the presence of inhibitors of H^+ ATPase and antibodies specifically directed against H^+ ATPase. Significantly, stable imposed gradients of H^+ resulted in much lower ATP concentrations than those obtained in cells growing normally with similar naturally formed gradients, leading to the suggestion that such ΔpH measurements fail to monitor the true energetic capacity of the system. Guffanti and Krulwich (18) speculate that ATP synthesis under alkaline conditions utilizes H^+ that is made available by some localized pathway between proton pumps and the ATP synthase, so that at least some of the protons avoid complete equilibrium with the protons in the highly alkaline medium. ATP synthesis was not observed where $\Delta\mu_H^+$ was reduced all the way to zero - this is interpreted as a requirement for a small bulk $\Delta\mu_H^+$ to retain protons within the localized pathways.

Thus there seems to be two different bioenergetic strategies for coping with the stress of high pH. In one case, alkaliphilic and alkalitolerant bacteria are assumed to have a significant component of the $\Delta\mu_{Na}^+$ system, operating at high pH, perhaps with a $\Delta\mu_H^+$ system operating at lower pH values (this might particularly be the case for alkalitolerant organisms) (Figure 1a). The other case supposes a cell with a 'normal' $\Delta\mu_H^+$ system responsible for most solute transport and ATP synthesis with the important modification compared with 'neutrophiles' that some at least of the protons extruded by respiration do not equilibrate with the bulk phase, but enter a localized pathway leading to the ATP synthase (Figure 1b). Such cells, however, still have a requirement for Na^+ in pH homeostasis and there may be some Na^+/solute symporter activity.

Although the picture is beginning to become more clear, alkaliphiles continue to pose a challenge for the physiologist and will continue to provide an experimental system that enables us to test the dogmas surrounding the nature of energy transduction.

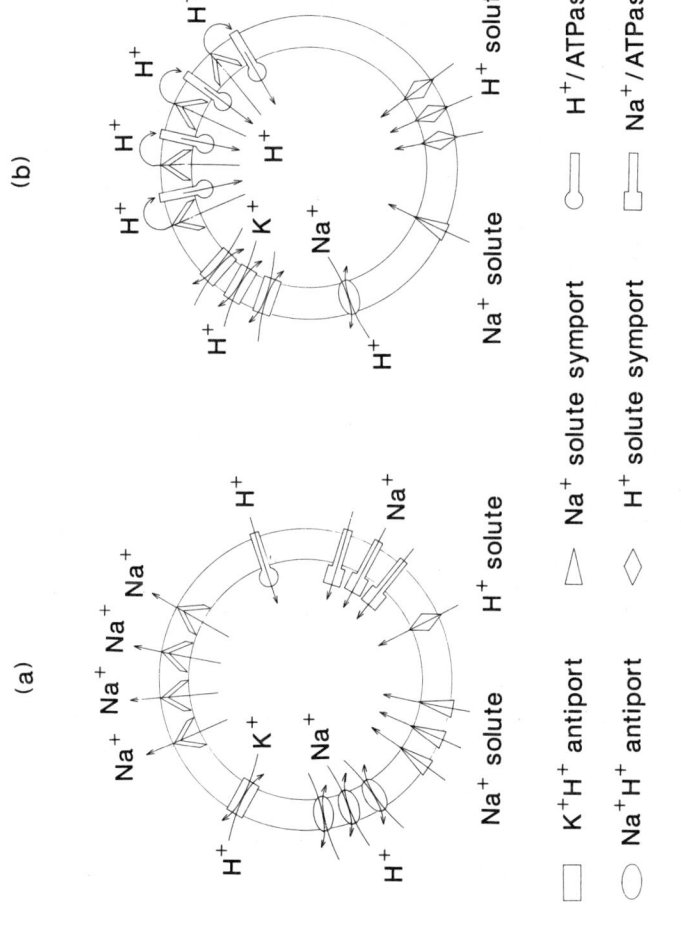

Figure 1. Ion movements in alkaliphiles.

2. **Alkaliphilic enzymes for industry**

With one or two notable exceptions, the main industrial applications for alkaliphilic enzymes are in the detergent industry. Detergent enzymes account for approximately 25% of total worldwide enzyme production and these necessarily alkaliphilic or alkalitolerant proteins represent a good example of a successful commercial exploitation of biology by industry. Companies with a significant interest and commercial stake in detergent enzymes include Gist-brocades and Novo Industries. The commercial interest can be gauged by the penetration of the market by detergent enzymes, some 75% of the market in Europe, 50% in Japan and greater than 30% in the USA (19). As the trend continues towards lower and lower washing temperatures, one can expect the penetration of enzyme detergents to increase.

Proteases and amylases are the most widely used enzymes. By using a protease and an amylase it is possible to hydrolyse proteins and polysaccharides which tend to cause adherence of dirt to clothes, and of course, are in themselves difficult materials to remove by simple washing procedures. Detergents usually have a pH in solution between 8 and 10.5. For an enzyme to be useful as a detergent additive it must be active in solution at alkaline pH values and stable in the presence of detergent additives such as bleaching agents, bleach activators, anticorrosion agents, perfumes, stabilizers and so on. Furthermore, for an enzyme to be useful in a detergent composition it must exhibit long term stability in the detergent product, preferably additionally in liquid detergent formulations.

In particular, alkaline proteases produced by cultivation of various strains of Bacillus sp. are widely used in detergent compositions. These proteases are usually extracellular and can be isolated readily from spent culture filtrates.

Proteases are divided into several groups on the basis of their pH optima, sensitivity to thiol reagents, and sensitivity to diisopropylphosphofluoridate (DFP, which is a serine-directed reagent) and dependence on metals for activities. Alkaline DFP proteases comprise the largest group and have the widest taxonomic distribution. These proteolytic enzymes, classified generically as serine proteases are widely used in detergent compositions. Not all of these are produced by alkaliphiles - the most widely studied serine proteases are the so-called subtilisins produced by 'neutrophilic' B. subtilis strains (20,21). More recently introduced serine proteases include Alcalase^R, Esperase^R, Savinase^R, Maxatase^R. The protease in Alcalase^R is produced from B. licheniformis, those in Esperase^R, Maxatase^R and Savinase^R from alkaliphilic Bacillus isolates. Several other enzymes produced by Gram-positive, Gram-negative bacteria and fungi have possibilities (22-25) for the future. Such enzymes are produced in a variety of different formulations for different purposes. In addition, in recent years, use has been made of such enzymes in the hide-dehairing process (20) where dehairing is carried out at pH values between 8 and 10. In the old dehairing process, the hides were placed in a bath containing calcium hydroxide and sodium sulphide at pH around 12. This process had the disadvantages of being detrimental to the hairs (which sometimes have commercial value) and causing swelling of the skins, resulting in difficulties in further processing. Specially formulated products include Rapidermase^R and Novo Unhairing Enzyme^R.

The inclusion of α amylase into detergents also assists in the removal of food residues from soiled fabrics, and for removing starch residues from food products derived from potatoes or spaghetti. Termanyl^R produced by B. lichiniformis and Maxamyl^R derived from Bacillus spp. are commercially available thermostable and alkalitolerant α amylases that have been widely used. Use is also made in the desizing and softening of denims (e.g. Rapidase^R).

The development of detergent enzymes has centred mainly on enzymes able to solubilise soilings, but in recent years alkaline cellulases have been introduced. These are believed to modify the structure of cellulose fibrils by removing microfibrils that develop during usage and washing and thus restoring the fabric to its original structure. Examples include the Bacillus-derived alkaline cellulase containing detergent (Attack^R) (30) recently introduced in Japan which has already gained around 40% of the total share of laundry detergent in the Tokyo area, and the Novo detergent cellulase SP 227 (19).

The trend towards lower and lower washing temperatures led by the demand to save energy has also enhanced the demand for detergent lipases. There is clearly a major market in the area and work is actively progressing to find appropriate alkalitolerant and detergent resistant lipases. Preliminary studies have already been carried out with some lipases e.g. Novo mucor lipase SP 225 (19).

A variety of other alkaliphilic enzymes have been investigated over the years (see 1,20), although most of these have not yet been exploited commercially. In particular, Horikoshi and co-workers have investigated many alkaliphilic enzymes from Bacillus spp. including xylanases (26-29) a variety of cellulases (30,31) penicillinase (32) glucosidase (33) mannosidase and mannanase (34,35) amylases (36) and cyclodextrin transferases (1,37). The genes for a number of these enzymes have now been cloned and analysed (34-48).

The work with cyclodextrin transferase, and pencillinase deserve particular mention. β-cyclodextrins are important in the food, cosmetic and pharmaceutical industries, being used for stabilizing volatile materials. B. macerans enzymes have been widely used for the production of β-cyclodextrins. Cyclodextrin glycosyltransferases (CGTs) catalyse the conversion of starch to cyclodextrins. These compounds comprise rings containing 6-8 glucose units linked by α, 1-4 bonds. A number of more efficient GGTs were found to be produced by an alkaliphilic Bacillus sp. (37). These have enhanced pH and temperature stability. A process based on an enzyme from this Bacillus now underpins the Japanese cyclodextrin industry. The enzyme has now been cloned and sequenced (38,42) and shows considerable homology with B. macerans CGT (42).

The work by Horikoshi and colleagues on penicillinase has gone in an unexpected direction. In the course of an analysis of alkaliphilic penicillinase it was noted that most of the penicillinase from an alkaliphilic Bacillus strain cloned into E. coli was released into the culture medium by E. coli strains bearing penicillinase gene-carrying plasmid vectors based on PMB9 (49-51). E. coli periplasmic proteins were also excreted. This proved to be due to the mutational activation of a dormant kil gene in PMB9 by the DNA of the alkaliphilic Bacillus (52). The kil gene is responsible for colicin E1 release by cell lysis, but is not expressed in PMB9 because it lacks a promoter. The weakly activated kil gene produced by insertion of promoter-like DNA from the alkaliphilic Bacillus (the Ex promoter) forms the basis of secretion vectors that have been used for the extracellular production of xylanases, cellulases (53,54), human growth hormone (55) and human immunoglobulin (56). This particular development represents a good example of how 'pure' research may lead in unexpected commercial directions. Other excretion promoting plasmids have since been discovered (57,58).

ALKALIPHILES AND THE FUTURE

Given constantly changing formulations for detergents, there will always be a need for improved enzyme additives. Improvements may be brought about in either of two ways:

(1) To date, we have been remarkably conservative in our choice of environments as source material. Enrichments made on soil samples at high pH will almost always yield Bacillus sp. and indeed one could expect any 'normal' environment to yield similar results. As a consequence, new organisms are hard to find in these environments and this is reflected in the number of Bacillus-derived enzymes that have been studied and brought to commercial production. Naturally-occurring highly alkaline environments such as the lakes of the East African Rift Valley have only recently been considered as potentially valuable sources of new alkaliphiles. These lakes are amongst the most productive environments on earth (59) and contain high densities of all kinds of microorganisms. Taxonomic studies of the bacteria in these lakes are in their infancy, but the preliminary results indicate a wide range of

taxa producing alkaliphilic (or alkali-tolerant) enzymes, and thus probably, significantly different types of proteases, amylases etc that might be valuable for detergent formulations of the future (and indeed biotransformations in general).

(2) Given a sufficient range of different alkaliphilic enzymes, it should be possible to produce detailed models of the enzymes themselves and begin to ask questions about the biochemical basis of alkaliphily. The sequence information obtained by Horikoshi and co-workers provides valuable starting points. Appropriate computer graphics, hardware and software already exists. These techniques are already of value to the enzyme designer in facilitating specific site-directed alterations (60,61). The state of the art of protein engineering makes it difficult to forsee the effects of a specific mutation in many cases - if enough well characterised enzymes were available, better predictions of enzyme properties would be possible.

REFERENCES

1. Horikoshi, K. and Akiba, T., *Alkalophilic Microorganisms*, Berlin : Springer. 1982.

2. Collins, M.D., Lund, B.M., Farrow, J.A.E. and Schliefer, K.H., Chemotaxonomic study of an alkalophilic bacterium. *Exiguobacterium aurantiacum* gen. nov., sp. nov. *J. Gen. Microbiol.*, 1983, 129, 2037-2042.

3. Grant, W.D. and Tindall, B.J., The isolation of alkalophilic bacteria. In *Microbial Growth and Survival in Extremes of Environment*, eds. G.W. Gould and J.G.L. Corry. London, Academic Press 1980, pp.27-36.

4. Tindall, B.J., Ross, H.N.M. and Grant, W.D., Natronobacterium gen. nov. and Natronococcus gen. nov., two new genera of haloalkaliphilic archaebacteria. Syst. Appl. Microbiol. 1984, 5, 41-57.

5. Warakit, S., Boone, D., Mah, R.A., Abdel-Sansil, M.G. and El-halwagi, M.M., Methanobacterium alkaliphilum sp. nov., an H_2-utilizing methanogen that grows at high pH values. Int. J. Syst. Bact. 1986, 36, 380-382.

6. Shiba, H. and Horikoshi, K., Isolation and characterization of novel anaerobic, halophilic eubacteria from hypersaline environments of Western America and Kenya. Abstracts of "The Microbiology of Extreme Environments and its Biotechnological Potential" FEMS Symposium 1988.

7. Krulwich, T.A. and Guffanti, A.A., Physiology of acidophilic and alkalophilic bacteria. Adv. Microb. Physiol. 1983, 24, 173-214.

8. Booth, I.R., Regulation of cytoplasmic pH in bacteria. Microbial. Rev. 1985, 49, 359-378.

9. Krulwich, T.A., Bioenergetic problems of alkalophilic bacteria. In. Membranes and Transport, ed.A.N. Martonosa New York, Plenum Press 1982, pp.75-81.

10. Guffanti, A.A., Firchs, R.T., Schneier, M., Chiu, E. and Krulwich, T.A., A transmembrane electrical potential generated by respiration is not equivalent to a diffusion potential of the same magnitude for ATP synthesis by Bacillus firmus RAB. J. Biol. Chem. 1984, 259, 2971-2975.

11. Dibrov, P., Lazarova, R.L., Skulachev, V.P. and Verkhjovskaya. M.L., The sodium cycle II. Na^+-coupled oxidative phosphorylation in Vibrio alginolyticus cells. Acta Biochim. Biophys. 1986, 850, 458-465.

12. Crider, B.P., Carper, S.W. and Lancaster, J.R., Electron transfer driven ATP synthesis in Methanococcus voltae is not dependent on a proton electrochemical gradient. Proc. Nat. Acad. Sci. USA. 1985, 82, 6793-6796.

13. Hirota, N. and Imae, Y., Na$^+$-driven flagellar motors of an alkalophilic Bacillus strain YN1. J. Biol. Chem. 1983, 258, 10577-10581.

14. Dibrov, P.A., Kostyrko, V.A., Lazarova, R.L., Skulachev, V.P. and Smirnova, I., The sodium cycle 1 Na$^+$-dependent motility and modes of membrane energizaiton in the marine alkalotolerant Vibrio alginolyticus. Biochim. Biophys. Acta 1986, 850, 449- 457.

15. Tokuda, H. and Unemoto, T., Na$^+$ is translocated at NaOH : quinone oxidoreductase segment in the respiratory chain of Vibrio alginolyticus. J. Biol. Chem. 1984, 259, 7785-7790.

16. Skulachev, V.P., Eur. J. Biochem. 1985, 151, 199-208.

17. Grant, W.D., The enigma of the alkaliphile. Microbiol. Sci. 1987, 4, 251-255.

18. Guffanti, A.A. and Krulwich, T.A., ATP synthesis is driven by an imposed Δ pH or $\Delta \mu^-_H$ but not by an imposed Δ pNa$^+$ or $\Delta \mu^-_{Na}$ in alkalophilic Bacillus firmus of 4 at high pH. J. Biol. Chem. 1988, 263, 14748-14752.

19. Christensen, P.N., Thomsen, K. and Branner, S., Development of detergent enzymes. Proceedings of the 2nd World Conference on Detergents, Montreaux, Switzerland, eds. A.R. Baldwin, Pub. Amer. Oil. Chem. Soc. 1987, pp.181-186.

20. Sharp, R.J. and Minster, M.J., Biotechnological implications for microorganisms from extreme environments. In Microbes in Extreme Environments, eds. R.A. Herbert and G.A. Codd. London. Academic Press 1986, pp.215-297.

21. Markland, F.S. and G.L. Smith., Subtilisins : primary structure and physical properties. In The Enzymes, Vol 3, ed. P.D. Boyer, London Academic Press, 1971, pp.561-608.

22. International Patent Application WO88/03947.

23. European Patent Application 0104554.
24. International Patent Application W088/03948.
25. International Patent Application W088/03946.
26. Okazaki, W., Akiba, T., Horikoshi, K. and Akahoshi, R., Purification and characterization of xylanases from alkaliphilic thermophilic Bacillus spp. Agric. Biol. Chem. 1985, 49, 2033-2039.
27. Honda, H., Kudo, T., Ikura, Y. and Horikoshi, K., Two types of xylanases of alkalophilic Bacillus sp No. C-125. Can. J. Microbiol 1985, 31, 538-542.
28. Ohkoshi, A., Kudo, T., Mase, T. and Horikoshi, K., Purification of three types of xylanases from an alkalophilic Aeromonas sp. Agric. Biol. Chem. 1985, 49, 3037-3038.
29. Okazaki, W., Akiba, T., Horikoshi, K. and Akahoshi, R., Production and properties of two types of xylanases from alkalophilic thermophilic Bacillus spp. Appl. Microbiol. Biotechnol. 1984, 19, 335-340.
30. Fukumori, F., Kudo, T. and Horikoshi, K., Purification and properties of a cellulase from alkalophilic Bacillus sp. No. 1139. J. Gen. Microbiol. 1985, 131, 3339-3345.
31. Horikoshi, K., Nakao, M., Kurono, Y. and Sashihara, N., Cellulases of an alkalophilic Bacillus strain isolated from soil. Can. J. Microbiol. 1984, 30, 774-779.
32. Sunaga, T., Akiba, T. and Horikoshi, K., Production of penicillinase by an alkalophilic Bacillus. Agric. Biol. Chem. 1976, 40, 1363-1367.
33. Yamamoto, M. and Horikoshi, K., Cloning and expression of an alkalophilic Bacillus oligo-1,6-glucosidase gene in Escherichia coli. J. Jpn Soc. Starch. Sci. 1987, 34, 300-304.
34. Akino, T., Nakamuru, N. and Horikoshi, K., Production of β-mannosidase and β-mannanase by an alkalophilic Bacillus sp. Appl. Microbiol. Biotechnol. 1987, 26, 322-327.

35. Akino, T., Nakamuru, N. and Horikoshi, K., Characterization of three β-mannanases of an alkalophilic Bacillus sp. Agric. Biol. Chem. 1988, 52, 773-779.

36. Hayashi, T., Akiba, T. and Horikoshi, K., Production and purification of new maltohexaose-forming amylases from alkalophilic Bacillus sp. H-167. Agric. Biol. Chem. 1988, 52, 443-448.

37. Nakamuru, N. and Horikoshi, K., Characterization and some actual conditions of a cyclodextrin glycosyltransferase-producing alkalophilic Bacillus sp. Agric. Biol. Chem. 1976, 40, 753-751.

38. Kaneko, T., Hamamoto, T. and Horikoshi, K., Molecular cloning and nucleotide sequence of the cyclomaltodextrin glucotransferase gene from the alkalophilic Bacillus sp. strain 38-2. J. Gen. Microbiol. 1988, 134, 97-105.

39. Sashihara, N., Kudo, T. and Horikoshi, K., Molecular cloning and expression of cellulose genes of alkalophilic Bacillus sp. strain N4 in Escherichia coli. J. Bact. 1984, 158, 503-506.

40. Fukumori, F., Kudo, T., Narahashi, Y. and Horikoshi, K., Molecular cloning and nucleotide sequence of the alkaline cellulase gene from the alkalophilic Bacillus sp. strain 1139. J. Gen. Microbiol. 1986, 132, 2329-2335.

41. Honda, H., Kudo, T. and Horikoshi, K., Molecular cloning and expression of the xylanase gene of alkalophilic Bacillus sp. strain C-125 in Escherichia coli. J. Bact. 1985, 161, 784-785.

42. Hamamoto, T., Kaneko, T. and Horikoshi, K., Nucleotide sequence of the cyclomaltodextrin glucotransferase (CGTase) gene from alkalophilic Bacillus sp. strain No. 38-2. Agric. Biol. Chem. 1987, 51, 2019-2022.

43. Fukumori, F., Kudo, T. and Horikoshi, K., Truncation analysis of an alkaline cellulose from an alkalophilic Bacillus sp. FEMS Microbiol. Lett. 1987, 40, 311-314.

44. Kato, C., Kudo, T. and Horikoshi, K., Gene expression and production of Bacillus No 170 penicillinase in Escherichia coli and Bacillus subtilis. Agric. Biol. Chem. 1984, 48, 397-401.

45. Kudo, T., Oshikoshi, A. and Horikoshi, K., Molecular cloning and expression of xylanase gene of alkalophilic Aeromonas sp. No. 212 in Escherichia coli. J. Gen. Microbiol. 1985, 131, 2825-2830.

46. Fukumori, F., Sashihara, N., Kudo, T. and Horikoshi, K., Nucleotide sequences of two cellulose genes from alkalophilic Bacillus sp. strain N-4 and their strong homology. J. Bact. 1986, 168, 479-485.

47. Hamamoto, T., Honda, H., Kudo, T. and Horikoshi, K., Nucleotide sequence of the xylanase A gene of alkalophilic Bacillus sp. strain C-125. Agric. Biol. Chem. 1987, 51, 953-955.

48. Fukumori, F., Ohishi, K., Kudo, T. and Horikoshi, K., Tandem location of the cellulase genes in the chromosome of Bacillus sp. strain N4. FEMS Microbiol. Lett. 1987, 48, 65-68.

49. Kato, C., Kudo, T., Watanabe, K. and Horikoshi, K., Extracellular production of Bacillus penicillanase by Escherichia coli carrying pEAP2. Eur. J. Appl. Microbiol. Biotechnol. 1983, 18, 339-342.

50. Horikoshi, K., Genetic application of alkalophilic microorganisms. In Microbes in Extreme Environments, eds. R.A. Herbert and G.A. Codd. London, Academic Press 1986, pp.297-316.

51. Kudo, T., Kato, C. and Horikoshi, K., Excretion of the penicillinase of an alkalophilic Bacillus sp. through the Escherichia coli outer membrane. J. Bact. 1983, 156, 949-951.

52. Kobayashi, T., Kato, C., Kudo, T. and Horikoshi, K., Excretion of the penicillinase of an alkalophilic Bacillus sp. through the Escherichia coli outer membrane is caused by insertional activation of the kil gene in plasmid pMB9. J. Bact. 1986, 106, 728-732.

53. Kato, C., Kobayashi, T., Kudo, T. and Horikoshi, K., Construction of an excretion vector : extracellular production of Aeromonas xylanase and Bacillus cellulases by Escherichia coli. FEMS Microbiol. Lett. 1986, 36, 31-34.

54. Kato, C., Ohkoshi, A., Kudo, T. and Horikoshi, K., Extracellular production of xylanase L in Escherichia coli using excretion vector pEAP2. Agric. Biol. Chem. 1986, 50, 1067-1068.

55. Kato, C., Kobayashi, T., Kudo, T., Furusato, T., Murakami, Y., Tanaka, T., Baba, H., Oishi, T., Ohtsuka, E., Ikahara, M., Yanagida, T., Kato, H., Moriyama, S. and Horikoshi, K., Construction of an excretion vector and extracellular production of human growth hormone from Escherichia coli. Gene 1987, 54, 197-202.

56. Kitai, K., Kudo, T., Nakamuru, S., Masegi, T., Ichikawa, Y. and Horikoshi, K., Extracellular production of human immunoglobulin GFc region (hIgG-Fc) by Escherichia coli. App. Microbiol. Biotechnol. 1988, 28, 52-56.

57. Honda, H., Kudo, T. and Horikoshi, K., Extracellular production of alkaline xylanase of alkalophilic Bacillus spp. by Escherichia coli carrying pCX311. J. Ferment. Technol. 1986, 64, 373-377.

58. Honda, H., Kudo, T. and Horikoshi, K., Selective excretion of alkaline xylanase by Escherichia coli carrying pCX311. Agric. Biol. Chem. 1985, 49, 3011-3015.

59. Grant, W.D. and Tindall, B.J., The alkaline, saline environment. In Microbes in Extreme Environments eds. R.A. Herbert and G.A. Codd London Academic Press, 1986, pp.22-54.

60. Pantoliano, M.W., Ladner, R.C., Bryan, P.N. Pollence, M.L., Wood, J.F. and T.L. Poulos., Protein engineering of subtilisin BPN' : enhanced stabilization through the introduction of two cysteines to form a disulfide bond. Biochemistry 1987, 26, 2077-2082.

61. Nedkov, P., Oberthur, W. and Braunitzer, G., Determination of the complete amino acid sequence of subtilisin DY and the comparison with the primary structures of the subtilisins BPN', Carsberg and Amylosacchariticus. Biol. Chem. Hoppe-Seyler 1985, 366, 421-430.

MODIFICATION OF IMMUNE RESPONSE BY EXTREME HALOPHILIC BACTERIA

C. RUIZ, M. MONTEOLIVA-SANCHEZ AND A. RAMOS-CORMENZANA
Department of Microbiology. Faculty of Pharmacy.
University of Granada. 18001 Granada. Spain.

ABSTRACT
Several species of Halobacterium and Halococcus morrhuae presented marked modificatory effects of humoral or cellular responses and these effects enhance or suppress the response in experimental animals depending on the genus and the inoculation model.

INTRODUCTION
The cell wall components from various bacteria have been found to be potent biological response modificators. Thus, this property of microorganisms may be attributable to superficial cellular fraction (1).
 The halobacteria are extreme halophiles that differ significantly from most prokaryotes in possesing cell walls which contain no muramic acid. The cell wall composition in these bacteria may or may not be family characteristic, and at least two types are reported. One of sulfated glycoprotein subunits in the Halobacterium genus and the other a sulfated heteropolysaccharides in the Halococcus genus. The presence of glycoproteins on the cell surface is of special interest, because shuch glycoproteins are common in eukaryotic cells but little known in prokaryotic cells (2).
 The object of the present paper was to investigate the possible immunomodulation activity of halobacteria in the consideration to their differential structures in relation to other prokaryotes.

MATERIALS AND METHODS
Animals
Female Swiss mice (6 to 8 weeks old) were purchased from the University of Granada laboratories.
Microorganisms and culture conditions
The strains studied were Halobacterium salinarium CCM 2084, H. saccharovorum ATCC 29252, H. mediterranei ATCC 32500 and Halococcus morrhuae CCM 537.
 A solid medium was used as described Rodriguez-Valera et al. (3) for growth of bacteria.

Antibody forming cell assay (PFC)
For this assay were used the methods described by Cunningham
and Szenberg (4). Experiments were carried out in three
differents intraperitoneal (i.p.) treatments: (i) 2 days before
(ii) the same day, or (iii) 2 days after i.p. immunization with
1×9^{10} sheep red blood cells (SRBC).
Phagocytosis "in vivo" (clearance test)
The clearance test was performed following the methods of
Liany-Takasaki et al. (5) at 24 and 72 hours after of
treatments.
Delayed-type hypersensitivity (DTH)
This assay was performed as described Ruiz et al. (6).
Statistics
For analysis of results the Student's t test was used.

RESULTS

The immunomodulators effects obtained with the four extreme
halophilic bacteria studied on humoral response by PFC assay
are presented in Figure 1. The PFC response was clearly
different between the several species of the genus
Halobacterium and the specie Halococcus morrhuae tested.
H. saccharovorum and H. mediterranei caused increase in the
number of PFC per spleen when they were injected at the same
time, 2 days before or 2 days after SRBC immunization. However,
H. salinarium was lightly potentiator inoculated at the same
day of immunization and caused suppression when was injected
2 days after. In contrast, the effect on PFC response obtained
with H. morrhuae was suppressor in all models assayed.

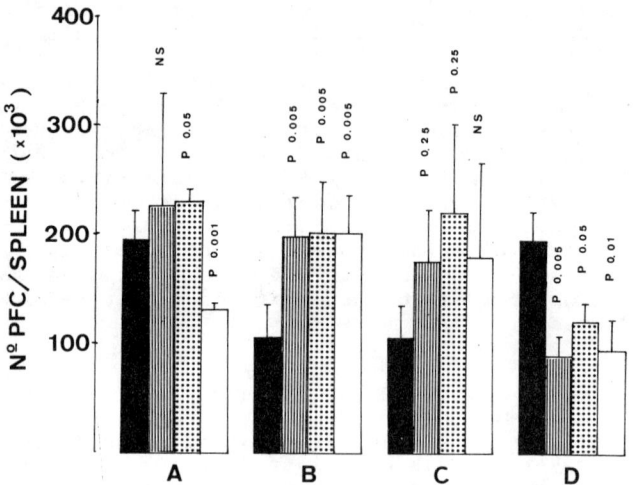

Figure 1. Modulation of the PFC response to SRBC in mice by:
(A), H.salinarium; (B), H.saccharovorum; (C), H.mediterranei;
(D), H.morrhuae. Model of inoculation i.p. in treated mice:
▨ , 2 days before; ▨ , the same day; ☐ , 2 days after i.p.
immunization;■, control mice. NS: not significative data.

H. saccharovorum and H. morrhuae showed the best immunomodulator response, and by this we analysed in these strains other immunology parameters as phagocytosis "in vivo" and delayed-type hypersensitivity. As showed table 1 the phagocytosis assay hasn't significance with any strains. However the delayed-type hypersensitivity was significatively modified with both strains tested. H. morrhuae enhance the DTH response to SRBC, while H. saccharovorum suppressed this response.

TABLE 1
Effects of H. saccharovorum and H. morrhuae on the DTH and phagocytosis in mice

	H.saccharovorum		H.morrhuae	
	DTH^a　Phagocytosis^b		DTH　Phagocytosis	
Control	0.77 ± 0.21	0.15^c　0.09^d	0.77 ± 0.21	0.15^c　0.09^d
Treatment	0.33 ± 0.17 $P < 0.025$	0.23　0.12	2.20 ± 0.57 $P < 0.0001$	0.21　0.10

a Foot swelling (0.1 mm units).
b Clearance rate of bacteria per minute.
c,d, 24 and 72 hours respectively.

DISCUSSION

Our results show that the inoculation of mice with extreme halophilic bacteria caused modification of immune response against unrelated antigens. The experiments show clearly that H. saccharovorum possess immunomodulating properties different to Halococcus morrhuae. This effect is possible to attribute to their structural differences in relation with their cell walls (7), although these microorganisms lack of typical immunomodulators structure, as murein, which is present in mayority of prokaryotes organisms. It is reasonable to suppose that the modification of the subunits that formed the cell wall sulfated glycoproteins in Halobacterium and sulfated heteropolysaccharides in Halococcus, caused a immunomodulation pattern different. However, it is dificult to atribute their immunomodulating properties only to cell wall. A bacteria represents a complex association of antigens, mitogens and adjuvant when the bacterial components interact with cells of the immune response system (8). The literature shows many microorganisms and microbial products may markely influence antibody formation, as well as cell-metiated immunity, both "in vitro" and "in vivo". Indeed, many microbial products have been used as immunologic enhancing agents for non-specific purposes. Is widely accepted that in many cases the immunomodulator mechanisms of several microbial cell component are potent stimulators of soluble mediators formation or release (1).

Further studies are needed to know the implication of different structures of halobacteria in the immunomodulating process and the immunomodulators mechanisms by these bacteria.

REFERENCES

1. Yamamura, Y., Kotani, S., Azuma, I., Koda, A. and Shiba, T., eds., Immunomodulation by microbial products and related synthetic commpounds. Excepta Medica, Amsterdam, 1982.
2. Kushner, D.J., The Halobacteriaceae. In The Bacteria vol.VIII. eds., C.R. Woese and R.S. Wolfe, Academic Press, New York, 1985, pp. 171-206.
3. Rodriguez-Valera, F., Ruiz-Berraquero, F. and Ramos-Cormenzana, A., Isolation of extremely halophilic bacteria able to grow in defined inorganic media with single carbon sources. J. Gen. Microbiol., 1980, 119, 535-8.
4. Cunningham, A.J. and Szenberg, A., Further improvements in the plaque technique for detecting single antibody forming cells. Immunol., 1968, 14, 559-60.
5. Liang-Takasak, C.J., Saxen, H., Makela, P.H. and Leive, L., Complement activation by polysaccharide of lipopolysaccharide: An important virulence determinant of Salmonellae. Infect. Immun., 1983, 41, 563-9.
6. Ruiz, C., Ruiz-Bravo, A., Alvarez de Cienfuegos, G. and Ramos-Cormenzana, A., Immunomodulation by myxospores of Myxococcus xanthus. J. Gen. Microbiol., 1985, 131, 2035-9.
7. Kandler, O., Cell wall structures and their phylogenetic implication. Zbl. Bakt. Hyg., 1982, I Abt Orig C3, 149-60.
8. Warren, H.S., Vogel, F.R. and Chedid, L.A., Current status of immunological adjuvant. Ann. Rev. Immunol., 1986, 4, 369-88.

ISOLATION AND CHARACTERIZATION OF NOVEL ANAEROBIC, HALOPHILIC EUBACTERIA
FROM HYPERSALINE ENVIRONMENTS OF WESTERN AMERICA AND KENYA

HIROTAKA SHIBA[*] AND KOKI HORIKOSHI[+]
The Superbugs Project, Research Development Corporation of Japan,
2-28-8, Honkomagome, Bunkyo-ku, Tokyo 113, Japan
* Present adress: Institute for Fundamental Research, Suntory Limited,
1-1-1, Wakayamadai, Shimamoto-cho, Mishima-gun, Osaka 618, Japan
+ Present address: Tokyo Institute of Technology, 2-12-1, O-okayama,
Meguro-ku, Tokyo 152, Japan and The Institute of Physical and Chemical
Research, 2-1, Hirosawa, Wako-shi, Saitama 351-01, Japan

ABSTRACT

Anaerobic halophiles which can grow at more than 20 % NaCl, were isolated from salty soil, mud, and salt piles in hypersaline environments of California and Nevada, USA and from soda lakes in Kenya. Several different types of fermentative eubacteria were found including neutrophiles, alkalophiles and an extremely halotolerant bacterium. These bacteria from aerobic surface sediments were similar to previously known strictly anaerobic halophiles isolated from the anaerobic bottom sediments in hypersaline lakes. One new type of alkalophile was discovered in samples from both California and Kenya. This extreme alkalophile is a facultatively anaerobic, rod-shaped bacterium and chemotaxonomic data (the G+C content of DNA, the cellular fatty acid composition and DNA-DNA hybridization) indicate that this does not belong to any previously reported genus.

INTRODUCTION

Most studies of the bacteria which grow in extreme environments of high salt concentration have focused on aerobic halophiles and studies of anaerobic halophiles have been few. Now, only four species of strictly anaerobic, moderately halophilic, chemo-organotrophs are known and available for study (reviewed by Oren [1]); Haloanaerobium praevalens, Halobacteroides halobius, Sporohalobacter lortetii, and Sporohalobacter marismortui. These four strains were isolated from the deep-bottom sediments or the salt flat of hypersaline lakes (the Great Salt Lake, USA and the Dead Sea, Israel), whose condition was completely anaerobic. Here, we succeeded in the isolation of strictly anaerobic halophiles from samples of aerobic surface sediments in an evaporating pond and soda lakes of the USA and Kenya.

In alkaline saline environments, the existence of alkalophilic anoxygenic phototrophic bacteria of the genus Ectothiorhodospira, haloalkalophilic archaebacteria of the genera Natronobacterium and Natronococcus, and haloalkalophilic methanogens is known [2], but anaerobic fermentative bacteria have not yet been reported. In this communication, we report the isolation of a new type of a moderate halophile, which is a facultatively anaerobic, obligately alkalophilic, eubacterium found in soda lakes of both America and Kenya.

MATERIALS AND METHODS

Hypersaline soil, mud, and water from California and Nevada were sampled in March, 1987 and salty samples from soda lakes in Kenya were generously supplied by Drs. W. D. Grant and T. Hamamoto. Most of the methods used in this study have been previously described [3]. Fatty acid composition was measured by the method of Komagata and Suzuki [4]. DNA-DNA hybridization was performed by the nitrocellulose membrane method [5] with some modification.

RESULTS AND DISCUSSION

Isolation of Strictly Anaerobic Halophiles from Aerobic Hypersaline Environments

Three strains of strict anaerobes (strains SS-11, SS-15 and SS-21) were isolated from 0 - 10 cm depth of surface sediment of hypersaline muds in the closed lagoon at the south east rim of Salton Sea, California, using a screening medium containing 23.4 % NaCl. Two alkalophilic strains (strains M-20 and KY-402) were isolated from the surface sediments in Big Soda Lake, Nevada, and from Trona of Lake Magadi, Kenya, using a screening medium at pH 10, containing 20 % NaCl. Growth properties and DNA base compositions of the five isolates are listed in TABLE 1.

TABLE 1
Characteristics of isolated anaerobic halophiles

	SS-11	SS-15	SS-21	M-20	KY-402
Sampling place	Salton Sea USA	Salton Sea USA	Salton Sea USA	Big Soda Lake USA	Lake Magadi Kenya
G+C content	30.3 %	30.5 %	34.9 %	31.1 %	31.9 %
Growth NaCl conc.	3-25 %	3-25 %	10-30 %	3-20 %	3-25 %
(Optimum)	5-10 %	3-10 %	12-20 %	5-10 %	5-10 %
Growth pH	6.5-9.5	6.5-9.0	6.5-9.0	7.0-10.0	6.0-10.5
(Optimum)	7.0-9.0	7.0-8.5	7.0-8.0	8.5-9.5	8.5-10.0

All isolates were Gram-negative, non-sporeforming, motile, rod-shaped moderate halophiles, similar to previously described species. Fatty acid glycerol esters were found in their cell membranes. Accordingly, these strains belong to the eubacteria. Strain SS-21 tolerated more than 30 % NaCl in the medium. Taxonomic properties including morphological and physiological data, fermentation products from glucose, the G+C content of DNA, membrane fatty acid composition, and DNA-DNA hybridization data, suggest that strains SS-11 and SS-15 are members of the genus Haloanaerobium [6], but the other three isolates differ from previously described halophilic anaerobes [1].

The existence of these anaerobic halophiles in the hypersaline surface sediment samples confirms that strict anaerobic halophiles are universally distributed in hypersaline environments, as are aerobic halophiles. Halophilic methanogens have been detected in such environments, for example, Oremland et al. [7] described a moderately halophilic, alkalophilic methanogen from Big Soda Lake where strain M-20 was isolated. Also haloalkalophilic methanogens were isolated from the same Salton Sea samples that we used (Nakatsugawa, N., personal communication). Zeikus et al. [6] have reported that in the Great Salt Lake, Haloanaerobium praevalens has a role as a producer of substrates for methanogenesis. So, anaerobic heterotrophic bacteria such as our isolates may be participating in the recycling of nutrients in their environment together with halophilic methanogens.

Isolation of Novel Facultatively Anaerobic Haloalkalophiles

Three halophilic, obligately alkalophilic eubacteria were isolated from salt piles at Owens Lake, USA and from soda lakes in Kenya, listed in TABLE 2.

TABLE 2
Characteristics of facultatively anaerobic haloalkalophiles

	M-12	KY-284	KY-362
Sampling place	Owens Lake USA	Lake Magadi Kenya	Lake Nakuru Kenya
G+C content	37.2 %	37.2 %	37.3 %
Growth NaCl conc.	0.1-25 %	0.1-25 %	0.1-25 %
(Optimum)	3-5 %	3-10 %	3-10 %
Growth pH	8.0-11.0	8.5-11.5	8.5-11.0
(Optimum)	9.0-10.0	9.0-10.0	9.0-10.0

These bacteria were Gram-negative, rod-shaped and exhibited motility by means of peritrichous flagella. Spore formation was not detected. These three isolates were facultative anaerobes and lacked catalase and oxidase activities, cytochromes and quinones. So far, no similar type of halophile has been reported. Cellular fatty acid components of these three isolates were similar and the major fatty acids were n-C14:0, n-C16:0, iso-C15:0 and anteiso-C16:1, a pattern which resembles that of genus Bacillus. As shown in TABLE 3, DNA-DNA hybridization data indicated that these three isolates constitute one group which exibits low homology with the differ from two types of alkalophilic Bacillus tested. This type of haloalkalophilic anaerobe was considered to be a new genus. More varieties of anaerobic halophiles may be living in such extreme environments.

TABLE 3
DNA-DNA homology of anaerobic haloalkalophiles

Unlabeled DNA	^{32}P-labeled DNA			
	M-12	KY-284	KY-362	Bacteroides fragilis DSM 2151
M-12	100	48	44	<1
KY-284	49	100	90	1
KY-362	51	96	100	<1
Bacillus alcalophilus DSM 485	6	6	7	
Bacillus alcalophilus DSM 497	6	6	7	
Bacteroides fragilis DSM 2151				100

ACKNOWLEDGMENTS

We thank R.H. Doi and F.E. Robinson (University of California, Davis) for valuable advice about sampling, K. Uematsu (Suntory Limited) for arrangement of sampling, and also thank W.D. Grant (University of Leicester) and T. Hamamoto (The Institute of Physical and Chemical Research) for use of samples from Kenya, K. Suzuki (The Institute of Physical and Chemical Research) and N. Amano (Suntory Limited) for stimulating discussion, H. Nagayama (The Superbugs Project) for helpful assistance and W. Bellamy (The Superbugs Project) for his critical reading of this manuscript.

REFERENCES

1. Oren, A., The ecology and taxonomy of anaerobic halophilic eubacteria. FEMS Microbiol. Rev., 1986, **39**, 23-29.
2. Grant, W.D. and Tindall, B.J., The alkaline saline environment. In Microbes in Extreme Environments, eds., Herbert, R.A. and Codd, G.A., Academic Press, London, 1986, pp. 25-54.
3. Shiba, H., Yamamoto, H. and Horikoshi, K., Isolation of strictly anaerobic halophiles from the aerobic surface sediments of hypersaline environments in California and Nevada. in preparation.
4. Komagata, K. and Suzuki, K., Lipid and cell-wall analysis in bacterial systematics. In Methods in Microbiology, Vol. 19, eds., Colwell, R.R. and Grigorova, R., Academic Press, London, 1987, pp. 161-207.
5. Suzuki, K., Kaneko, T. and Komagata, K., Deoxyribonucleic acid homologies among coryneform bacteria. Int. J. System. Bacteriol., 1981, **31**, 131-138.
6. Zeikus, J.G., Hegge, P.W., Thompson, T.E., Phelps, T.J. and Langworthy, T.A., Isolation and description of Haloanaerobium praevalens gen. nov. and sp. nov., an obligately anaerobic halophile common to Great Salt Lake sediments. Curr. Microbiol., 1983, **9**, 225-233.
7. Oremland, R.S., Marsh, L. and DesMarais, D.J., Methanogenesis in Big Soda Lake, Nevada: an alkaline, moderately hypersaline desert lake. Appl. Environ. Microbiol., 1982, **43**, 462-468.

THE POTENTIAL USE OF HALOPHILIC EUBACTERIA FOR THE PRODUCTION OF ORGANIC CHEMICALS AND ENZYME PROTECTIVE AGENTS

E.A. GALINSKI
Institut für Mikrobiologie, Meckenheimer Allee 168, 5300 Bonn 1
Federal Republic of Germany

ABSTRACT

Screening of a large number of isolates (including halophilic actinomycetes) has permitted the selection of strains which are potentially useful for the mass-production of organic compounds (compatible solutes), and has led to the discovery of a number of yet unknown osmotica. Actinomycete strain A5-1, which is a potent producer of the novel amino acid "ectoine" and a second not yet identified amino compound (Y), has been studied in greater depth. Investigations into the effect of physiological parameters on solute composition have shown that it is possible to influence the relative proportion of compatible solutes and thus direct the biosynthesis towards a desired product. A standard extraction procedure followed by chromatographic separation was applied to process bacterial cell mass and has enabled us to routinely recover a compatible solute fraction of 16% of the cells' dry weight. Studies on salt protecting and temperature stabilizing effects, which have so far been confined to readily available compounds like betaine, proline and glycerol only, may therefore be widened to encompass ectoine and further novel compatible solutes, which seem to have a potential technological application.

INTRODUCTION

Concentrated salt solutions are naturally occurring extreme environments, being found as coastal lagoons, salt and soda lakes and solar evaporation ponds [1]. To cope with the physiological stress of low water activity (high osmolality) of the surrounding medium, the majority of halophilic and halotolerant eubacteria synthesize organic osmotica which belong to three typical classes of compounds: SUGARS, AMINO ACIDS and QUATERNARY NITROGEN COMPOUNDS (betaines). These have been named "compatible solutes" on the basis that they do not interfere with enzyme action [2]. The notably high cytoplasmic concentration (2-3 mol/kg water) of organic compounds makes halophilic organisms potentially useful for the biotechnological production of chemicals, namely chiral molecules, the bioproduction of which strongly competes with chemical methods of synthesis. Besides this, the prospect of detecting novel substances produced in quantity bears great technological potential. A novel cyclic amino acid has been isolated and identified from extremely halophilic species of the bacterial genus *Ectothiorhodospira* [3]. This hitherto unknown compound (Fig.1) has since been found in a variety of different organisms and without doubt plays a major role in haloadaptation.

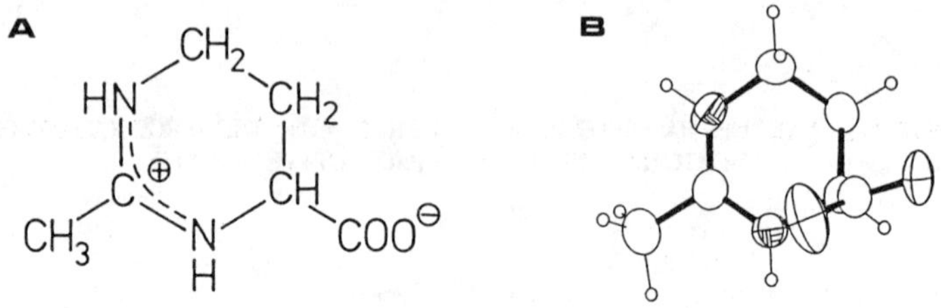

Figure 1. Structural formula (A) and ORTEP-plot (B) of 1,4,5,6-Tetrahydro-2-methyl-4-pyrimidine carboxylic acid. This novel cyclic amino acid was given the trivial name "ectoine" due to its discovery in the bacterial genus *Ectothiorhodospira*.

In addition to their function as osmotica, compatible solutes are required to maintain the enzymes' hydration shell in a low water environment and to counterbalance the effect of elevated ionic strength within the cytoplasm. In view of this second physiological function compatible solutes may also find a technological application as stabilizers and protective agents in enzyme technology.

MATERIALS AND METHODS

Screening and Culture Conditions

In contrast to the screening methods applied by other researchers, we laid emphasis on using a basic mineral salt medium (10% NaCl) supplemented with 0.1 g/l yeast extract and a variety of carbon sources, including polymeric naturally occurring materials like starch, cellulose and chitin [4]. Thus all halophilic/halotolerant organisms which accumulate organic osmotica from media constituents but are unable to synthesize compatible solutes were excluded from our screening project. The fermentation medium for halophilic actinomycetes was essentially the same except that yeast extract was omitted and NH_4HCO_3 raised to 3 g/l. The fermentation vessel of a 13 l working volume was specially modified by BIOENGINEERING (Wald, Switzerland) to meet our requirements in view of increased material resistance and has been described elsewhere [4]. Actinomycete strain A5-1 was chosen for the production of ectoine and served as a model organism to evaluate halo-fermentation techniques and processing procedures.

Recovery and Identification of Compatible Solutes

For screening purposes a minimum of 200 mg of cell material (wet weight) was extracted following a modification of the Bligh & Dyer technique [5], desalted on an ion retardation column (BIORAD AG 11A8) und subsequently analyzed using a combination of HPLC methods [6]. Potent producers of yet unknown organic compounds were fermented to yield a minimum of 10 g of cell material, which was processed accordingly and subjected to ^{13}C NMR analysis (BRUKER WH-90 at 22.628 MHz and 90.02 MHz for the decoupling channel).

RESULTS AND DISCUSSION

The Spectrum of Compatible Solutes

Our screening programme, which - admittedly - has only covered a rather small range of moderately halophilic microorganisms mainly from hypersaline marine environments, aimed at the isolation of halotolerant eubacteria which produce an interesting range of compounds and grow well on simple synthetic media. The spectrum of organic compounds present in high cytoplasmic concentrations (Table 1.) does not include polyols, which have been reported mainly from halophilic algae and osmophilic yeast and fungi. Our results do not support the concept proposed by others [7] that glycine betaine is the main compatible solute of halophilic eubacteria. Nor do we believe there is a need to postulate a hypotonic cytoplasm [8]. These apparently conflicting views may be explained by our finding that betaine is easily taken up from complex media constituents (or possibly synthesized from precursors supplied with the medium) and by the failure to detect unusual or unidentified compounds. Our preliminary screening programme has shown that ectoine is in fact the most abundant organic osmoticum, which was - besides other compounds - detected in the vast majority of the strains examined. As yet trehalose, alanine, proline and ectoine seem to be promising candidates for a possible biotechnological production.

TABLE 1
Spectrum of compatible solutes found in moderately halophilic eubacteria; preliminary result based on 15 strains from culture collections and approx. 150 isolates from mediterranean salinas

Sugars	Amino Acids	Betaines	Unknown
Glucose Sucrose Trehalose	Glutamate Alanine Proline Ectoine	Glycine Betaine	X, Y, Z α, β, γ

The rather limited range of osmotica is almost equalled by a number of yet unidentified compounds which are under present investigation. The prospect of discovering even more novel compounds makes it worthwhile to systematically extend our screening project towards more unusual saline environments and further systematic bacterial groups. Strain A5-1 was chosen as a model organism to optimise production and processing strategies since actinomycetes are widely used in industrial fermentation processes and since this particular strain produces a range of interesting compounds (trehalose, ectoine and substance Y).

The Relative Proportion of Compatible Solutes

Of the physiological growth parameter examined, salinity and temperature seem to have the most prominent effect on the amount and relative proportion of the compatible solutes. As can be seen in Fig.2 (A) an increase in salinity from 5% to 15% will double the quantity of solutes produced. It is, therefore, likely that other additional compounds (possibly salts) will also contribute towards osmotic equilibrium. Assuming a cytoplasmic volume of approx. 0.2 ml/g wet weight, one would estimate a cytoplasmic concentration of up to 2 moles osmotica per kg cell water (growth in 15% NaCl).

We have so far been able to produce 20 g/l of bacterial cell mass (wet weight) and to gain approximately 1 g/l of compatible solutes, equivalent to 16% of the bacterial dry weight. The economical feasibility of the fermentation process can still be improved and will certainly profit from the use of overproducing or leaky mutants.

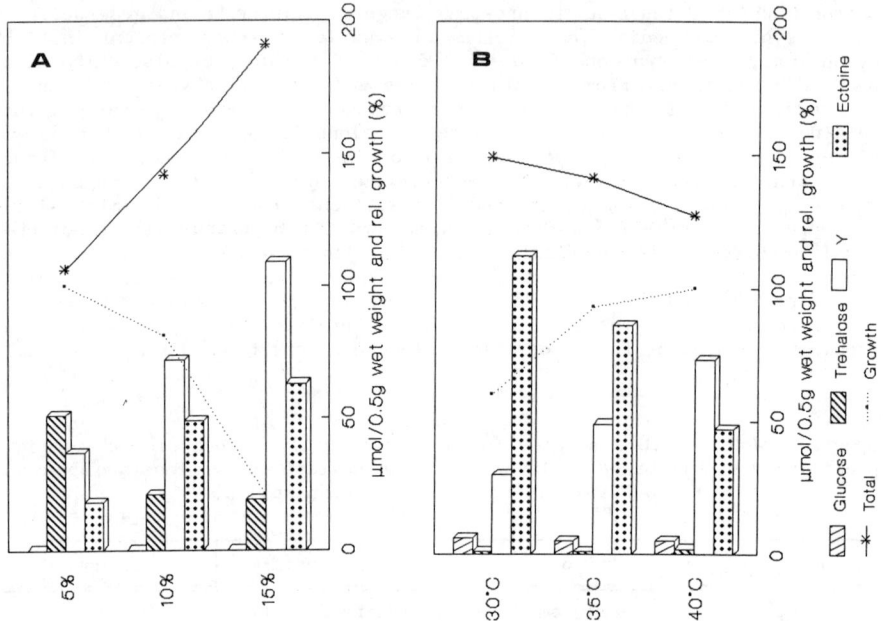

Figure 2. Effect of salt concentration (A) and growth temperature (B) on the compatible solute spectrum of actinomycete strain A5-1

Figure 2 (B) reveals a striking correlation between growth temperature and the ratio of the two predominant compatible solutes ectoine and Y. The underlying principle of this remarkable phenomenon has still to be investigated in detail. We believe the organisms adaptation may possibly reflect a temperature stabilizing effect of the novel unidentified compound Y. On the basis of these findings it seems, therefore, possible to channel the bioproduction of compatible solutes towards certain products by varying the physiological growth conditions.

Enzyme Protection

It has been shown in a number of cases, mainly with betaine as a solute, that compatible osmotica are able to fully or partly relieve salt inhibition on certain enzymes (Table 2). Although numerous theories have been put forward to explain this protective action, the principle behind it is by no means clear. Our own investigations have also given evidence for a protective function against heat denaturation superior to that of glycerol [9]. A systematic study of this class of compounds may, therefore, provide a significant contribution towards the understanding of enzyme protective mechanisms and without doubt meets biotechnological demands.

TABLE 2

Enzymes which gain partial salt protection through the presence of betaine

Enzyme	Organism	Reference
Malic enzyme, Pyruvate kinase	Hordeum vulgare	[10]
Glucose-6-P dehydrogenase	Aphanothece halophytica	[11]
Glutamine synthetase	Synechocystis DUN52	[12]
Isocitrate lyase	Ectothiorhodospira halochloris	[6]

REFERENCES

1. Trüper, H.G. and Galinski, E.A., Concentrated brines as habitat for microorganisms. Experientia, 1986, 42, 1182-1187.

2. Brown, A.D., Microbial water stress. Bact. Rev., 1976, 40, 803-846.

3. Galinski, E.A., Pfeiffer, H.P. and Trüper, H.G., 1,4,5,6-Tetrahydro-2-methyl-4-pyrimidinecarboxylic acid, a novel cyclic amino acid from halophilic phototrophic bacteria of the genus Ectothiorhodospira. Eur. J. Biochem., 1985, 149, 135-139.

4. Galinski, E.A., Halo-fermentation, a novel low water process for the production of organic chemicals and enzyme protective agents. In Bioreactors and Biotransformations, ed., G.W. Moody and P.B. Baker, Elsevier Applied Science Publishers, London, 1987, pp. 201-212.

5. Bligh, E.G. and Dyer, W.J., A rapid method of lipid extraction and purification. Can. J. Biochem. Physiol., 1959, 37, 911-917.

6. Galinski, E.A., Salzadaption durch kompatible Solute bei halophilen phototrophen Bakterien. Ph.D.thesis, University of Bonn, April 1986.

7. Imhoff, J.F. and Rodriguez-Valera F., Betaine is the main compatible solute of halophilic eubacteria. J. Bac., 1984, 160, 478-479.

8. Vreeland, R.H., Mechanisms of halotolerance in microorganisms. CRC Crit. Rev. Microbiol., 1987, 14, 311-357.

9. Jürgens, E., Der Einfluß von Ionen und Glycinbetain auf die Isocitrat-dehydrogenase von Ectothiorhodospira abdelmalekii. Diploma thesis, University of Bonn, January 1988.

10. Pollard, A. and Wyn Jones, R.G., Enzyme activities in concentrated solutions of betaine and other solutes. Planta, 1979, 144, 291-298.

11. Pavlicek, K.A. and Yopp, J.H., Betaine as a compatible solute in the complete relief of salt inhibition of glucose-6-phosphate dehydrogenase from a halophilic blue-green alga. Plant Physiol., 1982, 69, 58 S

12. Warr, S.R.C., Reed, R.H. and Stewart, W.D.P., Osmotic adjustment of cyanobacteria: the effects of NaCl, KCl, sucrose and glycine betaine on glutamine synthetase activity in a marine and a halotolerant strain. J. Gen. Microbiol., 1984, 130, 2169-2175.

CYCLODEXTRIN PRODUCTION BY EXTREMOPHILIC BACILLI

E.G. AFRIKIAN
Institute of Microbiology, Academy of Sciences
of the Armenian SSR, 378510, Abovian City, USSR

Valuable and numerous fields of application of cyclodextrins, their derivatives and inclusion complexes attract significant attention to the production of these substances. Some countries e.g. Japan and Hungary have organised large-scale production of cyclodextrins, mainly β-cyclodextrin for the use in medicine, agriculture and industry (1).

K. Horikoshi has developed cyclodextrin production using alkalophilic strains of bacilli (2). The great advantage of this technology is that the process requires no organic solvents. The immobilisation technique permits the development of a highly economic and intensive procedure for cyclodextrin production by continuous bioconversion. In the aerobic spore-forming bacteria cultures of Bacillus macerans, B.circulans, B.megaterium, B.subtilis and B.stearothermophilus have been listed as cyclodextrin producers (3). A great number of strains of the genus Bacillus isolated from different ecological and geographical regions have been tested for cyclodextrin formation. Table 1 summarises the results of our researches on distribution of cyclodextrin producing bacilli, including extremophilic forms. Except psychrophiles, cyclodextrin producers have been detected in all groups investigated. The strains studied were identified by their morphological, physiological and biochemical features. A number of cyclodextrin-forming strains occur in B.circulans-polymyxa group, so the use of this property for differentation of this group from B.macerans is of doubtful significance.

TABLE 1
Distribution of cyclodextrin producing strains among aerobic sporeforming bacteria

Groups, species	Number of strains tested	Number of CD producers
B. macerans	7	6
B. circulans-polymyxa	29	3
B. megaterium	26	2
Thermophiles	130	2
Psychrophiles	20	0
Alkalophiles	151	10

We were able to obtain strains selectively producing β-cyclodextrin from all the above-mentioned groups. The detailed study of thermophilic strains proved that the majority belong to B. stearothermophilus. The alkalophile group of extremophilic bacilli are one of the most efficient sources for obtaining β-cyclodextrin; some cultures can form 60% β-cyclodextrin from the starch used. The biosynthesis of cyclodextrins is species-specific, and with well defined cultures it is possible to select the production of the cyclodextrin desired.

The main stages of enzymatic and microbiological production of cyclodextrins (Fig. 1) includes the use of α-amylase for starch after its bioconversion. Thermophilic strains of bacteria permit 10% or more starch concentration to be used.

B. stearothermophilus strains possess high activity of both intracellular and extracellular CGT-ase enzyme from thermophilic cultures is thermostable with optimum within the limits of 65-70°C permitting high temperature processing.

Different methods have been used for immobilisation of microbial cells and enzymes in cyclodextrins production. The

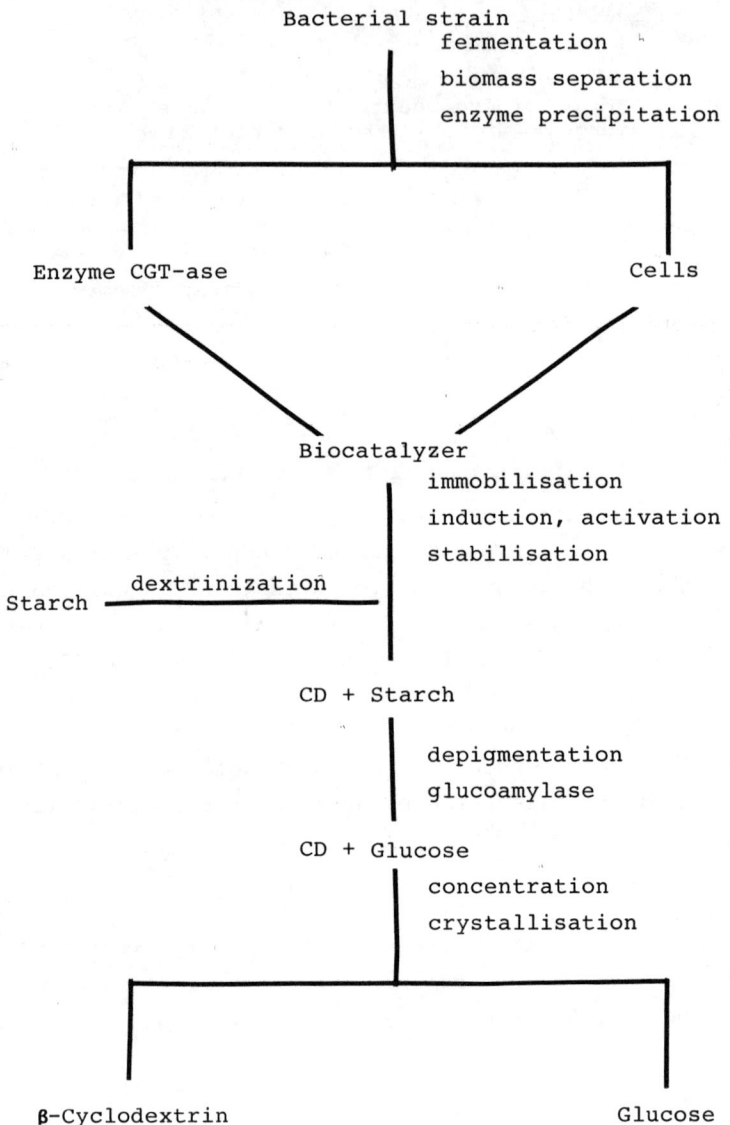

Figure 1. Main stages for β-cyclodextrin (CD) preparation.

efficiency of some methods, including some developed by us, are presented in Table 2.

TABLE 2

Efficience of different immobilisation system
for cyclodextrin production
(strain B.stearothermophilus, 55°, pH 6.5, SV $0.6h^{-1}$)

Immobilisation system	Yield (in per cent) of	
	α-CD	β-CD
Intact cells	8.4	49
Immobilised cells with		
Ca-alginate 1%	1.7	53
" 2%	1.5	53
" 5%	0	0
PAAG	0	0
Cross-linked (no carrier) SV=1.0	8.0	44
CGT-ase native	8.1	47
Immobilised CGT-ase with starch	2.0	31
Starch + CD	2.0	42
Inulin	2.0	17
Inulin + CD	3.2	32
Ca-alginate 2%	4.3	51
" 5%	0	0

REFERENCES

1. Horikoshi, K. and Akiba, T. Alkalophilic microorganisms Springer-Verlag, Berlin, Heidelberg, New York, 1982, pp 213.

2. Horikoshi, K. Enzymology and molecular genetics of CD-forming enzymes. 4th International Symposium on cyclodextrins, Munich, April 22-28, 1988, p A-1.

3. Szejtli, J. Cyclodextrins and their inclusion complexes. Akademiai Kiado. Budapest, 1982, pp 289.

ALKALINE PHOSPHATASE FROM THERMUS RUBER

D. COSSAR, R.J. SHARP
Division of Biotechnology, Public Health Laboratory Service,
CAMR, Porton Down, Salisbury, SP4 OJG, U.K.

Thermus strains isolated from Icelandic hot springs were isolated and examined for the production of Alkaline phosphatase. From over 100 strains screened strain 16107 was selected for further study. This strain was originally isolated from Hvergerthi, it has an optimum growth temperature of 60°C and does not grow at 70°C or 37°C. It has an optimum growth pH of 8.0-9.0. Alkaline phosphatase is produced constitutively throughout the growth of the culture and does not seem to be affected by phosphate level up to 1.0% (w/v) KH_2PO_4. The enzyme is cytoplasmic and may be membrane associated. In crude extracts alkaline phosphatase had a pH optimum at 8.9 and a temperature optimum at 67°C. The enzyme has been purified on the basis of cation exchange (CM-Sepharose) chromatography and gel permeation HPLC. Enzyme activity is apparently irreversibly lost on exposure to anion exchange resins in 20mM Tris-HCl buffer (pH 7.5). The use of 20mM triethanolamine buffer (pH 7.5) did not result in such a loss of activity. Similarly, $(NH_4)_2SO_4$ precipitation did not prove to be an effective purification procedure which may reflect a membrane association of the enzyme. The inclusion of an anion exchange step (DEAE-cellulose) allows purification of an enzyme which has activity against o-nitrophenyl phosphate at pH 9.0 (0.1M Diethanolamine) but does not appear to be alkaline phosphatase. The enzyme is monomeric with a native molecular weight of 40-50,000 Daltons. The final preparation gave only one peak on RP-HPLC and had a specific activity of approximately 40 $\mu mol.min^{-1}.mg\ protein^{-1}$ (100 x purification) at an overall yield of 5%.

LOSS OF PIGMENTATION IN THERMUS SP.

D. COSSAR, R.J. SHARP
Division of Biotechnology, Public Health Laboratory Service,
CAMR, Porton Down, Salisbury, SP4 OJG, U.K.

Thermus sp. isolated from geothermally heated pools generally have a bright yellow pigment or in the case of Thermus ruber pink or orange. Strains isolated from man made environments such as hot water systems, etc. are much less pigmented, and often appear as white colonies on agar plates.

Thermus aquaticus strain YS045, isolated from hot springs near St Mary's bay at the edge of Lake Yellowstone in the USA was studied to determine the effect of light on pigmentation. The strain has an optimum growth temperature of 70°C and a maximum of 80°C. Cultures grown under lactose limitation in a chemostat with reduced light intensity were observed to behave differently than when grown under continual illumination. Exposure of the chemostat to an alternate light and dark regime showed concomitant changes in pigmentation with an overall trend selecting for strains with reduced pigment content.

Thermus pigmentation is due to the production of carotenoid pigments which are assumed to offer the cells protection against the effects of light. Extracts of cells in 90% (v/v) acetone show characteristic spectra with two main peaks at 480 and 450 nm with a shoulder at 426 nm.

It is evident that the production of pigment is in response to the presence of light. Whether this is a direct response to light or a secondary response resulting from production of oxygen radicals etc. remains to be confirmed. In batch culture on complex media it is difficult to observe this phenomenon since pigment production does not seriously hinder growth of the organism. However, in the highly selective environment of the nutrient limited chemostat, this is not the case and pigment production represents a significant drain on carbon and energy. Further, cells which do not produce so much pigment will be able to divert more of their resources to growth and hence will predominate.

CHARACTERISATION OF AN EXCISION REPAIR DEFICIENT MUTANT OF BACILLUS CALDOTENAX

P. RILEY, T. ATKINSON, A. VIVIAN, R.J. SHARP
Division of Biotechnology, Public Health Laboratory Serivce, CAMR, Porton Down, Salisbury, Wiltshire SP4 OJG, U.K.

Two spontaneous mutants defective in their ability to remove mitomycin-C induced damage from their DNA (i.e. exhibit sensitivity to mitomycin-C relative to the wild-type), have been isolated from populations of Bacillus caldotenax. These mutants have been characterised with respect to a variety of DNA damaging agents: UV 8-methoxypsoralen, Methane sulphonic acid (MMS), N-methyl-N'-nitro-N-nitrosoguanidine (MNNG) and γ-radiation provided by a Cobalt-60 source. One mutant, BT1-40, exhibited sensitivity to a number of these DNA damaging agents relative to the wild-type, whereas the other mutant, BT1-41, exhibited no increased sensitivity to any of the DNA damaging agents used. BT1-40 was chosen for further study. It was shown to be capable of undergoing transformation and transduction suggesting an intact recombination function. However, BT1-40 was unable to reactivate UV damaged bacteriophage DNA or remove UV induced pyrimidine dimers from its DNA. Thus, it was concluded that BT1-40 was an excision repair deficient mutant of the wild-type B. caldotenax.

Comparison of the mutation rate occurring in the wild-type and BT1-40 following γ-radiation treatment showed BT1-40 exhibited five times the mutation rate of the wild-type. A study of the spontaneous mutation rate with the same two strains in chemostats over 500 h also demonstrated that BT1-40 generated spontaneous mutants at a faster rate than the wild-type.

PRELIMINARY TAXONOMIC STUDIES ON 1,000 ISOLATES OF THERMOPHILIC BACILLI

D. WHITE[1], F.G. PRIEST[2], R.J. SHARP[1]

[1] Division of Biotechnology, PHLS Centre for Applied Microbiology & Research, Porton Down, Salisbury, Wiltshire, SP4 OJG, U.K.

[2] Department of Microbiology, Heriot-Watt University, Edinburgh, U.K.

Thermophilic Bacilli are an extremely heterogenous group of microorganisms whose DNA composition varies from 42-71% G+C. Only B. stearothermophilus, B.coagulans, B.acidocaldarius and B.schlegelii are included in the 10th edition of Bergey's Manual although it is apparent that there are many more species yet to be clearly described and characterised.

Approximately 1,000 strains of thermophilic bacilli, isolated from different geographical and geophysical environments have been characterised using forty-four character states. This preliminary classification will be used as the basis of a more extensive study of 300 strains. UPGMA analysis gave 14 clusters at the 84-90% level. Thirteen of these formed two large groupings at the 78% level of similarity with 7 and 6 clusters. Groupings containing reference strains were present in six of the groups. These groups held strains containing B. stearothermophilus, B. caldotenax, B.thermodenitrificans, B.licheniformis and B.pallidus. The position of the remaining 8 clusters remains to be determined. Representatives of all 14 clusters will be examined in a more extensive study to include a greater number of characteristics.

AMINO ACID UTILIZATION BY STRAINS OF THE GENUS THERMUS

G. HOLTOM[1], D. COSSAR[2], R. SHARP[2], R.A.D. WILLIAMS[1]
[1]Department of Biochemistry, London Hospital Medical College
Turner Street, London E1 2AD, U.K.

[2] Microbial Technology Laboratory, CAMR, Porton Down,
Salisbury, Wilts. SP4 ORJ

Strains of Thermus grow on tryptic casein hydrolysate (0.3%) yeast extract (0.1%) with mineral salts at 65-70°C. The partial protein hydrolysate and yeast extract may be replaced by an amino acid mixture and a vitamin cocktail, indicating no requirement for peptides. For most strains glutamate is the major amino acid utilized, and almost all are capable of growth on glutamate alone in the presence of vitamins and mineral salts.

Chemostat culture shows that for all strains tested the presence of leucine, isoleucine and valine is also required for good growth. In addition, one strain also required phenylalanine and another as yet undetermined complex mixture of other amino acids.

Washed whole cells of Thermus strains take up $U^{14}C$ glutamate rapidly at 65°C, one strain, incapable of growth on glutamate, appears to have a deficiency in the uptake mechanism. In the presence of tryptone, carbon from glutamate shows unrestricted incorporation into cellular protein, lipid and nucleic acid.

All strains tested had an NADH-dependent glutamate dehydrogenase (E.C. 1.4.1.3) as the main glutamate producing system. One strain however, that could not use glutamate as the sole growth substrate had a low activity, 5-15% that of other strains. Aminotransferase activities were detected in cell free extracts, including glutamate/oxaloacetate (E.C. 2.6.1.1) glutamate/pyruvate (E.C. 2.6.1.2) and glutamine/oxoglutarate (E.C. 1.4.1.3) (NADH dependent) aminotransferases.

A NEW SEQUENCE-SPECIFIC ENDONUCLEASE FROM THE GENUS THERMUS

N.D.H. RAVEN[1], R.MULLINGS[2], P.EASTLAKE[3], R.A.D. WILLIAMS[1]

[1]Department of Biochemistry, London Hospital Medical College
Turner Street, London E1 2AD , U.K.

[2]Department of Biochemistry, University of Bristol, U.K.

[3]Northumbria Biological Ltd, Cramlington, U.K.

In the course of an investigation into the taxonomy of Thermus by DNA:DNA hybridisation, large numbers of purified chromosomal DNA samples were prepared. It was considered that isolates whose DNA's were digested by the known restriction endonucleases from the genus Thermus (Taq I, Taq II, Taq XI, Tth 111 I and Tth 111 II) would be good candidates for an initial screen for new restriction endonucleases in Thermus. We describe the isolation and characterisation of a new type II restriction endonuclease, selected for study on this basis.

Thermus sp. strain A1 was found to possess a type II restriction enzyme, designated Tsp AI. The enzyme was separated from contaminating activities by a four-stage chromatographic procedure. Purified enzyme was then used to digest a number of DNA substrates of known sequences. From the fragment patterns generated and the DNA sequencing results we conclude that the recognition site of TspAI is
5' CA(C/G)TG NN 3'.

ISOLATION AND CHARACTERISATION OF THERMOPHILIC AEROBIC BACTERIA, PRODUCING THERMOSTABLE ENZYMES

M. KAMBOUROVA, E. EMANUILOVA
Institute of Microbiology, Bulgarian Academy of Sciences
1113 Sofia, Bulgaria

The aim of this work was to obtain thermophilic bacteria, producing thermostable extracellular hydrolases. A number of thermophilic bacterial strains were isolated from samples of soil, water and decaying organic materials, collected from different Bulgarian hot springs. Some of their properties were: hydrolysis of different substrates, growth characteristics as well as some morphological properties were investigated using lavsan nuclear filter technique (1). To determine thermostability of their amylase, pullulanase, protease and lipase, the strains were cultivated as described previously (2). Cells were then centrifuged and supernatants were treated at temperatures between 50-100°C for 10 or 20 minutes. The residual enzyme activities were compared with non-treated samples by an agar diffusion method. The following substrates were used : 0.5% insoluble starch, 15% skimmed milk, 1% Tween 80 and 1% pullulan. Forty-four thermophilic aerobic bacterial strains were isolated, of which thirty-one were extreme thermophiles. The Gram-positive bacilli predominated, and most had these enzyme activities. Thermophily did not always correlate with the heat stability of the enzymes studied. The proteolytic activity in the supernatants of five of the strains remained unaffected after 10 minutes treatment at 90°C. The same was true for the amylolytic activity of another four strains. Pullulanase activity in supernatants of two of the strains, and lipase activity of another two, remained after treatment at 70°C for 10 minutes.

The preliminary studies have shown that the enzyme producers isolated are promising for further more detailed investigations in order to obtain and characterise their thermostable enzymes.

REFERENCES

1. Kambourova, M. and Emanuilova, E. (1983) Comp.rend.de l'Acad.Bulg.Sci. 36: 655-658

2. Kambourova, M. and Emanuilova, E. (1985) Acta Microbiol.Bulg., 17: 3-9

METABOLISM OF CARBAMOYLPHOSPHATE IN EXTREME THERMOPHILES

M. VAN DE CASTEELE[1], C. LEGRAIN[2], N. GLANSDORFF[1,2]
A. PEIRARD[2,3].
[1] Microbiologie, Vrije Universiteit Brussel
[2] Research Institute of CERIA-COOVI, Brussels
[3] Microbiologie, Universite Libre de Bruxelles, Belgium

Life at high temperature might require mechanisms that would protect from decomposition thermolabile metabolic intermediates. In this respect the arginine and pyrimidine biosynthetic pathways appear interesting because they both depend on carbamoylphosphate, a highly thermolabile precursor.

Comparison of the enzymes involved in the synthesis and utilisation of carbamoylphosphate in extreme thermophilic bacteria with those of well characterised mesophilic bacteria may contribute to our understanding of molecular evolution. Two organisms have been considered: Thermus aquaticus, an extreme thermophilic eubacterium and Sulfolobus solfataricus, an extreme thermoacidophilic archaebacterium.

The carbamoyltransferases of T.aquaticus are similar to their mesophilic counterparts as far as their sizes, kinetics and regulatory properties are concerned, but are quite different as regards their thermal stability. The ornithine carbamoyltransferase of T.aquaticus has been purified. The enzyme consists of a trimer of identical subunits, as in the mesophilic microorganisms analysed so far. This enzyme is reversibly inactivated when submitted to a "cryoscopic shock", suggesting the involvement of hydrophobic interactions in its thermostabilisation.

The aspartate- and ornithine carbamoyltransferases of Sulfolobus are highly thermostable enzymes, apparently devoid of regulatory properties. Their molecular weights are considerably higher than those of Thermus or the mesophilic transferases. Ornithine carbamoyltransferase activity can be resolved into isoenzymes by ion exchange chromatography. These isoenzymes are built up by the association of two types of subunits(MW 38,000 and 28,000). A subunit of MW 38,000 is also found in the ACTase complex. It is presently being investigated whether the same subunit is common to the two enzyme and whether it plays a role in making the synthesis and the utilisation of carbamoylphosphate compatible with high temperatures.

CELLULOLYTIC ANAEROBIC BACTERIA FROM HOT SPRINGS

B.E. JONES
Royal Gist-brocades N.V.,
P.O. Box 1, 2600 MA Delft, The Netherlands

Samples of mud and water from hot springs in Italy (Naples, Tuscany), Iceland, Java, U.S.A. (Colorado, California) were examined for thermophilic, anaerobic, cellulolytic bacteria. An attempt was made to correlate the isolates with the type of spring from which they were isolated and compare them with strains from unheated soils collected randomly world-wide. No cellulolytic anaerobes were isolated from highly acidic springs (solfataras).

The strains isolated at neutral pH were divided into 2 groups on the basis of thermophily :-

(a) moderate thermophiles with optimum temperature 50°-55°C
(b) extreme thermophiles with optimum temperature 70°-75°C

These were compared on the basis of their fermentation products from cellulose.

The moderate thermophiles were all ethanologenic with either a mixed acid (formic, lactic and/or acetic) or an acetogenic fermentation pattern. The extreme thermophiles were either acidigens (predominantly lactic or acetic acid) or had an ethanologenic mixed acid (lactic and/or acetic, no formic) fermentation pattern.

The extreme thermophilic ethanologens tended to be derived from the cooler springs.

THERMOPHILIC METHYLOTROPHIC BACILLI: POSSIBLE BIOTECHNOLOGICAL POTENTIAL

A.G. BROOKE, M.M. ATTWOOD, D.W. TEMPEST
Department of Microbiology, University of Sheffield,
Sheffield, S10 2TN, U.K.

We have isolated a number of thermotolerant spore-forming methylotrophic bacilli in pure cultures (1). These novel methanol utilisers may be of considerable commercial interest. However, in order to assess their biotechnological potential a detailed knowledge of the regulation of metabolic fluxes from one carbon substrates is needed.

Using methanol-limited chemostat cultures of strain Ts1 under conditions of optimal biomass production, the organism displayed a low maintenance energy requirement of 0.81 mmol methanol h^{-1} (g cells)$^{-1}$ and growth yield of 17.0 g cells (mol methanol)$^{-1}$ which is amongst the highest so far reported for methylotrophic bacteria (2). The culture was also able to accelerate the rate of methanol consumption substantially and instantaneously upon relief of the growth limitation (i.e. following a methanol pulse) without any concomitant change in growth rate. Hence Ts1 is able, at least transiently, to extensively dissociate methanol catabolism from anabolism. This raised the question as to how this strain would respond to steady-state cultural conditions in which the rate of anabolism is constrained by the availability of an anabolic substrate, whilst the catabolic substrate (methanol) remains in excess.

K^+ limited chemostat cultures resulted in substantial increases in q methanol, 12.1-51.0; qO_2, 9.45-31.3 and qCO_2, 2.97-15.6 mmol.h^{-1} (g dry wt.)$^{-1}$. Only 42.5% of the input methanol was recovered as biomass, CO_2 or residual methanol. However, when taking into account TOC analysis of culture supernatants, >90% carbon recoveries were obtained.

Thus a significant proportion of methanol-carbon was converted into overflow metabolites which were excreted into the culture medium. Preliminary data from GC/mass spectroscopy suggest 3 products. Studies are currently underway to identify these. Similar responses were seen under N, Mg, PO_4 and SO_4-limited growth.

REFERENCES

1. Dijkhuizen, L., Arfman, N., Attwood, M.M., Brooke, A.G., Harder, W. and Watling, E.M. (1988) FEMS LEtts. 52, 205-208

2. Brooke, A.G., Watling, E.M., Attwood, M.M. and Tempest, D.W. (1988) Arch Microbiol. (Submitted for publication)

METHANOL OXIDISING SYSTEM OF THERMOPHILIC <u>BACILLUS</u>

A. NETRUSOV [1], M. GUETTLER [2], R.S. HANSON [2]
[2] Gray Freshwater Biological Institute, University of Minnesota,
P.O Box 100, Navarre, MN 55392, U.S.A.

[1] Microbiology Department, Moscow University,
Moscow, 119899, USSR

A gram-positive endospore-forming rod (MGA3) that grows rapidly in a minimal salts medium with methanol was isolated. Growth was optimal at 50-53°C with a doubling time of 84 min in batch culture. This bacterium requires vitamin B_{12} for growth and pH 7.2 is optimal. Of 19 substrates tested, methanol and mannitol were preferred. Strain MGA3 forms endospores when cultures are grown at 50°C and then switched to 37°C, but does not sporulate at 50°C. The spores are oval and subterminal in swollen sporangia; dipicolinic acid (DPA) can be readily detected in spore containing cultures. Cell extracts had a high hexulose phosphate synthase activity, but hydroxypyruvate reductase activity was not detected; this suggests that methyl carbon is assimilated by the ruMP cycle of formaldehyde fixation rather than the serine pathway. The G+C% content in DNA was 44 mol% as determined by the thermal melting point method. The bacterium oxidises methanol by soluble NAD^+-dependent alcohol dehydrogenase, which has a pH optimum at 9.0, temperature optimum at 65°C and can oxidise methanol, ethanol and propanol, but not formaldehyde. This is the first NAD^+-dependent bacterial methanol dehydrogenase described.

Mn2+ RECOVERY FROM A LOW GRADE ORE BY MICROORGANISMS

B. PAPONETTI[1], L. TORO[1], C. ABBRUZZESE[2], A. MARABINI[2]
M.Y. DUARTE[3]

[1] Department of Chemistry, Chemical Engineering and Materials
University of L'Aquila, L'Aquila, Italy

[2] Institute of Mineral Processing, (C.N.R.), Rome, Italy

[3] Facultad de Ciencias Exactas, Naturales y Agrimensura
Universidad del Nordeste, Corrientes, Argentina

Manganese is a metal of particular industrial importance in the production of steel(s) and alloy(s). The traditional processes for extracting manganese from minerals containing manganese dioxide make use of reducing agents such as sulphur dioxide which can become a dangerous pollutant in high concentrations. The recovery of manganese by bioleaching of different microorganisms can be an interesting alternative process for the future. In the present paper a laboratory scale bioleaching process for extracting manganese from low grade domestic resources is described.

Strains of microorganisms with improved technical properties emerged after subculturing for two years in batches with increasing contents of manganese dioxide pulp. In particular autotrophic species as Thiobacillus ferrooxidans, mixotrophic species and heterotrophic species as Aspergillus niger were used. Mutants of Thiobacillus ferrooxidans cultivated on sulphides in the presence of manganese dioxide produce sulphur dioxide as an intermediate during the oxidation of the sulphide to sulphate. The recovery of manganese catalysed by Aspergillus niger follows the production of reducing agents such as oxalic and other organic acids. The dissolution of manganese catalysed by the presence of different microorganisms was detected in the liquors by means of an atomic absorption spectrophotometer. The same analyses were carried out with the samples pretreated with 4% SDS and 0.1% HCl to detect the manganese associated with the cells of the microorganisms, and that reprecipitated in form of Mn(OH)2, respectively. Using polarography, it was possible to determine the oxidation state of the manganese in the liquors. The consumption of the ore and the growth of the microorganisms was detected by scanning electron microscopy. The effect of the biotreatment on the elemental chemical composition of the ore was investigated using I.R. spectrophotometry, Petri-plates and microanalyser connected to the S.E.M. Research in progress is evaluating the effect on the solubilisation of manganese dioxide of several reducing agents, both organic acids and enzymatic proteins present in different microorganisms.

ACIDOPHILIC BACTERIA

M.E. SIMAS MARQUES [1], M.L. QUINTA[1], C.M. RANGEL[2]
[1] Laboratorio de Microbiologia Industrial, LNETI
[2] Corrosao e Proteccao Materials, LNETI
Azinhaga dos Lameiros, 1699, Lisboa Codex, Portugal

The scope of this paper concerns acidophilic bacteria growth on metal and alloys and iron and sulfur rich ecologic niches. Acidophilic mesophilic aerobic autotrophic or heterotrophic gram-negative bacilli have been isolated from numerous of these niches and their isolation and identification is of great interest to biomining and biocorrosion. This study reports on the acidophilic bacteria isolated on solid media from 70.30 cupro-nickel tubing used in cooling water heat exchangers (4). The tubes are subjected to a treatment that deposits a protective iron based film using ferrous sulphate injection. The coating is non-adherent mostly lepidocrocite (γ-FeOOH). The recent introduction of solid media with agarose (3) has allowed detailed studies of metabolism of some established cases considered as mixotrophy. Evidence of mixed cultures that include autotrophic and heterotrophic microorganisms was obtained. Two types of acidophilic bacteria were observed in 9K BS medium. The conducted tests revealed the presence of :-

1. <u>Thiobacillus ferrooxidans</u> (T) small, gram-negative bacilli, oxidising iron and sulphur with typical colonies being observed in solid media with agarose.
2. <u>Heterotrophic</u> (C) gram-negative bacilli, growing in media of pH 3-6.5, having carbon as an organic source with better growth being observed in glucose supplemented with yeast extract.

Cultural media used were: 9K BL (5); 9K BL + 0.5% agarose NA; Acidi (2); Acid:Acidi without glucose; TSB (OXOID CM 129) The direct cause-effect of the presence of these bacteria in the Cu-Ni tubing corrosion cannot be established at the present but the acidogenic character of the bacteria might be the basis of the localised attack suffered by the alloy(1) Further studies are needed to identify the heterotrophic bacilli.

REFERENCES
1. Clarke, C.K. (1988) <u>Corrosion</u>, 88: St Louis, Missouri
2. Harrison, A.P. Jr. (1981) <u>Int.J.Syst.Bacteriol.</u>, 31: 327-332
3. Harrison, A.P. Jr. (1984) <u>Ann.Rev.Microbiol.</u>, 38: 265-292
4. Rangel, C.M., Quinta, M.L., Simas, M.E., Sousa, F.(1988) 1st EFC Workshop on Microbial Corrosion, Sintra, Portugal
5. Silverman, M.P., Lundgren, D.G. (1959). <u>J.Bacteriol.</u> 77: 642-647

A NEW SOLID MEDIUM FOR THE ISOLATION AND PHENOTYPIC CHARACTERISATION OF THIOBACILLUS FERROOXIDANS

E. BIANCHI, P. VALENTI, P. VISCA, N, ORSI
Istituto di Microbiologia, Universita di Roma "La Sapienza"
00100 Roma, Italy

The growth of seven Thiobacillus ferrooxidans strains was tested on solid media prepared according to published procedures and a series of new formulations (TSMs, Thiobacillus solid media). These media differed in the gelling agent, and in the concentration of phosphate and ferrous ions. It was determined that established formulations provided qualitatively and quantitatively unsatisfactory results whereas, among the new media, TSM1 produced quantitative yields of T.ferrooxidans colonies. TSM1 can be readily prepared by combining three separately sterilised solutions, each at 60°C These contain the following components (g./L final concentrations). Solution A: $(NH_4)_2SO_4$, 3.0; $MgSO_4.7H_2O$, 0.5; KCl, 0.1; K_2HPO_4, 0.05; $Ca(NO_3)_2$ $4H_2O$, 0.015; all dissolved in 600ml distilled water acidified to pH 2.5 with H_2SO_4 and autoclaved at 121°C for 15 min. Solution B: $FeSO_4.7H_2O$, 22.0 in 150ml distilled water acidified to pH 2.5 and filter-sterilised. Solution C: Agarose (Biorad, high Mr 162-0001), 5.0; in 250ml distilled water at pH 7.0 sterilised at 121°C for 15 min.

The TSMI is recommended for the titration of cell numbers of T.ferrooxidans in liquid cultures and for the isolation of single clones. The morphology of colonies formed by different T.ferrooxidans strains was similar. We have also successfully used TSMI for the evaluation of the minimal inhibitory concentrations of several heavy metals for T.ferrooxidans, to determine selective markers able to discriminate between genetically distinct strains. A solid medium suitable for the growth of T.ferrooxidans is a fundamental advance in the standardisation of the biological procedures required for a better phenotypic and genetic characterisation of this chemolithotrophic acidophile.

PURIFICATION, PROPERTIES AND CLONING IN E. COLI OF THE STRUCTURAL GENE OF THE S-LAYER COMPONENT FROM THERMUS THERMOPHILUS

J. BERENGUER, M.L. FARALDO, J.R. CASTON, M.A. DE PEDRO
Centro de Biologia Molecular CSIC-UAM
Campus de Cantoblanco, 28049 Madrid, Spain

Thermus thermophilus has on its surface a crystalline layer of which the main component is a 100 kd protein called by us P100. We have purified this protein by exchange chromatography after digestion of whole membranes with lysozyme and solubilisation in neutral detergents. The purified protein is self-associated, and forms trimers able to resist high temperatures and strong detergents without disaggregation. As in native membranes, the stability is dependent on the presence of Ca^{2+}, being able to bind 12 molecules per mol of protein with an apparent affinity of 50×10^{-6} M, as shown by Scatchard plot with radioactive calcium. The aminoacid composition is similar to other S-layer subunits, especially that of Deinococcus radiodurans, with one half of the protein composed by hydrophobic aminoacids, a quarter of acidic, amino acids and the virtual absence of cysteine, proline and tryptophan.

We have also cloned the protein from a genetic library constructed in pUC9/E.coli using antiserum and monospecific antibodies as screening system. The complete gene was detected in a 5.8 kb fragment, and expressed in E.coli lonA mutants from pLac and from its own promotor. Nevertheless, the level of expression, even from pLac, is relatively low, probably because of a down-promotor region with strong secondary structure, as suggested by the preliminary sequence of this region.

MALATE DEHYDROGENASE FROM CHLOROFLEXUS AURANTIACUS

A.K. ROLSTAD, E. HOWLAND, B. SYNSTAD, R. SIREVÅG
Department of Biology, Division of Molecular Cell Biology,
University of Oslo, Norway

Malate dehydrogenase (MDH) from the thermophilic, green non-sulfur bacterium Chloroflexus aurantiacus was purified by a two-step procedure involving affinity chromatography and gel filtration. The enzyme consists of identical subunits with a molecular weight of approximately 35,000 which in its active form at 55°C form tetramers. At lower temperatures, inactive dimers and trimers exist. Antibodies against the purified enzyme were produced and immunotitration and ELISA-tests showed immunochemical homology between MDH from C. aurantiacus and MDH from several other bacteria. The amino acid composition was similar to that of other MDHs. The N-terminal amino acid sequence was enriched with hydrophobic amino acids, which showed a high degree of functional similarity to amino acids in the N-terminal end of both E.coli and Thermus MDH. A genomic library was prepared from Chloroflexus, using lambda replacement vector EMBL-3.

The library was screened with a synthetic oligonucleotide (15 bases) deduced from part of the N-terminal amino acid sequence. Of the 4,000 plaques screened, 12 possible clones were obtained. These clones will be characterised by restriction endonuclease digestion. The nucleic acid sequence will be determined by the Sanger method.

REFERENCE

1. Rolstad, A.K., Howland, E. and Sirevag, R. (1988) J. Bacteriol. $\underline{170}$: 2947-2953

STABILITY OF THERMOPHILIC PLASMID DNA IN HETEROLOGOUS HOSTS

D.J. HARDMAN, M.F. TUITE
The Institute for Biotechnological Studies,
University of Kent, Canterbury, Kent CT2 7PD, U.K.

A series of recombinant plasmids have been constructed to investigate the stability of plasmids containing DNA from thermophilic bacterium Bacillus stearothermophilus, in heterologous hosts. A 3Kb cryptic plasmid (pMT400) from B.stearothermophilus HR17 was cloned into the Escherichia coli vectors pBR325, pBR328 and pAT153 and into the E.coli - Saccharomyces cerevisiae shuttle vector pMA3a and transformed into the heterologous hosts. Subsequent isolation and analysis of the plasmid DNA from the E.coli transformants demonstrated structural rearrangement of the thermophilic DNA in all the chimeric plasmids (pUKC1, pUKC2, pUKC4 and pUKC300). These rearrangements occurred at the time of construction or transformation and for pUKC1, 2 and 300, which represented pMT400 cloned into the SalI sites of pBR325, pBR328 and pMA3a respectively, the same deletion/rearrangement occurred whereas when the chimer pMT400::pAT153 was transformed into E.coli the structural rearrangement of the thermophilic DNA was considerably greater.

The chimeric plasmids pUKC1 and 2 could be used to retransform E.coli cells, and stability studies in shake-flask and continuous culture under selective conditions over 120 generations indicated a high degree of structural stability. Under non-selective conditions the chimeric plasmids demonstrated segregational instability which was seen to be associated with a depressed copy number.

Under non- or slow-growth conditions further deletion events were observed in pUKC1 and 2 which resulted in the complete loss of the thermophilic moiety from pUKC1 and loss of all but 200bp in pUKC2.

When transformed into S.cerevisiae, pUKC300 showed both structural and segregational stability over 75 generations of non-selective growth, and retained the copy number of the parent plasmid pMA3a (approximately 150 per cell) under selective growth.

THERMOTROPIC BEHAVIOUR OF THE POLAR LIPIDS OF B.STEAROTHERMOPHILUS

A.S. JURADO, M.S. COSTA, V.M.C. MADEIRA
Centro de Biologia Celular, Departamento de Zoologia,
Universidade de Coimbra, 3049, Coimbra Codex, Portugal

B.stearothermophilus grows over the temperature range of 45 to 70°C in a complex medium containing 115µM Ca^{2+} and 95µM Mg^{2+}. Addition of Ca^{2+} (2.5 to 10mM) stimulates growth at sub and supraoptimal temperatures extending the range of the maximal temperature for growth (1). Phospholipids of B.stearothermophilus were identified as diphosphatidylglycerol, phosphatidylglycerol, phosphatidylethanolamine, two aminoacyl phospholipids and a phosphoglycolipid. The most striking change with temperature is the increase of the phosphoglycolipid concentration, barely detectable at 45°C and reaching about 12% of the total phospholipid at 71°C (2). The phosphoglycolipid may be involved on the regulation of membrane thermostability, providing a putative site for divalent cation interaction, with subsequent membrane stabilisation at high temperatures. Membrane lipid fatty acids also show some significant changes with temperature, including the ratio of straight to branched fatty acyl chains and the amount of unsaturated fatty acids (3). To study the thermotropic behaviour of the polar lipids of B.stearothermophilus, we are using a fluorescent probe for fluidity, 1,3-di(2-pyrenyl) propane [2 Py(3)2 Py] able to form intramolecular excimers. Preliminary studies indicate that this probe locates well inside the membranes and the excimerisation rate, evaluated as the excimer to monomer fluorescence intensity ratio (I'/I), is very sensitive to lipid phase transitions (4). Liposomes were prepared with the membrane polar lipids of cultures of B.stearothermophilus grown at different temperatures. Thermograms were obtained in the range of zero to 70°C. At all temperatures of growth there is a broad phase transition beginning at about 12°C. The thermal profiles obtained with cultures grown in the basal medium at 55 and 69°C (i.e. below and above the optimal temperature range, respectively) are very similar. The addition of Ca^{2+} to the growth medium of cultures grown at 55°C, increases significantly the fluidity of the polar lipids as revealed by a shift of the thermal profile to higher I'/I ratios over the entire temperature range. However, Ca^{2+} has no apparent effect on the thermotropic behaviour of the polar lipids of cultures grown at 69°C. These observations suggest a direct effect of Ca^{2+} on the thermostabilisation of the membrane at high temperatures. However, at low temperatures, Ca^{2+} may induce alterations on the lipid composition leading to growth improvement.

REFERENCES

1. Jurado, A.S., Santana, A.C., da Costa, M.S., and Madeira, V.M.C. (1987) <u>J.Gen.Microbiol.</u>, 133: 507-513
2. Jurado, A.S., Santana, A.C., da Costa, M.S. and Madeira, V.M.C. (1987) Symposium on Membrane Lipids, Sintra
3. Ornelas Soares, A., Jurado, A.S. and Madeira, V.M.C. (1988) III Congresso Luso-Espanhol de Bioquimica. Santiago de Compostela.
4. Jurado, A.S., Almeida, L.M., Madeira, V.M.C. (1988) III Congresso Luso-Espanhol de Bioquimica. Santiago de Compostela.

ACKNOWLEDGEMENTS

This work was supported by INIC

BACILLUS SCHLEGELII AND HYDROGENOBACTER SPP. COMPARED ECOLOGY AND TAXONOMY OF TWO THERMOPHILIC, HYDROGEN AND/OR SULFUR OXIDISING AEROBES FROM GEOTHERMAL AREAS

M. ARAGNO, F. BONJOUR
Laboratoire de Microbiologie, Universite, P.O. Box 2,
CH-2007 Neuchatel, Switzerland

Two neutrophilic, aerobic hydrogen oxidising bacteria growing at similar temperatures, Bacillus schlegelii and Hydrogenobacter spp. show different ecological and evolutionary behaviours. B.schlegelii (1) is a spore-forming, aerobic, facultative autotroph utilising molecular hydrogen and/or thiosulfate as electron and energy sources. It grows optimally at 70°C, with a maximum above 75°C. Originally isolated from cold environments (lake sediment, glacier ice), it was later found in air samples in a non-geothermal area, and finally in geothermal areas: Italy, Iceland (Geysir and Heimaey's Eldfell vulcano)(2), and Antarctica (Erebus vulcano)(3). All the strains studied from any of these environments showed a high DNA:DNA homology (>75%), so that the species appears homogenous. The genus Hydrogenobacter (4) comprises non-sporeforming, obligately chemolithoautotrophs utilising hydrogen and/or thiosulfate as electron and energy sources (5). They grow optimally at 70-75°C, with a maximum above 80-85°C. They were only isolated from warm, geothermal environments. Although all Hydrogenobacter are phenotypically very similar, we found, amoung 18 strains isolated from four distant locations (Iceland, Italy, Kamchatka, Japan), five clear-cut homology groups, with high intra-group homologies (>80%) and low intergroup homologies (>20%). These groups coincided with the geographical origin of the strains, with the exception of Italy, where two groups were found. The latter differed, however, by their maximal temperature (80, resp. 86°C). In geothermal environments, the sporeformers were generally found in ponds with strongly variable temperature and water level (6). By contrast, we mainly found Hydrogenobacter in hot springs and ponds with fairly stable parameters.

The genotypic homogeneity of the species B.schlegelii and its allochtonous occurrence in cold habitats can be explained by the survival of airborne spores allowing transport in any location, and thus either competition or genetical exchange. This is not the case with Hydrogenobacter, whose potential habitats are so distant from one another that there is practically no chance for this non-sporeformer to mix genetically and for individuals from different geographical areas to enter in competition: such geographical isolation would then lead to genetical divergence, whereas the phenotypical properties would be kept by the extreme environmental pressure of their habitat.

REFERENCES

1. Schenk, A., Aragno, M. (1979) J.Gen.Microbiol. 115: 333-341
2. Bonjour, F., Graber, A., Aragno, M. (1988) Microb.Ecol. (in press).
3. Hudson, A., Daniel, R.M., Morgan, H.W. (1988) FEMS Microbiol.Lett. 51: 57-60
4. Kawasumi, J.K., Igarashi, Y., Kodama, T., Minoda Y. (1984) Int.J.Syst.Bacteriol. 34: 5-10
5. Bonjour, F., Aragno, M. (1986) FEMS Microbiol.Lett. 35: 11-15
6. Bonjour, F., Aragno, M. (1984) Arch.Microbiol. 139: 397-401

COLD ADAPTATION OF BIOPOLYMER-DEGRADING BACTERIA FROM PERMANENTLY COLD ENVIRONMENTS IS THERE A BIOTECHNOLOGICAL POTENTIAL?

W. REICHARDT
Institut fur Meereskunde an der Universitat Kiel
D-2300 Kiel, W. Germany

Psychrophilic (1) protein- and polysaccharide-degrading bacteria with growth temperature optima between 4° and 12°C are most abundant in Antarctic marine sediments. Over 600 aerobic psychrophilic isolates from fauna-rich, bioturbated environments of the Antarctic sea floor (100 - 2500 m water depth) were screened for extremely cold-adapted enzyme activities involved in the degradation of detrital biopolymers (2). Conversion of glucose to CO_2 showed temperature optima in the range of the growth temperature maxima of the isolates (around 20°C). A few hydrolase activities also revealed relatively low temperature optima with 25°C and 28°C found for chitobiase and alkaline phosphatase, respectively, from several of the psychrophilic isolates. These hydrolases could become most easily available for potential applications in biotechnological processes that require unusually low temperatures.

On the other hand, certain hydrolases involved in the extracellular depolymerisation of scleroprotein and structural polysaccharides (chitin) failed to show a mode of cold adaptation at the activity level that would be comparable to that of the above mentioned hydrolases (3). Despite extreme cold adaptation of growth, extracellular degradation of particulate biopolymers was cold adapted mainly at the level of enzyme production. Furthermore, chitobiase as a largely cell-bound hydrolase gives an example for the existence of dual strategies of cold adaptation, viz, low temperature optima at the activity level as well as enhanced enzyme production at temperatures close to growth temperature optima (between 5° and 10°C). Finally, hydrolytic enzymes with extremely low temperature optima (20-25°C) are not only produced by psychrophiles, but also by members of the widespread group of psychrotrophs (1) as shown for a fresh water isolate of <u>Cytophaga johnsonae</u>.

REFERENCES

1. Morita, R.Y. (1975) <u>Bact.Revs.</u> 39: 144-167

2. Reichardt, W. (1988) <u>Microb.Ecol.</u> 15: 311-321

3. Reichardt, W. (1988) <u>Mar.Ecol.Prog.Ser.</u> 40: 127-135

MICROBIAL PROTEIN THERMOSTABILITY CORRELATES DIRECTLY WITH RESISTANCE TO DENATURATION IN BIPHASIC ORGANIC:AQUEOUS SOLVENTS

R.K. OWUSU, D.A. COWAN
Department of Biochemistry,
University College London, Gower Street,
London WC1E 6BT, U.K.

The averaged thermostabilities of crude protein extracts from a range of mesophilic, thermophilic and extremely thermophilic microorganisms (growing over a temperature range of 25°C to 90°C) show a strong positive correlation with resistance to denaturation in a range of biphasic organic: aqueous solvent systems. The correlation was also observed when similar experiments were carried out using a range of purified proteolytic enzymes.

Protein stability was greater in biphasic systems where the organic component showed lowest aqueous miscibility (high log P). Where solvents of similar log P were used, stability correlated inversely with interfacial tension. Organic solvent stability was further reduced by the presence of gas-liquid interfaces with certain organic solvents, but unaffected with others.

These findings suggest that thermostable enzymes from extremely thermophilic organisms should be applicable in many types of organic solvent biocatalysis.

PARTIAL PURIFICATION OF INTRACELLULAR ENZYMES BY HEAT-PRECIPITATION OF CELL DEBRIS: A MODEL STUDY

Y. TAKESAWA[1], D. COWAN, M. HOARE[2], J. BONNERJEA[2]
Department of Biochemistry, University College London,
Gower Street, London WC1E 6BT

[1]Shell Research Ltd., 4052-2 Nakatsu Aikawacho,
Aikougan Kanagawa, Japan

[2]Department of Chemical and Biochemical Engineering,
University College London, Torrington Place, London WC1E 7JE

Heat denaturation has been used successfully as a means of partially purifying thermostable proteins, whether intrinsic to the source organism or cloned from thermophiles into mesophilic hosts. We have studied aspects of this process by exogenous addition to a yeast cell homogenate of a partially purified thermophilic β-galactosidase.

Purification of a soluble intracellular enzyme requires the initial removal of insoluble debris (cell wall fragments, membranes, organelles, etc.) and soluble macromolecules. While much of the former can be removed by a low-speed centrifugation step, the remaining small-particle fraction requires high-speed centrifugation. This is time-consuming and expensive for large-scale processing.

Incubation at temperatures above 60°C yielded large aggregates of cell debris and denatured protein. A low-speed centrifugation yielded a supernatant of greater than 95% clarity. The rate of aggregate formation was rapid, more than 85% being formed during the first 10 minutes of the incubation at 70°C. After incubation at 90°C for 0.5h, over 90% of the soluble protein fraction was removed. At temperatures below 100°C, the activity of an exogenous T.aquaticus β-galactosidase preparation was retained in the supernatant and purification factors of 35 to 50-fold were obtained.

INVESTIGATION OF THE ENZYMES RELATED TO THE ENERGY METABOLISM OF SULFOLOBUS ACIDOCALDARIUS, A THERMOACIDOPHILIC ARCHAEBACTERIUM.

T. WAKAGI, T. YAMAUCHI, H. WAKAO, H. EGUCHI, T. OSHIMA
Department of Life Sciences, Tokyo Institute of Technology, Nagatsuta, Yokohama 227, Japan

Sulfolobus acidocaldarius grows optimally at pH 2-3, 75°C. The organism has a novel pathway of energy metabolism, such as glycolysis (like Entner-Doudoroff type) and oxidative phosphorylation (a strange respiratory chain and a H-ATP synthase different from FoFl). In the present report we have examined [1] the components of respiratory chain and [2] isocitrate dehydrogenase of the organism.

[1] The components of respiratory chain.
Sulfolobus membranes consumed oxygen upon addition of succinate or NADH. Purified NADH dehydrogenase was a dimeric flavoprotein with an Mr of 95,000. The enzyme reacted with DCIP, ferricyanide, benzoquinone, naphthoquinone and "Caldariellaquinone", a major benzothiophenquinone found in Sulfolobus, as an electron acceptor. Cytochrome c is lacking in the organism, while a type a cytochrome bound to membranes appeared to oxidise cytochrome c from other sources. Cytochrome a was solubilised with sucrose-monocaprate, and separated by ion-exchange chromatography. Two heme-a-containing fractions (A and B) were separated. Fraction A and B appeared to correspond to cytochrome aa_3 and $a1$ (and/or b. with contamination of aa_3), respectively. Both showed oxidase activity sensitive to cyanide and azide. These results suggest that the respiratory chain of Sulfolobus is composed at least of NADH dehydrogenase, caldariellaquinone and cytochrome aa_3.

[2] Isocitrate dehydrogenase
Isocitrate dehydrogenase (IDH) was similar to dimeric NADP-dependent IDHs from other bacteria, except for several points such as remarkable stability, amino acid composition and pI. Higher content of polar amino acid and lower cysteine content were observed. The V_{max}/K_m value for NADP was about 280-fold larger than that for NAD.

PURIFICATION AND CHARACTERISATION OF THE CO DEHYDROGENASE FROM METHANOTHRIX SOEHNGENII

M.S.M. JETTEN, A.J.M. STAMS, A.J.B. ZEHNDER
Department of Microbiology, Agricultural University Wageningen,
H.van Suchtelenweg 4, NL 6703 CT, Wageningen,
The Netherlands

Acetate is quantitatively the most important intermediate in the anaerobic degradation of biopolymers under methanogenic conditions; seventy per cent of the methane is derived from acetate (1,2). The most abundant acetate-cleaving methanogenic bacterium in applied methanogenic systems is Methanothrix, an organism which can only use acetate as an energy substrate (3). Carbon monoxide dehydrogenase plays an important role in the methanogenic cleavage of acetate. The enzyme, which was found to be present in activities of 5 U/mg in cell extracts, was purified to homogeneity (4). In contrast with the carbon monoxide dehydrogenase from most other anaerobic bacteria, the purified enzyme from Methanothrix soehngenii was remarkably stable towards oxygen and it was only slightly inhibited by cyanide (4,5).

The native molecular weight of carbon monoxide dehydrogenase determined by gel filtration was 190,000. The enzyme is composed of subunits with molecular weights of 77,400 and 19,400 in a $\alpha_2 \beta_2$ oligomeric structure. The enzyme from Methanothrix soehngenii contains 1.8 g-atoms of Ni and 18.7 g-atoms of Fe and it constitutes 4% of the soluble cell protein. Analysis of enzyme kinetic properties revealed an apparent Km of 0.7 mM for CO and 65µM for methylviologen. At the optimum pH of 9.0 the Vmax was 140 U/mg. A high degree of thermostability was found.

REFERENCES

1. Smith, P.H. and Mah, R.A. (1966) Appl.Microbiol. 14: 368-371
2. Gujer, W. and Zehnder, A.J.B. (1983) Wat.Sci.Tech. 15: 127-167
3. Dubourguier, H.C., Prensier, G., Semain, E. and Albagnac, G. (1985) in Energy from biomass, (Palz, W., Coombs, J. and Hall, D.O., eds.) pp542-551, Elsevier Applied Science Publishers, Amsterdam
4. Kohler, H.P.E. and Zehnder, A.J.B. (1984) FEMS Microbiol.Lett. 21: 287-292
5. Diekert, G.B., Fuchs, G. and THauer, R.K. (1985) In Microbial gas metabolism, mechanistic, metabolic and biotechnological aspects (Poole, R.K. and Daw, C.S. eds) pp 115-130, Society for General Microbiology

GROWTH OF A THERMOPHILIC BUTYRATE-DEGRADING BACTERIUM IN AXENIC CULTURE WITH BUTYRATE AS SOLE CARBON AND ENERGY SOURCE

B.K. AHRING,　P. WESTERMANN,　R.A. MAH
School of Public Health,
University of California, Los Angeles, U.S.A.

A thermophilic butyrate-degrading bacterium was isolated together with <u>Methanobacterium thermoautotrophicum</u> and the physiology of this coculture has previously been described. By means of a palladium-catalyst it was possible to cultivate the butyrate-degrading bacterium in axenic culture under simultaneous reduction of ethylene to ethane.

In the poster presentation we report on the growth characteristics of the thermophilic butyrate-degrader when grown in axenic culture with butyrate as sole carbon and energy source.

PROPERTIES OF ENZYMES AND ELECTRON CARRIERS ISOLATED FROM THE THERMOPHILIC SULFATE-REDUCING BACTERIUM DESULFOVIBRIO THERMOPHILUS

G. FAUQUE [1], M. CZECHOWSKI [2], A.R. LINO [3], Y. BERLIER [1], D.V. DERVARTANIAN [2], I. MOURA [3], P.A. LESPINAT [1], L. KANG [2], J. LAMPREIA [3], A.V. XAVIER [3], H.D. PECK Jr [2], J.J.G. MOURA [3], J. LEGALL [1,2].

[1]A.R.B.S., C.E.N. Cadarache, 13108 Saint-Paul-lez-Durance Cedex, France

[2]Department of Biochemistry, University of Georgia, Athens, GA 30602, U.S.A.

[3]Centro de Quimica Estrutural, Complexo 1, 1096 Lisboa, Portugal

Desulfovibrio thermophilus DSM 1276 is a non-spore-forming Gram-negative sulfate-reducing bacterium isolated in a stratal water of a petroleum deposit in the Caspian Sea at a temperature of 84°C. The optimum and maximum temperatures for growth of this microorganism are 65 and 85°C respectively. We have isolated and characterised enzymes and electron carriers involved in the metabolism of hydrogen and sulfur by this sulfate-reducer.

A nickel-three iron-containing hydrogenase has been partially purified. This protein, with a low specific activity both in the hydrogen production and the proton-deuterium exchange reaction is the first hydrogenase so far isolated from a thermophilic sulfate-reducer. The adenylyl sulfate (APS) reductase, a non-heme iron flavoprotein, is membrane-bound in contrast to most of the APS and sulfite reductases from Desulfovibrio which are cytoplasmic. The dissimilatory sulfite reductase, a tetrameric enzyme with a molecular weight of 175 kDa, presents properties similar to those of desulfofuscidin found in Thermodesulfobacterium commune. A protein containing molybdenum and cobalt has also been purified but its physiological function is still unknown.

Two electron carriers have also been isolated and characterised from D.thermophilus: a tetrahemic cytochrome C3 with a molecular weight of 13 kDa and a three-iron-containing ferredoxin with an absorbance ratio A 397 nm/A 277 nm equal to 0.73 and a molecular mass of around 16 kDa.

BIOSYNTHESIS OF ACETATE FROM METHANOL BY METHANO-SARCINA BARKERI AS MONITORED BY IN VIVO ^{13}C NMR

[1]H. SANTOS, [1]P. FARELEIRA, [2]R. TOCI, [2]J. LE GALL, [1]A.V. XAVIER

[1]Centro de Quimica Estrutural, Av Rovisco Pais, 1096, Lisbon, Portugal
[2]Section d'Enzymologie et Biochimie Bacterienne, ARBS, CEN Cadarache, 13108 St Paul-lez-Durance Cedex France

Methanogens are strict anaerobic organisms that possess unique biochemistry. The elucidation of the mechanism of methanogenesis has been the aim of intense research studies (1). However, often the in vitro approach has been preferred and a direct connection between the in vivo and the in vitro mechanisms remain to be demonstrated in many cases. In vivo NMR has already proved useful to investigate cellular processes in a non-invasive way (2). However, due to its intrinsic insensitivity, one of the most common problems arising when trying to use in vivo NMR is related with the small intracellular pools of the intermediate compounds. In fact, when ^{13}C methanol was fed to a cell suspension of M.barkeri, and the cell metabolism was monitored in a series of ^{13}C NMR spectra, only resonances due to the substrate itself and to the final products, CO_2, HCO_3 and CH_4 could be detected with a useful signal to noise ratio. The difficulty in observing the intermediate metabolites has been overcome by carrying out the exchange between extracellular and intracellular acetate leads to the scavenging of ^{13}C labelled compounds from the intracellular space, allowing their detection. Under these experimental conditions, ^{13}C resonances due to the three isotopomers of acetate ($^{13}CH_3$$^{13}COO-$, $^{13}CH_3COO-$, $CH_3$$^{13}COO-$) could be observed, in addition to other resonances in the carboxylate and methylene regions. Cyanide, an inhibitor of CO dehydrogenase, inhibited the synthesis of all three acetate isotopomers, although methanogenesis was not affected. Iodobutane, a corrinoid antagonist, completely inhibited the synthesis of $^{13}CH_3$$^{13}COO-$ and $^{13}CH_3COO-$, although the exchange between $^{13}CO_2$ and the carboxyl group of acetate, as well as the methanogenesis, were not affected. These inhibitory effects were reversed by light, implying that a corrinoid is involved in the transfer of the methyl group of methanol to the synthesis of acetate, but probably not in the formation of methane. These results constitute a direct demonstration of the synthesis of acetate as a precursor of cell carbon in vivo and support the involvement of the acetyl-CoA pathway (3) in the assimilation of methanol by M.barkeri.

REFERENCES

1. Keltjens, J.T. and van der Drift, C. (1986) <u>FEMS microbiol.Rev.</u> 39: 259-303
2. Shulman, R.G. (1988) <u>Trends Biochem.Sci.</u> 13: 37-39
3. Wood, H.G., Ragsdale, S.W. and Pezacka, E. (1986) <u>FEMS Microbiol.Rev.</u> 39: 345-362

ACKNOWLEDGEMENT

This work was supported by JNICT, grant no 832.86.178)

REDOX PROPERTIES OF TWO B_{12} CORRINOID PROTEINS ISOLATED FROM METHANOSARCINA DSM 800 AND DSM 2905 IN THE METHYLATED AND AQUO FORMS

A.R. LINO[1], J.J.G. MOURA[1], A.V. XAVIER[1], J.LEGALL[2], I.MOURA[1]
[1] Centro de Quimica Estrutural (UNL and FCL), Lisboa, Portugal

[2] Department of Biochemistry, University of Georgia, Athens, GA, U.S.A.

Previously, we reported (1,2) the isolation and the characterisation of the B_{12} proteins isolated from the methanogenic bacteria Methanosarcina barkeri (DSM 800, DSM 804 and DSM 2905) by u.v./visible, E.P.R. and N.M.R. spectroscopies.

A native methylated form of the protein is isolated from M. barkeri (DSM 2905) while the proteins from M.barkeri (DSM 800 and 804) were isolated in the aquo form. The N.M.R. spectra showed that factor III is the cofactor bound to both proteins, which suggest that this cofactor could be conserved in Methanosarcina species. The E.P.R. spectra shows a typical signal of a stable Co(II) complex. Triplets are observed from the N-hyperfine interaction of the co-ordinated benzimidazol base.

We report the redox potentials of the proteins isolated from M.barkeri (DSM 800 and 2905) when bound to different axial ligands (aquocobalamine and methylcobalamine).The metabolic role attributed to these proteins is still controversial. The redox potentials of the different forms may help to elucidate the possible physiological role of these proteins.

REFERENCES

1. Lino, A.R. et al. (1987) Recueil des Travaux Chimiques des Pays-Bas, 106: 350

2. Lino, A.R. et al. (1987) Ciencia Biologica, 12: 5A 80

ACKNOWLEDGEMENTS

This work is supported by INIC and JNICT

ISOLATION AND CHARACTERISATION OF TWO NOVEL METHANOGENS, A NEW HALOALKALOPHILIC METHANOGEN AND A NEW ALKALOPHILIC METHANOSARCINA

N. NAKATSUGAWA, K. HORIKOSHI
The Superbugs, ERATO program,
Research Development Corporation of Japan, 2-28-8 Honkomagome,
Bunkyo-ku, Tokyo 113, Japan

There is currently great interest in methanogenic archaebacteria from extreme environments. We have embarked on research attempting to isolate super-methanogens which can grow in extreme environments, for use in industrial methane production and to develop a system for genetic engineering of methanogens. We have isolated two new methanogens, strain NY-218 and strain NY-728. Strain NY-218 is a halophilic and alkalophilic methanogen which was isolated from a hypersaline, alkaline environment in California, USA. The isolate was a motile coccus with a flagellum. Cells stained Gram-negative and were 0.5 - 1.5µm in diameter. The optimum NaCl concentration, pH and temperature for growth and methanogenesis were 2.5 - 3.0M, 8.0 - 8.5, 30-40°C, respectively. Methanogenesis occurred at more than 3.5 - 4.0 M NaCl concentration. Methanol, methylamine, dimethylamine and trimethylamine were used for growth and methanogenesis, but the last supported the most rapid growth. The DNA base composition was 42 mol% guanine plus cytosine. Strain NY-218 is different from previously reported moderately or extremely halophilic methanogens in its ability to grow at high pH in the presence of flagella.

Strain NY-728 is an alkalophilic methanogen belonging to the genus Methanosarcina, which was isolated from Japanese lake sediment Cells were stained Gram-positive. The optimum pH and temperature for growth and methanogenesis were 8.1 - 8.7 and 34-42°C. Acetate, H_2/CO_2, methanol and methylamine were used for growth and methanogenesis. Methanol supported the most rapid growth. The DNA base composition was 52 mol% guanine plus cytosine. Strain NY-728 is the first reported alkalophilic Methanosarcina.

ISOLATION AND IDENTIFICATION OF BACTERIA LIVING IN ENVIRONMENTS SEVERELY CONTAMINATED WITH HEAVY METALS

L. DIELS, L. HOOYBERGHS, A. RYNGAERT, M. MERGEAY
Laboratory of Microbial Genetics and Biotechnology
SCK/CEN, B-2400 Mol, Belgium

Many different bacteria have been isolated from environments strongly polluted with heavy metals and divided in different groups according to their metal resistances. Members of each group seemed to contain practically the same protein pattern with only a few exceptions. Some groups contained plasmids, others did not. The different isolated groups were resistant to some of the following metals: Cd, Zn, Cu, Co, Cr, Ni, Bi, Tl, Ce, Se, Ag, As, Te, Sb, Hg. The mechanisms for these resistances involve genes on plasmids, transposons or on the chromosome.

Previously we isolated an Alcaligenes eutrophus strain CH34 bearing two large plasmids coding for resistance to Cd^{++}, Co^{++}, Zn^{++}, Ni^{++}, Hg^{++}, Cu^{++}, Pb^{++} and CrO_4^{--}. Now it seems that CH34 is representative of a remarkable group of Alcaligenes eutrophus strains isolated from 5 different sites polluted with heavy metal (industrial sites in Belgium or mines in Zaire). All these A.eutrophus strains bore two large plasmids and hybridised with a 9kb DNA fragment from CH34 which codes for Cd, Co and Zn resistance (ccz^+). The strains were also comparable at the level of their protein pattern (on SDS-PAGE). All these soils were polluted with zinc which seems to be the driving force in the selection of these strains. This proves the existence of at least one universal mechanism for resistance against Cd, Co and Zn, which is dispersed all over the world like antibiotic resistance genes.

Preliminary results showed efflux systems for Cd and Ni ions (at low concentrations). Accumulation of Cd and Zn ions started at high ion concentrations. Total Cd elimination was obtained with some formation of Cd carbonates and diphosphates. Questions arise about a possible specific role of A.eutrophus in soils with high content of heavy metals.

EFFECT OF EXTERNAL SALINITY CHANGES ON AMINOACIDS AND IONS COMPOSITION OF DELEYA HALOPHILA

A. DEL MORAL, M.J. VALDERRAMA, M.R. FERRER, E. QUESADA, F. PERAN, A. RAMOS-CORMENZANA
Department of Microbiology, Faculty of Pharmacy, University of Granada, 18001, Granada, Spain

It has been reported that bacterial adaptation to osmotic stress involves changes in the concentrations of some aminoacids and/or cations. We show in this communication the preliminary results about the changes in the total composition in some aminoacids and cations of Deleya halophila in response to external salinity changes.

MH medium was utilised at 3, 7.5 and 20% (w/v). All cultures were incubated in an orbital shaker, harvested in the early exponential, stationary and late stationary phases by centrifugation at 10,000g for 15 min at 4°C. Cells were washed twice with cold saline containing NaCl at a concentration equal to that in the growth medium, and then lyophilised. Cells were lysed by sonication and analysed for Na^+ and K^+ in a Beckman autoanalyser. The protein content was determined by the Lowry method. Amino acid analyses were performed by HPLC using a Beckman model 342.

Growth in the medium with higher salt concentration did result in a small increase in internal K^+ and a significantly larger increase in the internal Na^+. It has been reported that the only cation required for growth by Deleya halophila was Na^+. Our results show that this cation is accumulated by the microorganism when grown at high salt concentrations.

Our preliminary results show that all bacterial samples examined contained aspartate, glutamine, serine, glycine, alanine, isoleucine, leucine and lysine. The concentration of glycine, aspartic and glutamine increased as the external salt concentration increased, glycine being the aminoacid that registered the most significant increase.

DOES THE HIGH Mg^{2+} CONTENT INHIBIT THE $CaCO_3$ PRECIPITATION BY DELEYA HALOPHILA?

A. RIVADENEYRA[1], R. DELGADO[2], E. QUESADA[1]
A. RAMOS-CORMENZANA[1]
[1] Department of Microbiology, Faculty of Pharmacy, Granada, Spain
[2] Department of Edaphology, Faculty of Pharmacy, Granada, Spain

Deleya halophila is a moderately halophilic species isolated from a saline soil located near Alicante (Spain). This microorganism has its optimal growth at 7.5% (w/v) total salts, being capable of growing between 2 and 30% (w/v) total salts.

Previously, we have studied the influence of salt concentration on the calcium carbonate precipitation by this bacterium. Our results indicated the optimal salt concentration for this process was 2.5% (w/v) while at a high salt concentration (20% w/v) calcium carbonate precipitation was strongly inhibited. These results confirmed the observations made by Billy (1980) who affirms that an excess of salt influences calcium carbonate crystal formation negatively. Nevertheless it is not known whether the inhibitory effect of high salt concentration is due to the high Mg2+ content of the salt solution used, since a negative effect of magnesium ions has been reported.

In this work we have studied the calcium carbonate precipitation by 27 strains of Deleya halophila and by other marine species of the genus Deleya, using media with different total salts (2.5, 7.5 and 20% w/v), but without magnesium. The experiments have been carried out in duplicate, at 22 and 32°C. Deleya halophila could become a useful tool for these type of studies since this bacterium is capable of growing in media containing NaCl as the sole salt.

All the strains tested were capable of calcium carbonate precipitation at the three different salt concentrations and two incubation temperatures studied (except those strains incapable of growing in these conditions). The amount of crystals formed was very high at all salt concentrations tested, although we could note a slight diminution at 20% (w/v) total salts. Comparison of these results with those previously obtained using media containing magnesium suggests that the inhibitory effect of this ion is the predominant cause of reduction in precipitation of calcium carbonate.

HALOADAPTATION AND MEMBRANE LIPID COMPOSITION IN A HALOPHILIC VIBRIO sp. HX

R.L. ADAMS, N.J. RUSSELL
Department of Biochemistry, University College Cardiff,
P.O. Box 78, Cardiff CF1 1XL, Wales, U.K.

The moderately halophilic bacterium Vibrio sp.HX requires 0.5M NaCl for growth and at 30°C grows in up to 4.5M NaCl with a broad optimum between 1.0 and 2.0M NaCl, which is shifted slightly to higher salinity by growth at 37°C. The major lipids in Vibrio HX are phosphatidylglycerol, phosphatidylethanolamine and diphosphatidylglycerol (cardiolipin) together with smaller amounts of lysophosphatidylethanolamine and phosphatidylserine. The major fatty acids are 18:1, 16:0, 16:1 and C17 and C19 cyclopropane.

Growth of Vibrio HX adapts rapidly to as much as a fourfold shift-up in salinity. Lags in growth are seen only when the shift-up is greater than threefold. The final NaCl concentration is most influential in determining the growth rate, although it is also influenced by the magnitude of the shift-up. Following a shift-up, the proportion of phosphatidylglycerol (the major anionic lipid) relative to phosphatidylethanolamine (the major zwitterionic lipid) increases rapidly, i.e. in less than one generation time - to the composition of isotonic cultures growing at the higher salinity. These results suggest that Vibrio HX is a particularly versatile moderate halophile which can cope rapidly with sudden and large changes in salinity, and that regulation of membrane lipid composition is an integral part of phenotypic haloadaptation.

THE GENE OF A PUTATIVE DNA-BINDING PROTEIN FROM HALOBACTERIUM HALOBIUM

G. BALDACCI
Institut de recherches sur le cancer,
C.N.R.S.   ER 272, Laboratoire 11A
7, Rue Guy Moquet, 94802, Villejuif Cedex, France

In the course of the study of the structural features of genes in Halobacterium halobium, we have isolated a SalI-SmaI DNA fragment, 850 bp long, which hybridises to RNA. We have determined the nucleotide sequence of this region and the size (about 500-600 nucleotides) of the corresponding transcript. The hybridisation of a single-stranded probe allowed the identification of the coding strand. In addition in vitro nuclease S1 protection experiments and the results of hybridisation to RNA of small DNA probes allowed the boundaries of DNA/RNA hybrids to be defined. Taken together, our data indicate the existence in this region of H.halobium DNA of a transcribed region potentially coding for a 163 aminoacid long peptide. Computer analysis of the predicted protein shows clustered analogies with protamines, histones and other DNA-binding proteins. At present, we are trying to express this region in E.coli in order to obtain large amounts of the peptide in order to study its possible interaction with nucleic acids.

SCREENING FOR COMPATIBLE SOLUTES OF HALOPHILIC EUBACTERIA

A. WOHLFARTH, J. SEVERIN, E.A. GALINSKI
Institut fur Mikrobiologie,
Rheinische Friedrich-Wilhelms-Universitat Bonn,
Meckenheimer Allee 168, D-5300 Bonn 1

Marine and halophilic microorganisms have only recently become the objects of investigation in the search for bioactive agents (1). Among these, compatible solutes comprise a diverse group of highly soluble substances such as amino acids, sugars, polyols and betaines. These substances serve to compensate the osmotic pressure in osmotically stressed micororganisms. They accumulate to high intracellular concentrations, and seem not to affect metabolism. Indications of enzyme protection and stabilising effects have been detected (2).

Little is known about the spectrum of compatible solutes in marine and halophilic eubacteria, and the mechanism of halotolerance has still not been fully explained (3). Studies of certain halophilic bacteria (3) have shown a deficiency of compatible solutes, i.e. too little to cause osmotic balance across the membrane. These organisms seem to be hypoosmotic with respect to the medium.

In bacteria from marine environments the intracellular spectrum of compatible solutes depends on growth conditions, growth phase and the composition of the medium. Proline, glutamate, trehalose and glucose were detected as well as glycinebetaine, which usually accululated when the organisms were grown on a yeast extract medium. More important, however, is the detection of hitherto unknown osmotica, such as Ectoin (4). This substance, recently described in the anaerobic phototrophic bacterium Ectothiorhodospira halochloris, was found to be one of the most important osmotically active compounds in halophilic eubacteria, e.g. Halomonas elongata, Paracoccus halodenitrificans, Planococcus halophilus, Flavobacterium halmephilum, Micrococcus halobius. Our results show that, taking further new substances into account, halophilic micororganisms are indeed able to balance the osmotic pressure exerted by the environment.

REFERENCES

1. Okami,Y. (1986)   Microb.Ecol. 12: 65-87
2. Jurgens, E. (1988) Diplomarbeit, Universitat Bonn
3. Vreeland, R.H. (1987)   Crit.Rev.Microbiol. 14: 311-357
4. Galinski, E.A., Pfeiffer, H.P., Truper, H.G. (1985) Eur.J.Biochem. 149: 135-139

ENHANCEMENT OF PROLINE PRODUCTION OF BACILLUS SUBTILIS DURING SALT STRESS

E. MÜLLER, H.G. TRÜPER
Institut für Mikrobiologie, Rheinische Friedrich-Wilhelms Universität Bonn, Meckenheimer Allee 168, D-5300 Bonn 1

Halotolerant and halophilic microorganisms accumulate great amounts of compatible solutes within the cytoplasm to withstand the high osmotic pressure caused by extreme salinity. Suitable organic compatible solutes are amino acids, polyols, betaines and sugars(1). Of the amino acids, the predominant compatible solutes are proline, glutamic acid, glutamine, alanine and γ-aminobutyric acid. The concentration of proline in gram-positive bacteria is about ten times that in gram-negative organisms(2).

Compatible solutes have become the topic of investigation in several laboratories and in industry because these substances are accumulated intracellularly by halophiles to molar concentrations. The risk of contamination of halophilic or halotolerant fermentation processes is greatly reduced due to the lowered water activity.

In Australia and Israel pilot plants have been set up with the aim of producing compatible solutes and β-carotene from halophilic algae(3). The probable industrial potential of proline as a compatible solute is obvious because it has been shown that even low cytoplasmic concentrations of proline result in high fermentation yields of this amino acid(4)

Our own investigations dealing with the halotolerant Bacillus subtilis have shown that the cytoplasmic concentration of proline varied to a great extent. The composition of the fermentation media has a great effect on growth and proline production of Bacillus subtilis. In addition, proline production may be increased by the choice of a suitable fermentation process, and by the optimization of fermentation parameters. This enhances the proline production of Bacillus subtilis by several orders of magnitude.

Literature:

1. Galinski, E.A. (1986) Dissertation, Bonn
2. Tempest, D.W., Meers, J.L., Brown, C.M. (1970) J.Gen. Microbiol. 64: 171-185
3. Ben-Amotz, A., Avron, M. (1983) Ann.Rev.Microbiol. 37: 95-119
4. Sugiura, M., Kisumi, M. (1985) Appl. Environm. Microbiol. 49: 782-786

ANTIBIOTIC-RESISTANT MODERATELY HALOPHILIC GRAM-NEGATIVE MOTILE RODS FROM HYPERSALINE WATERS

J. QUEVEDO-SARMIENTO, A. DEL MORAL, M. RITA FERRER, A. RAMOS-CORMENZANA
Department of Microbiology, Faculty of Pharmacy, University of Granada, 18001 Granada, Spain

The investigation of bacteria of faecal origin has dominated studies of antibiotic-resistant bacteria in nature because of the association of these bacteria with disease. To date, little work has been done on antibiotic resistance patterns of non-coliform bacteria that occur in less polluted environments, although there is evidence that indigenous bacteria in marine and freshwater habitats can possess the same kinds of antibiotic resistance patterns as coliform populations.

In a previous study, we isolated and characterised a group of motile, Gram-negative, moderately halophilic bacteria which predominated in hypersaline waters from the ponds of an inland saltern. They were related to species of the genera Vibrio, Deleya, Alteromonas and Pseudomonas. The aim of the present study was to provide detailed descriptive information about the antibiotic resistance patterns of moderately halophilic bacteria isolated from hypersaline waters and their taxonomic groupings.

The incidence of resistance to 14 antibiotics was determined in 152 moderately halophilic bacteria. All the strains examined were multiple antibiotic resistant.

Results obtained were quite unexpected because the incidence of multiple antibiotic-resistance in moderately halophilic bacteria from hypersaline water is higher than that found for multiple antibiotic resistant coliforms isolated from such different habitats as: sewage and faecal samples, sediments, river, lakes and drinking water. However, the current findings should be treated with care, because the interpretation of antibiotic resistance tests is founded on detailed knowledge of numerous isolates of medical and veterinary significance.

POLYOL ACCUMULATION BY YEASTS AND A YEAST-LIKE FUNGUS UNDER OSMOTIC STRESS

M.F. NOBRE, M. S. DA COSTA
Centro de Biologia Celular, Departamento de Zoologia,
Universidade de Coimbra, 3049 Coimbra Codex, Portugal

Osmotolerant eucaryotic microorganisms produce and accumulate polyhydroxy alcohols (polyols or sugar alcohols) in response to osmotic stress. Arabitol, erythritol, glycerol and mannitol are the most important polyols accumulated by these microorganisms as compatible solutes. In this work we studied the effect of the exogenous carbon source on the osmoregulation of three yeasts (Debaryomyces hansenii, Candida famata and Pichia farinosa and a yeast-like fungus (Geotrichum sp.) when grown under conditions of low water activity imposed by high salt concentration.

Growth of D.hansenii in medium containing 1M NaCl and 1%(w/v) glucose as carbon source, resulted in the accumulation of glycerol during exponential phase and arabitol during stationary phase. When meso - erythritol was used as the sole carbon source, the yeast accumulated large amounts of this polyol during exponential phase of growth, instead of glycerol. The yeasts C.famata and P.farinosa, showed the same behaviour under the same conditions of salt stress. Erythritol replaced intracellular glycerol as compatible solute causing a decrease in the intracellular concentration of this polyol when a complete medium (YEPD) or a minimal medium (YNB) containing 1M NaCl were used.

Geotrichum sp. is a yeast-like fungus that accumulates arabitol during exponential phase of growth, and mannitol during stationary phase, in glucose media containing high concentrations of salt or sugars. Substitution of glucose by glycerol as carbon source did not cause alterations in the accumulation of arabitol as osmoregulatory solute.

These data showed that, unlike the yeasts studied in which exogenous polyol (erythritol) could accumulate as compatible solute, in Geotrichum sp. exogenously supplied glycerol cannot accumulate under water stress and arabitol seems to be the only compatible solute in this fungus.

REFERENCES

1. da Costa, M.S. and Niederpruem, D.J. (1980) Temporal accumulation of mannitol and arabitol in Geotrichum candidum. Arch.Microbiol. 126: 57-64
2. da Costa, M.S. and Niederpruem, D.J. (1982) Arabitol accumulation in Geotrichum candidum. Arch.Microbiol. 131: 283-286

3. Nobre, M.F. and da Costa, M.S. (1985) Factors favouring the accumulation of arabinitol in the yeast Debaryomyces hansenii. Can.J.Microbiol. 31: 467-471
4. Nobre, M.F. and da Costa, M.S. (1985) The accumulation of polyols by the yeast Debaryomyces hansenii in response to water stress. Can.J.Microbiol. 31: 1061-1064.

ISOLATION AND CHARACTERISATION OF BACTERIOPHAGES ACTIVE ON MODERATELY HALOPHILIC MICROORGANISMS

A.M. GARCIA DE LA PAZ, A. PEREZ MARTINEZ, C. CALVO SAINZ, A. RAMOS CORMENZANA
Departamento de Microbiologia, Facultad de Farmacia, Universidad de Granada, Granada, Espana

Lysogeny of Deleya halophila species has been studied after induction with Mitomycin C. Several bacteriophages have been isolated and influence of saline salts concentration on its replication and survival has been established.

On the other hand three bacteriophages active on moderately halophilic and/or halotolerant microorganisms have been isolated from the solar saltern located in La Mala (Granada, Spain). The influence of environmental factors such as salt concentration and temperature on the stability of these bacteriophages has been investigated.

IDENTIFICATION OF THE HALOCIN H4 GENE IN HALOFERAX MEDITERRANEI

B. GAMBIN[1], G. JUEZ[1], F. RODRIGUEZ-VALERA[1],
M. BETLACH[2], H.W. BOYER[2]

[1]Microbiology Division, University of Alicante, Spain

[2]Biochemistry and Biophysics Department,
University of California, San Fransisco, U.S.A.

The genetics of halobacteria and archaebacteria in general, have become the subject of considerable interest in recent years, particularly since the discovery of certain peculiarities of this third phylogenetic stem, such as the possesion of multiple repetetive sequences, of introns and transposable elements giving rise to the frequent genomic reorganisations observed in halobacteria.

However, very few genes have been described in halobacteria, amongst these the bacteriorhodopsin and halo-opsin genes. It would be of great interest to describe new genes in this group of bacteria especially if they can be used as genetic markers in future genetic experiments.

The aim of this work is to sequence and describe the gene which codifies for a bacteriocin produced by Haloferax mediterranei, the halocin H4. This halocin has been purified and sequenced, and two synthetic oligonucleotides (45 base pairs and 36 base pairs) have been made from the N-terminal fragment and a tryptic fragment of the amino acid sequence.

These oligonucleotides have been used as probes to hybridise against 300 clones from a genomic library of the producer strain. We have been able to identify a fragment of 2.1 Kb which gives strong hybridisation with probes. This fragment will be studied in more detail and sequenced in the near future, in order to describe the halocin H4 gene.

HALOCOCCUS SACCHAROLYTICUS SP.NOV., A NEW GROUP OF EXTREMELY HALOPHILIC ARCHAEBACTERIAL COCCI

A. VENTOSA[1], C.G. MONTERO[1], F. RODRIGUEZ-VALERA[2], M. KATES[3], N. MOLDOVEANU[3], F. RUIZ-BERRAQUERO[1]

[1]Department of Microbiology, Faculty of Pharmacy, University of Sevilla, Spain

[2]Department of Microbiology, Faculty of Medicine, University of Alicante, Spain

[3]Department of Biochemistry, University of Ottawa, Ontario, Canada K1N 9B4

In a recent study a large number of extremely halophilic non-alkalophilic cocci have been isolated, and a numerical taxonomic study of these bacteria has been carried out. In this study, four different phenons were obtained. Phenon A included all the reference strains and was considered to comprise members of the only named species of the genus Halococcus, H.morrhuae. Phenons B and C showed greater metabolic versatility than Phenon A. The four strains belonging to Phenon D were significantly different from the other phenons in many phenotypic features, especially in their acid production from sugars and their remarkably wide nutritional versatility.

In the present work we have studied in detail a representative of Phenon D with respect to its DNA base composition, polar lipid composition, ultrastructure, presence of plasmids etc. The guanine plus cytosine content of strain P-423 is 59.5 mol %, the lowest value found in strains belonging to the genus **Halococcus.** Phylogenetic differences have also been found between strains P-423 and H.morrhuae using DNA-rRNA hybridisation techniques. Since it seems justified that the strains belonging to Phenon D should be included in a new species, we propose to name them as H.saccharolyticus, and we also make a formal proposal with the description of the new species.

RESTRICTION ENZYME SCREENING ON BACTERIAL SPECIES OF AZORES AND MADEIRA MINERAL WATERS

J.M.B. VITOR, R.V. CORREIA DE SILVA
Genetic Engineering Laboratory, Faculty of Pharmacy,
Lisbon University, Av. dos Forces Armades,
1699 Lisbon, Portugal

We have isolated 37 bacterial strains from two glass-bottled mineral waters, 21 and 16 of them being recovered from the Lombadas (Azores) and Sao Vicente (Madeira) waters, respectively. Firstly, the isolates were subjected to appropriate growth conditions until purified, single colonies had been obtained on agar plates. Secondly, they were kept either as subcultures on agar slants in the refrigerator, or as suspensions in 50% glycerol-50% growth medium at -20°C until used.

The cells were submitted to an osmotic lysis procedure previously described, and their supernatants incubated at 37°C for 3 hours in the presence of either lambda DNA or their Bst E II fragments (BioLabs) in 6mM Tris (pH 7.5), 6mM 2-mercaptoethanol, and 6mM $MgCl_2$ buffer containing one of each 0, 50, 100 and 150 NaCl concentration. After the digestion products were resolved by agarose gel electrophoresis (AGE) those samples producing a clear DNA substrate cleavage were defined as restriction enzyme activity positive.

Restriction endonuclease activity was found to be associated with 8 strains. Others which, occurring in 5, 4 and 11 strains were considered as suspicious restriction enzyme, ligase-like and non-specific nuclease activities, respectively. Mainly, 4 positive samples were recovered from the Lombadas (Azores) mineral water, one of them growing well at 55°C but not surviving at 57°C. Since its restriction enzyme activity (observed at 37°C) was unable to digest DNA plasmids contained in these cells, this could be correlated in that a possible DNA methylation mechanism could afford some sort of cellular protection against adverse environmental growth conditions.

ACKNOWLEDGEMENTS

This work was supported by JNICT (grant no. 87272) and Faculty of Pharmacy of Lisbon University. We also thank Dr Dioguina Fonseca as well as the Portuguese BioMerieux for their collaborations on computer reading of bacterial profiles.

GROWTH CHARACTERISTICS AND SALT REQUIREMENT OF TWO NEW GROUPS OF MODERATELY HALOPHILIC MICROORGANISMS

M.J. VALDERRAMA, V. BEJAR, E. QUESADA, A. RAMOS-CORMENZANA
Department of Microbiology, Faculty of Farmacy,
Granada, Spain

The organisms used in this work were representative strains of Flavobacterium sp.and Acinetobacter sp., two new groups of moderately halophilic microorganisms previously described by us (1).

Growth response to salt was determined on 0.5, 2.5, 5, 7.5, 10, 15, 20, 25% total salts using a complex and a defined medium which composition were previously reported (2). Finally, we have determined salts and ions requirements by substituting NaCl at the optimal concentration by other salts in the defined medium.

Flavobacterium sp. presented its optimal growth at 5% (w/v) total salts at 22 and 32°C in complex medium and in defined medium. At 42% it exhibits a salt requirement of 7.5% for its optimal growth in complex medium, whereas it was unable to grow in defined medium at the same temperature. The bacterium required for growth Na^+ plus Mg^{2+} cations, however Cl^- anion could be substituted by SO_4^{2-}.

When Acinetobacter sp. was cultured in complex medium, as the incubation temperature was increased (22, 32, 42°) the salt concentration needed for optimal growth increased too (5,7.5, 10% w/v, respectively). Results in defined medium were similar: while a 5% (w/v) total salts permitted optimal growth at 22 and 32°C, a higher salt concentration (10%) is required at 42°C. This microorganism presented a sole and specific requirement for Na^+ cation, while Cl^- anion could be substituted by SO_4^{2-} or Br^-.

REFERENCES

1. Quesasa et al (1987) Syst.Appl.Microbiol. 9: 132

2. Quesada et al (2987) Curr.Microbiol. 16: 21

DEC 2 8 1989